MINERAL PROPERTY ECONOMICS
Volume 1: Economics Principles and Strategies

By

JOHN M. CAMPBELL, JR., Vice President
JOHN M. CAMPBELL, SR., President
ROBERT A. CAMPBELL, Vice President

Of

JOHN M. CAMPBELL AND CO.

MINERAL PROPERTY ECONOMICS

Volume 1

First Printing, July, 1978

Original Drafting By: Jerry Collins

Printed and Bound in U.S.A.

Published By: Campbell Petroleum Series
               121 Collier Drive
               Norman, Oklahoma 73069

DEDICATED TO ......

     Gwen Campbell the wife of one author and the mother of the other authors, who made this book possible.

ACKNOWLEDGMENT

This book is the most ambitious work on mineral economics ever attempted. It would not have been possible without the contribution of many persons.

The contributions began about 20 years ago with the publication of a predecessor to this book. Where possible we have attempted to acknowledge these by referencing the source. However, it literally has been impossible to know the source of all information presented, particularly that taken from our personal records. To all of you we say THANK YOU for your contributions to a better understanding of mineral economics.

A special thank you is due Shirley Ray and Joy Garrison. They prepared the final copy, edited and assisted in the layout. Their dedication is much appreciated.

Norman, Oklahoma
June, 1978

THE CAMPBELLS
John Sr., John Jr., Robert and Charles

TABLE OF CONTENTS

Volume 1

LIST OF FIGURES

Volume 1

1

ELEMENTS OF ECONOMIC EVALUATION

The evaluation of mineral properties is a subject that contains too many variables and is too complex to be defined in a simple manner. However, it might be described as an orderly, dispassionate, and planned analysis of all available information in order to determine the probable value of a property under consideration. Although this is not a complete definition, it does set up some general criteria in that it specifies that such an evaluation must be performed in a professional manner which, therefore, sets this type of evaluation apart from one based on whim, fancy, or wishful thinking.

It is necessary to make a distinction here between this type of evaluation and the informal kind that one so often performs from day to day. Every time a person makes a purchase, at least subconsciously, he makes some form of evaluation. In most cases the evaluation can hardly be called scientific. For example, a man buying a car will usually raise the hood and look at the motor, kick a tire or two, slam a door, and, maybe, get behind the wheel. If the car survives this "exhaustive" test, and if he likes the brand, the color scheme, and the upholstery, he will probably buy it. In other words, the price asked by the seller seems fair to this particular buyer since his evaluation has indicated that the probable value to him is equal to or greater than the asking price. On the other hand, another buyer might weigh the same set of "facts" and reach a negative decision.

If one pursues this line of reasoning, one may arrive at the following important conclusions:

1. The measure of success of an evaluation is not how technical it is but whether it accomplishes its intended purpose.

2. The relative importance and interpretation of the available data will vary somewhat with the person making the evaluation.

3. Value is not an absolute quality, but one which must be established in any particular case by mutual agreement of the opposite parties involved.

Item 1 does not imply that scientific methods are not desirable, or even needed. It simply points out that the evaluator must consider all facets of the problem, particularly those that are considered important by the recipient. Probably the most common fault of evaluations prepared by inexperienced engineers is their stress on engineering calculations with little, if any, regard for the economic limitations presented. In this, as well as in most other fields, engineering feasibility means nothing unless it is accompanied by economic feasibility. Engineering calculation, therefore, is a means of accomplishing an end result; i.e., of making a profit.

It must be stressed that true value can only be established by an "arms-length" transaction between two independent parties. The word "independent" signifies that the parties enter into the transaction with purely selfish motives and that they are independent economic entities. In other words, the value is a true one and not an artificial one.

It is important to understand this. Value is never fixed truly by a professional. It is only determined by a transaction. The professional must of necessity assume a role or viewpoint of potential parties to the proposed transaction.

This book is concerned with both economic and engineering factors which play a part in the study of a property. The engineering method is desirable, since it minimizes, as much as possible, the emotional element.

The engineering method has been described as an orderly, dispassionate, and planned manner of procedure in tackling a problem.

This procedure might be summarized as follows:

1. Study the problem to determine its various aspects and to understand its nature and scope.

2. Break down the problem into its numerous component parts and establish the relative importance of each and the relationships between them.

3. Summarize the known and the unknown elements.

4. Become acquainted with all pertinent facts that have been established by experience or research.

5. Solve the component parts separately and progressively fit them together in order to answer the original problem.

This procedure, then, consists of starting with known facts and established principles and enables one, by logical reasoning and resourcefulness, to proceed in an orderly manner, avoiding the pitfalls of wishful thinking or of personal and/or political considerations.

The fact that there is a wide latitude for interpretation raises a question, since it is probably true that no one can ever be truly unbiased. However, the problem here is to avoid any conscious bias. One logically wishes to please, and, for that reason, may be tempted to interpret data as he thinks a superviosr or a client would like to have it done. This raises a moral question.

The evaluator has an obvious responsibility to see that his employer's interests are protected at all times. The question, then, is - in the interpretation and presentation of data, are the figures presented the most probable ones, or are they biased strongly in the employer's favor with only a mere possibility of being correct? Since no sharp line can be drawn between them, decisions on this matter become an individual responsibility.

A good evaluation will establish a fairly narrow range of probable values. One can then choose the value in this range (distribution) that is consistent with the pruposes of the report.

In the long run a "fair" report usually serves the client's interest to the best advantage. A good test of fairness might be the following questions:

1. In the evaluator's private opinion do the results presented honestly fall within the limits of the most probable values?

2. Are all the significant facts presented in a clear manner without distortion or misrepresentation?

3. Can the facts presented be substantiated in the light of available data and present engineering and economic knowledge?

4. Can the report be used as an instrument for negotiation by all parties and as a basis for an agreement that is fair to these parties?

5. Can the recipient of the report use it as a basis for future decisions, subject to the obvious general limitations of an evaluation and those specifically presented in the report?

Unless these questions honestly can be answered in the affirmative, the report in question does not fall under the category of an evaluation in the sense used in this book. Excluded, therefore, are reports written to please the recipient or to provide unfair advantage to one party in an agreement, or any other report that ignores the precepts of the engineering method described above.

After studying the concepts reviewed in future pages, it should be evident that the only things separating a good evaluation from a poor one are the integrity, the professional status, and the good judgment of the evaluation engineer. Integrity is a necessary ingredient.

PURPOSE OF THE VALUATION

Mineral economic valuation is usually made for one of the following reasons:

(1) sale and exchange of property

(2) loans

(3) a guide for management decisions

(4) future exploration and development policy

(5) accounting and taxation

(6) settlement of estates

(7) a basis for unitization agreements

(8) lease bidding

(9) evaluation of governmental regulations

3

Because of these many uses, it is understandable that many approaches are possible, since, to a certain extent, an evaluation must be "tailor-made" to fit its own problem. In all cases, however, it must state the available reserves of minerals and their probable economic value. In most instances it is necessary to estimate the probable rate and economic value of future recovery as well. This means not only that the evaluation engineer must be competent in reservoir and production operations, but also that he be familiar with factors affecting mineral economics. This is particularly essential when one remembers that a property evaluation is, in effect, a look into future happenings over which he or his client can exercise minimal control. It would not be far wrong to say that any future event that affects world economics has at least some effect on a given evaluation. This includes wars, depressions, and political situations in different countries, all of which are highly unpredictable factors. It would therefore be an ideal situation if the evaluator also could be trained as an economist, or lacking this, training as a swami with fool-proof crystal ball. Since these are in short supply, it is necessary to rely on a sound analysis of past economic happenings and detailed study of probable economic conditions in the future.

In spite of the fact that prediction of the future is a precarious task, it is significant that such predictions by competent people are generally satisfactory. Virtually all of the investment capital outlay in our world structure is based on an evaluation of one type or another, and the success of the system speaks for itself. With the realization that control of the future is impossible, success in this prediction must inevitably stem from the semi-application of statistics and probability - making sure that the odds are in favor of the economics being as predicted.

The primary concern of a lending institution - either the company itself or creditors outside the company - is that there are sufficient mineral reserves to serve as collateral, and that they will be produced at an economic rate in a reasonable period of time. The valuation report should, therefore, stress these points. Furthermore, it is essential that the figures presented be conservative, yet realistic, in order to reduce the gamble of the lending agency. In arriving at such a position it is not recommended that a reserve figure be artibrarily cut to achieve this goal. Rather, it is more sound to weigh each factor during the course of the examination and to discard or to modify those that have dubious validity, or that appear to involve an undue amount of risk. For this reason, the figures presented can be described as the minimum probable results. Conservative estimates of a mineral prospects economic value calculated in this manner are likely to be more accurate than estimates obtained by just adjusting the values downward to reflect risk.

A valuation for management use should generally be more "middle-of-the-road" than that above, because management will add safety factors based on its own experience and judgment. Consequently, management depends on the report to be factual. This means that it must be based on the best available information (as should all reports) with its limitations clearly stated. However, it is generally a poor policy to attempt to compensate for these uncertainties to any great extent for if this were carried to the extreme, no investment would be attractive. A too liberal use of safety factors may completely distort the "facts" on which management must base a decision.

The primary purpose of the evaluation in the mind of most management is not merely that it presents facts, but that it affords a means of comparing the relative attractiveness of alternative investments or alternative approaches to the same investment. Although reserves and other engineering figures lend support to the conclusions, in the final analysis, they must be considered only the means to an end and, as such, of secondary importance to the financial consequences of a decision.

Among the most critical valuations are those which concern new acreage for future potential development, such as wildcat wells in the petroleum industry, because the major investment in mineral properties usually occurs during the drilling (or construction) and completion stage. After a mineral property has been completed its worth decreases with the age of the subject property; a valuation at depletion has little more than academic value to the subject property and serves primarily only as a tool for future work.

Unfortunately, the evaluations performed prior to the completion of development are the least accurate since the data available then are relatively meager and have a larger degree of uncertainty. In any evaluation the accuracy is a direct function of the quantity of reliable data available, and one must rely to a large extent on the formal application of the principles of probability and statistics. There is no doubt that many people have made money using inexact methods such as hunches, divining rods, and the like; but history also tells us that many more have been unsuccessful.

The primary factor to be considered is the cost of developing and producing relative to the value of the mineral, such as local marketing situations, nearness to existing commercial production, geophysical surveys such as seismic or magnetic, geological structure maps, and general regional characteristics of the potential zones. Other facts might also prevail but the above may never conscientiously be ignored. All major companies, for example, have data in their files on favorable mineral properties that have not yet been developed, for economic reasons alone.

The problem is to attach the proper weight to each of the factors and to use the principles of probability for drawing proper conclusions. The result of such an evaluation is not so much a summary of reserves or future revenue but a recommendation whether to develop a property. In doing so it is necessary to establish some arbitrary standards against which the findings may be compared.

The systematic appraisal of the many rather indefinite factors yields a result that is generally satisfactory and which at its worst is better than unsound methods. Unfortunately, too large a segment of all mineral industries often ignore this fact and let prejudice play too large a part. This common human frailty offers one of the prime pitfalls to an evaluation of this type.

Those valuations dealing with accounting, taxes, and estates are generally short because they are concerned only with the total reserves, their total value, and the value to each of the interests involved. Such valuations are often only a one-page summary suitable for routine company reference.

For example, unitization presents a special problem from the evaluation standpoint in the petroleum industry. The primary goal is to determine the producible hydrocarbons held by each operator in order to pro-rate the revenue from the unit. Although the formula is likely to vary slightly in each case, it usually involves some volumetric measures. It seldom occurs that the calculation is carred out to the point of determining reserves, since this involves some rather arbitrary assumptions.

Evaluations for the sale and exchange of property embody the general features of those discussed above. The only additional stress might be on taxation and the manner in which it affects values. Instances often arise in which a party might sell a property at less than its probable value and still show a profit because of a tax advantage. Taxation is at least an indirect consideration in all transfers of mineral property.

Economic values are affected significantly by the cost of acquiring mineral leases for development. Various forms of bonus bidding, royalty bidding and combinations of other bidding strategies where the owner of the lease shares the cost and risk alter the potential cash flows of a project, and, therefore, its economic value to the mineral company. Occasions often arise when the decision to bid and the amount of each bid at lease sales are retarded when investors must put up large front-end cash payments for high cost, high risk leases. Optimal development of mineral resources requires that leasing policies reflect the cost and risk of development.

Government regulations and policies exemplify non-economic and non-engineering forces determining the economic value of mineral leases. Selection of tax incentives (or disincentives), environmental regulation, expenditures on pilot projects and research-development of new mineral technology, and international law all exemplify governmental influences on economic value. Two mineral prospects, exactly alike in every engineering aspect, have different economic values depending upon the price producers are allowed to charge and the rate at which they can produce the mineral.

Mineral economics at one time were determined mainly by engineering and economic forces. Governmental inroads, however, have in recent years altered historical valuation procedures to the point where mineral leases considered to be uneconomical in the past are now profitable. Stripper wells shut-in several years ago are economical at free world prices, while oil receiving controlled prices is less valuable. Coal, similarly, benefitted from oil pricing controls in the early seventies as seen by the re-opening of many mines felt to be uneconomical before the major increase in oil prices. Governmental regulations and policies are often a more important force in determining the economic value of minerals than engineering forces at the present time.

RISK AND UNCERTAINTY

Traditionally, risk and uncertainty have been handled somewhat arbitrarily. A lending institution might divide evaluated reserves by a factor of 2-4 to arrive at a "safe-no risk" number for lending purposes. In other cases, the evaluator might consistently choose conservative values of each parameter contributing to the final answer. The goal - developing a minimum value that possesses little risk.

What it amounts to is that a person may arbitrarily choose a combination of values that suits his goals, biases, prejudices and needs. Too often one sees evaluation reports that are more nearly devised to justify a pre-conceived answer than to arrive at a reasonable answer.

This is not necessarily unethical. There is a risk in each number used. Professional expertise can usually determine the likely range of values for each parameter. The choice of any

6

number within this range is arbitrary. The average number in the range might be "comfortable" but not necessarily meaningful. A man standing with one foot in boiling water and one foot in ice water could be considered comfortable - on the average.

There are several factors which dictate using more formal methods for the assessment of risk.

1. The higher risk - higher cost - longer payout prospects being developed in the search for "big" reserves.

2. The tax rulings and balance-of-payments limitations which minimize the rate of internal capital return and the capability to finance new investments from internally generated capital.

3. The fewer prospects among which risk may be spread.

4. The growing percentage of drilling prospects which are on lands controlled by some governmental authority (both foreign and domestic).

There are several types of risk involved - (1) the risk in the numbers available; (2) the inherent time risks in the uncontrollable future that possess some degree of predictability (supply, demand, economic trends, technology changes, etc.); (3) the unpredictable future events (wars, political changes); (4) the risk in the always fallible human judgment.

Prior to about 1960 these risks could be accommodated by the simple exercise of good judgment. The number of prospects drilled and produced was large; the payout was fairly short (the high risk period); and the political climate was reasonably stable. The capital required for any one venture was usually small compared to total available capital and current cash flow. Maintenance of liquidity was an important but not necessarily critical concern. In this case the probability of success was favorable even though the laws controlling such probability were not formally applied.

This is no longer the case. The "rules of the game" have changed. The bulk of the reserves being found and developed are in areas removed from the market place and technology centers. The front end load to obtain development rights are large. This is followed by equally large costs to move and maintain men and materials. The capital costs for moving raw hydrocarbons to market introduces even further capital requirements. The results - fewer but higher cost risk ventures. The number of ventures compared to available capital is even smaller.

Risks (1) and (2) lend themselves to formal risk appraisal using available approaches. It is now possible for the evaluator to array his information in such a way that the decision maker may exercise his judgment in a more meaningful manner as he considers all risk factors. When properly used, formal risk assessment does not supplant judgment; it supplements it.

Time value of money concepts - as they are usually applied - do not assess risk in a formal manner. They serve primarily to assess the relative merit of alternatives. They may be used to incorporate the relative risk of future time into judgment considerations. But, in the final analysis, they really only reflect the cost of using money. Those who attempt to read much more than this into time value of money concepts do so erroneously.

There is no way that arithmetic will, or can, be used formally to predict risk categories (3) and (4). These are judgment areas for the decision maker.

FACTORS INFLUENCING THE VALUATION

The number and relative importance of the factors involved in any specific valuation must necessarily depend on the good judgment of the evaluator. In all cases these factors may be subdivided into two general classifications - engineering characteristics and economics. Nine factors that must be considered to some degree in any evaluation are discussed below:

General economics. The value of an income-producing property must certainly depend on the general health of our economic structure. The demand for commodities derived from the product depends on the relative prosperity according to the laws of supply and demand. This, in turn, controls prices and the relative value of the property. Mineral production offers no exception to these principles.

Marketing factors. These are a part of the general market picture that deals specifically with local marketing conditions existing under a given over-all economic picture. The local market could be poor despite a generally high level of prosperity and a generally heavy demand for mineral products. This localized situation might be largely dependent on transportation facilities, refining facilities and their distance from production, steady demand, proration, and competition with other minerals. This is one of the most important factors, for minerals in the ground that cannot be sold profitably possess no immediate value.

Role of property in company organization. Where an evaluation is being made for potential purchase or development of a property, it is necessary to consider its relationship to the overall company structure. An independent operator whose organization is geared primarily for drilling will generally handle complex production techniques less efficiently than one who has well-organized, experienced production personnel; and as a result, a property which requires extensive development will have less value for him. Therefore, we see that the value of a property might conceivably depend upon who is to operate it. Although great strides have been made in recent years, it is still true that the value of many leases could be improved by a revision of producing methods.

Probably the best way to evaluate this factor is to study the past history of the operator in order to determine whether available producing data reflect the true economic potential of the mineral deposit. Does he have experience personnel, maintain proper cost control on producing operations, and fit producing equipment to the needs of the individual lease? The final result of this study usually results in little more than a general opinion since there is no practical scale against which one may rate the findings. However, this opinion can, and should, be a factor in reaching a final value. Sale of a mineral property is often successful only because the buyer feels that he can do a better job of operation.

Drilling, development, and production costs. It is inherent that the investment necessary to derive a profit must be a consideration. After these are determined it might be found that a deal is too large, or too small, to fit into the company program even though it is economically favorable. These costs must also be balanced against the risks involved, for the greater the risk the less one would care to risk a large portion of his assets.

Value of money. The real value of a given amount of money varies with its tax position, the time necessary for its return, the cost of using it, its current and future purchasing power, and its availability; the "real value" of money to different parties varies with their circumstances.

Of these, taxation is probably the most important factor. Few, if any, current financial transactions are consummated without due examination of pertinent tax aspects. The growth of taxation has made it firmly entwined in all economic affairs and has raised tremendous moral and economic questions that are difficult, if not impossible, to resolve. Regardless of these issues, suppose it is true that if a corporation has $100,000 in yet untaxed money available for investment, it would pay $48,000 in taxes it it were not spent. Therefore, the money has a real value to him of only 52.0 cents on the dollar for this is the only real loss that would be personally sustained, the government sustaining the remainder. Obviously, this can be a dominant factor in the handling of money. The important thing to the investor is the real value returned versus the real value of the money invested.

In some instances the value of money is not involved, this being left to the recipient of the report. This is particularly true where the property is only part of a variety of interests about which the evaluation engineer has no knowledge, or at the request of the employer.

Fair market value. In the sale and exchange of property the results of a study must always be conditioned by current market practices. Any finding that is not generally consistent with this structure usually has little chance of acceptance unless strong evidence is presented. Even then the odds against success are high because it is impossible to "prove" the results obtained.

Because of all the extenuating circumstances, fair market value may best be described as that price at which the seller is willing to sell and can find a willing buyer, both being in the same tax bracket. The only exceptions to this are the somewhat arbitrary values sometimes fixed by the courts in condemnation proceedings and similar situations.

Reserves. Value must also depend on the amount of reserves available for collateral. This phase of the valuation, together with rate of production, embodies the largest amount of true engineering effort. Inasmuch as there is no positive way truly to measure them, the findings must stem from the rigorous application of established and proven engineering methods. This involves not only choice of method or methods among several available ones, but also checking the results to see that they are consistent with other data. For these reasons the accuracy of reserve figures is in direct proportion to the amount of data available.

Reserves are often shown as producing and nonproducing. The former encompass all zones open to production, while the latter include all other zones that will probably produce. Producing reserves are more positive because there are inherently more production and test data available to confirm the figures.

Nonproducing reserves may be further classified as "possible" or "probable," depending on the amount of data available. A zone on which logs, core analyses, geological sample descriptions, and drill stem tests are available can certainly be evaluated with greater accuracy than ones (usually older wells) where the only data are notations by the driller that some oil show was obtained, plus the drilling time log. These represent the extremes, but there are far too many old wells like the latter.

In the petroleum industry it is often necessary then to give little credit for a portion of the reserves behind the pipe where the supporting evidence is too meager. Rather, they are listed only as plus factors and noted as such in the report. Such reserves are often the main sources of disagreement between two valuations, since it is difficult to decide exactly what constitutes positive evidence of reserves. The owner usually desires that more credit be given to their value; whereas a buyer, lending agency, or other prospective investor will tend to discount them. Often the only solution is to assign a sliding arbitrary scale of values to various classes of reserves.

One can apply these same criteria to coal, oil shale and other energy minerals. One can sample deposits by core drilling and obtain a proven (documented) picture of in-place reserves. Said data can be extrapolated to include possible extensions of these reserves based on interpretive conclusions of the evaluator. They can be called "unproven," "possible," or whatever qualifier is deemed appropriate.

It must be emphasized that a reserve number is always qualified. When proposing a reserve it is subject to the phrase "based on current economic factors and current technology." A reserve of no commercial value today may develop value with a future price change or a technological advancement affecting economics. Over two-thirds of the oil discovered to date is uneconomical to produce as yet and is not listed as a commercial reserve. It is inevitable, however, that some combination of price and technology will ultimately add this to our proven reserves stockpile at some time in the future. We possess the technology to recover about one-half of this found but unproven reserve. Only price prevents its development.

The same situation applies for all minerals. Marginal and/or high sulfur coal deposits which are non-commercial by present day criteria will become commercial at some point in time.

Time value of money is also a factor in reserves numbers. It is popular to publish (and lament) that we only have x years reserves of say natural gas. The implication is that at the end of this time it will be all gone. Such a conclusion is utterly ridiculous. Reserves are an inventory the size of which is governed by economics. What is the point of investing money in an inventory that is not needed in a short enough period of time to provide a satisfactory return on the investment? The answer - none.

Stated simply, the inventory of any commodity or any mineral is determined by economics. If a survey shows a six months supply of refrigerators, does that mean none will be vailable thereafter? Of course not. More will be produced if it can be done profitably. The same is true of minerals at this point in time. Of course there are some differences ultimately since minerals are finite. But as yet we have not reached a point where that is significant to the point being made.

Rate of production. The rate of production governs the flow of profits and is therefore important. The problem is to estimate the future production potential of the reserve and then modify it in accordance with any expected regulation policy. The former may be accomplished rather accurately in the presence of ample data, but the latter is best handled with extreme care. Regulation is governed by political and economic factors that are difficult to predict. Consequently, the rate of production is usually assumed at some conservative figure that is based on allowables maintained during the past few years, if such years were considered normal ones. Production data during abnormal years in which major wars or world crises occurred would not fit this criterion.

The length of time necessary to recover the reserves and/or the orignial investment is important, since it costs money to use money, and the money returned will have a purchasing power different from that invested. It is necessary to compensate for each of these factors.

A study of economic history shows a continual inflation so that in order to be realistic, the future money should be adjusted to reflect the value of the money invested. It is often this sort of factor that makes payout time so important.

Salvage value. In the early stages of mineral production history the salvage value of capital equipment was negligible. Only prospects with large potentials were completed and the amount of equipment used was small. At present this is not so. Many properties are developed that have a relatively short life and/or require extensive lease-processing equipment to achieve maximum return. Salvage values of highly capital intensive mineral systems significantly alter economic values. In any event, salvage revenue becomes significant by reducing the amount of risk capital invested and should be included, after proper modification, in a valuation to account for the difference between present value and the value at the time of realization.

TOOLS OF THE EVALUATION

There is often a "feast or famine" aspect to the use of evaluation tools. The "famine" end of the scale is represented by the case where little factual data, of doubtful quality, are available; both the quantity and quality are insufficient for any sophisticated calculation. All one can do is inject judgment numbers based on experience and analogy. The net result is little more than a projection assuming that the entity being evaluated is typical for similar reservoirs in the area.

The "feast" end of the scale is encountered for that too seldom case where a sufficient amount of geological, formation and production test data are available to supply all of the parameters needed for applying more complex analytical procedures. Most cases fall somewhere in-between these extremes. In essence, the evaluation tools used must be compatible with the quality and quantity of data available.

There is always a tendency to over-compute our real knowledge. But ... what real precision results when we must superimpose ten judgment numbers on top of two or three fairly solid facts merely to allow our computer to print out an avalanche of numbers with many "significant" figures? The net result is meaningful only if we regard the results as an answer, not the answer.

We also tend to stress those numbers we know how to calculate - which might be the least important ones determining value. One of the important uses of a computer in evaluation is sensitivity analysis; the determination about which variables are the most critical to the establishment of value.

Never reduce evaluation to a series of purely number games. It may be fun (and maybe even impressive) but one might as well play Gusher or some similar game with play money. The results will be as meaningful in many instances.

NEED FOR TOTAL SYSTEM EVALUATION

Historically the evaluation of mineral properties has been based primarily on the value of a raw material produced at the surface. Little or no processing took place at the production site except basic separation and possibly some conditioning to meet purchase specifications. The ultimate refining and marketing operations were evaluated separately. This is no longer the situation necessarily. Consider the following examples

1. Offshore oil and gas production

2. Mineral production in less developed countries

3. Minerals produced as a feedstock for hydrocarbon fluids.

All are cases where the cost of exploration, exploitation and production may be small compared to cost of support services and/or the transportation, handling and finishing to produce a marketable product.

Consider first offshore production. One cannot use a standard reservoir type calculation alone to make the evaluation. How are you going to get the oil to shore? How do you handle it then? Will it go to an existing refinery without affecting capacity or is a modification of capacity, or even a new refinery needed? There are several alternative answers to each question. For each answer there is a corresponding money implication.

Natural gas is produced with the oil. Is there enough to justify a pipeline to shore. If not, what is its value for re-injection or as fuel on the platform since there will be no direct cash realization.

In other words, this is a capital intensive situation where all elements to final marketing must be considered. Oil on a platform is worth nothing unless someone will take it off your hands. Most offshore operations are inherently more attractive to integrated companies who can control the sequence of activities between exploration and marketing. This means that all costs and revenues from a mineral development project involve the same treasury. On a total project basis the costs include

- Exploration
- Platform and/or sub-sea facilities
- Pipeline or tanker transportation
- Land facilities at the end of the pipeline
- Additional refining cost (if any)
- Facilities for employees (in a less developed area)

The development cost of a large field in the North Sea has been as high as 5 billion dollars to date. Costs of 1-2 billion dollars are not uncommon. The actual cost of drilling and completion of wells, plus production equipment, is only a part of this. For one production complex the company had to build a pipeline to the Orkney Islands, a terminal there, and then transship to a new refinery built to handle the oil produced.

What must you do? Analyze the entire system to see if the reserve is large enough to serve as collateral and if the production rate will be high enough and long enough to provide satisfactory cash flow.

With these restrictions economically there are reservoirs containing several hundred million barrels of oil (proven) that are not commercial under present conditions. Such reservoirs might as well be "dry" for present purposes. They of course have a strategic value for the future.

Consider another example. You have found a large gas reserve in the jungle of a less developed country. There are no local roads, no developed towns or support services and there is a trivial local demand for natural gas. Is the reserve commercial? In addition to a reservoir study one must examine the processing, transportation and marketing situation. There are three basic alternatives; liquify the gas (LNG), produce methanol and/or construct some sort of petro-chemcial complex requiring a gas feedstock. In the short-term the reservoir is worthless without one of these. The choice among them will be made on the basis of marketing demand and price forecasts, capital requirements of the process alternatives, cash flow and profit projections, etc., etc. This includes tanker charges, receiving terminals for the product and a whole host of factors. Whatever the uncertainties of the reservoir study, on a relative basis it may be the most accurate of the various studies involved.

A similar problem is involved with oil shale or coal being produced for the manufacture of "synthetic" oil or gas. The plant for processing the ore is dependent on the size and quality of the raw material in the warehouse justifying that plant. These various elements may be evaluated separately but the final economic decision must look at the total sequential system.

In the interest of brevity we will forego the temptation to illustrate the principle by citing some real world examples. It should suffice to say that failure to look at the effect of the total system on the value of any element demonstrates one's incompetence.

SUMMARY

The remainder of this book discusses many of the individual factors affecting mineral property economics. In most cases each chapter could be expanded to a full book. Coverage detail is in some proportion to depth of knowledge needed by the minerals professionals to whom this book is directed. In many areas an overview only is provided. In such areas the minerals professional probably is not qualified to perform the primary tasks; they will be done by specialists. But, the recipients of this work must possess enough knowledge so that they can intelligently review and interpret the work of other specialists.

In other subject areas sufficient details are presented to enable the reader to perform the basic computations. Many of these will be on a computer in day-to-day activities. The information presented serves to illustrate the principles, and the limits for applications, of the subject method.

One may conclude that the evaluator must be an expert in the fields of reservoir engineering; process engineering, geology, economics, accounting, taxation, business management, and law. Though such an ideal combination seems unlikely to appear, one can at least become competent in the portions of these fields that have a direct relationship to evaluation and use these skills intelligently. Evaluation might be likened to cooking. Anyone can follow a recipe, but a good cook, like a good evaluator, must go beyond just that and exercise good judgment.

2

ECONOMIC EVALUATION NOMENCLATURE

The purpose of this chapter is to define briefly some evaluation terms in order to clarify discussions in ensuing chapters.

It should be borne in mind that many terms, particularly legal ones, might be defined in many ways, depending on the circumstances which vary between states and between countries. In the interest of brevity the definitions given are limited to the normal manner in which they would be applied to evaluations.

An understanding of such terms is essential in preparing an evaluation. This is particularly true where the interest of the party or parties being evaluated is subject to some liability or assigned interest. The gross value of this interest must, therefore, be corrected for such liability or assigned interest to arrive at a net value.

An essential part of any evaluation is the review of all agreements entered into prior to the time of the report. This is often tedious, involving detailed examination of legal documents and development of patterns in transactions which are sometimes complex. It normally involves determination of the original division of interest, subsequent assignments, and any current liabilities. It is a good practice not only to refer to the original documents but also to discuss them with the proper people. Sometimes assignments are not currently in the files for one reason or another. On the other hand verbal information is often faulty and summary sheets prepared on leases are subject to error. Cross-checking all sources minimizes error.

In some cases the evaluation is concerned with the value of the mineral alone, without regard to outstanding loans and other obligations. In such cases the report should contain a concise statement to this effect in order to avoid misinterpretation of the results. This is advisable even though the report is designed only for intra-company use, because one never knows what ultimate use the report might have. Such statement is a protective clause to prevent possible future embarrassment to the writer.

Not included are many engineering, economics, and tax terms whose definitions will become apparent during ensuing discussions.

Capital Assets. - Capital assets are the dollar value of assets that are tangible and which may be appraised by inspection. This includes buildings, machinery and equipment of all types which depreciate in value with age and usage. Also included in this category are negotiable instruments such as stocks, bonds and any unencumbered cash.

Intangible Assets. - Intangible assets are the class of assets that includes all types of minerals. No value may be established by direct inspection and the asset does not necessarily depreciate with time. It loses value only when produced and furthermore cannot be replaced.

The exact differentiation between a tangible (capital) asset and an intangible asset is somewhat arbitrary. From a practical standpoint, such differentiation depends on the current applicable tax provisions governing a specific type of property. Evaluations should not rely solely on past evaluations since the laws regarding intangible assets change constantly.

Depletion. - In the gneral sense, depletion describes the production of a wasting asset which, in turn, reduces in value as the supply decreases. This, then, is both physical and economic depletion. However, the tax aspects of depletion are those usually considered. For many tax purposes the minerals must not only be produced but also sold to be covered under depletion provisions.

The word depletion is often used as an abbreviation for "depletion allowance," i.e., that deduction which may be lawfully taken under provisions of the income tax statutes.

Ownership in Fee. - An owner in fee owns both surface and minerals under the property covered by a lease.

Surface Owner. - Ownership of the surface rights does not necessarily give a party any interest in the minerals unless this is specified. In the absence of such mineral rights the surface owner is not a party to the lease agreement. However, he is seriously affected by it for a portion of the land has to be appropriated for access roads and drilling locations.

The lease specifically gives the lessee the right to build the facilities necessary for drilling and production, but furthermore requires payment of damages.

In the absence of specific contract provisions the surface owner's only recourse is arbitration. Although he has no specific contract protection, history shows that he has fared quite well as a result both of oil company public relations and his nuisance value. Once a well has started the poorest clay soil supporting only scrub oak and weeds will immediately be capable of producing record crops. No livestock inadvertently killed by oil field equipment is ever anything but prime beef due to win a blue ribbon at the next exhibition.

There are exceptions undoubtedly, but one may safely conclude that most surface owners are amply compensated for damages and relatively few claims reach the courts. Except where land is very cheap, the operator can seldom justify purchasing or leasing surface rights.

Mineral Rights. - Mineral ownership legally may be separate from surface ownership except where otherwise provided. The mineral owner therefore has the right to recover said minerals from the premises.

Lease. - An oil lease is in reality not a lease in the general sense of the term. It is rather the delegation of exclusive rights to "capture" such minerals from the mineral owner (lessor) to a lessee. In return the lessee gets to keep and sell the greater part of these minerals as compensation for his efforts. The remainder, or royalty, belongs to the mineral owner - that portion to be recovered for him at no cost, by the lessee.

In other words, the entire minerals are recovered by the lessee, but a portion of the sale proceeds must be paid to the lessor without charge. In some instances the lease contract, however, does call for a small processing charge to be paid by lessor. For example, he might bear a portion of the cost of processing gas so that it can be sold.

The courts have ruled that the lessor retains title to his share at all times even though the operator handles it. This might not seem important at first glance as compared to the possible interpretation that the operator owns it all and just pays a share. However, the former interpretation has been held necessary in order that the royalty owner may have a legal right to be assured that the minerals are produced in an efficient manner. It is on this basis, therefore, that a royalty owner may question well spacing, production, and other practices.

Most leases give the right to capture all minerals below the surface at any depth, but in some areas the depth is specified. One lease might be signed for a depth to 10,000 feet and another for beyond 10,000 feet.

The common lease form provides for the term (usually one to five years) and often calls for a well to be started within a stated period of time. Once the well has begun to produce, the lease agreement continues in force until it is abandoned. If there is no production, the lease agreement expires at the stated time.

A lease bonus is usually paid at the time of the signing and specified delay rentals are paid annually in lieu of production. In some instances, such as leases on Indian lands, rentals are continued even though production is obtained.

The lease also contains the usual forfeiture clause in case the lessee fails to comply with all terms. It also contains a surrender clause which permits the lessee to be relieved of all obligations.

Conditions are specified under which wells are required when certain offset production exists which might presumably drain the property in question. Most current leases protect the lessee from any difficulties that might develop due to proration or state regulations, or where marketing becomes uneconomical.

The lessor's share of the total minerals recovered normally varies from one-eighth to one-fourth, with the former being more common. Except in unusual cases, royalties greater than one-eighth are only paid for minerals owned by units of a government. When this higher rate is paid it sometimes is further based on net barrels after payment of operating costs.

Lease Delay Rental. - This is annual rental paid to the lessor by the lessee in lieu of production during the life of the lease. In some cases, such rentals are paid even after production is obtained. A common rental price is one dollar per acre per year.

Lease Bonus. - A lease bonus is cash payment to the potential lessor by a potential lessee as an inducement to give a lease on a property. Where the bonus is large it is sometimes paid over a period of years, in which case it is known as a deferred bonus.

Royalty. - Royalty is the interest of a party owning minerals in the ground where another party (the working interest) has gained the right to capture such minerals under a lease agreement.

16

Such royalty interest is normally free of all costs of capture except for special treating costs that might be specified in the lease or assignment. This term is often used as an abbreviation for the term land-owner's royalty. There are many specific forms of royalty.

Overriding Royalty. - This is an additional royalty created out of the working interest and having the same term as this interest. It is said to be carved out if such royalty is assigned free of all operating and development expenses. It is said to be reserved if the lessee assigns the working interest and retains only a fractional share. The latter is the more common and usually results from the activities of land broker who obtains leases for the sole purpose of ultimately assigning them to a third party for development purposes. This form of royalty may also become a part of a farmout agreement.

Term Royalty. - This is a specific form of land-owner's royalty which is applicable only for a specified length of time. In most transactions where the holder of land-owner's royalty assigns all or part of his interest to another for a cash payment, a term is usually specified. If no minerals are produced in the meantime it reverts to the assignor.

Minimum Royalty. - This form occurs only where the rate paid depends on quality, quantity, and price of oil. It is the minimum payment under such an agreement.

Offset Royalty (Oil and Gas). - This is a misnomer and is really a payment from the lessee to the lessor in lieu of drilling an offset well. Said sum is theoretically equal to the revenue that would be received by the royalty owner from such an offset well if one is required in the lease agreement. It is in reality a form of compromise in those situations where the operator does not feel that said offset well would be profitable. Owing to its intangible nature the amount of such payment must be negotiated.

Barrel Royalty. - This results from a royalty assignment wherein the grantee gets not a certain percentage but a revenue equivalent to a certain number of barrels (or MMscf) per day.

Participating Royalty. - This is also a misnomer since it more nearly represents a quasi-portion of the working interest. It differs only in that the holder limits his liability by contributing only a fixed sum each month toward maintenance and operation. It is advantageous during flush production but often becomes a burden as production declines since the cost continues even though the revenue decreases. This arrangement is most prevalent in those cases where a promoter sells an interest in a well to a large number of small investors.

Assignment. - An assignment is that legal document whereby one party transfers all or a portion of his interest to another, subject to specific considerations.

Assignee (Grantee). - The assignee is the party to whom an assignment is made.

Assignor (Grantor). - The assignor is the party who makes the assignment.

Farm-out Agreement. - This is the name applied to a specific form of assignment wherein the lessee grants a conditional interest to a third party in consideration for the development of a prospect within a specified length of time on given acreage. It is usually undertaken where the lessee has leases on a relatively large block of acreage and does not wish to undertake the sole cost of developing it. In most instances this form of agreement is between a major company and the independent operator who cannot afford to acquire large acreage. Most farm-outs contain the following stipulations in oil and gas operations:

1. The grantee will drill a well in a diligent and workmanlike manner to a certain geological formation or to a specified depth, on a limited block of acreage, in a specified length of time.

2. Any geological or test information obtained from such drilling must be made available to the grantor.

3. If the well is a producer the grantee is assigned the well acreage subject to an overriding royalty to the grantor. After the return of a specified sum to the grantee, the override is usually discontinued and the grantor comes in for a share of the working interest. This last provision is sometimes optional to the grantor. On successful completion of the subject well the grantee is usually assigned all or a partial interest in certain undeveloped acreage in the vicinity of the well. Where the grantee is a drilling contractor, he is often given the contract to drill all future development wells on the given acreage under specified terms.

4. If the well is dry, the grantor retains all lease interest held prior to the agreement.

5. The grantor is free of all cost and liability incurred in drilling subject well.

This type of agreement is often an ideal solution to property development. It frees the independent of large and often prohibitive survey and leasing costs and yet permits use of his drilling equipment. On the other hand it is a means of developing and proving acreage by the lessee at relatively small cost. It is, therefore, a form of agreement that is often advantageous to both parties in that it allows each to perform that portion of the development most compatible with his current situation. It is sometimes used to develop acreage which the lessee cannot develop before expiration of the lease or where he feels that the chances of success are less favorable than elsewhere.

Farm-In Agreement. - This most often occurs after commercial production is found. The finder needs more capital to develop the property than is available. One or more partners are found to develop the mineral property. As assignment of interest is made in consideration of specific services provided.

Carried Interest. - A carried interest is an agreement between two or more partners in the working interest whereby one party (carried party) does not share in the working interest revenue until a certain amount of money has been recovered by the other party (carrying party). This type of arrangement may result when the carrying party advances all or part of the development costs of the carried party, or where land brokers use the device for a portion of their profits. It is also a provision in some farm-out agreements. Carried interest may apply to a single well or an entire lease.

Generally, the carried party does not participate in profits until the carrying party has recovered his investment or has been returned a specified amount of money. Some such agreements specify the amounts which may be subtracted as operating costs against the gross recovery. A land broker, for example, might come in for a one-fourth interest once the operator has recovered a specified amount or has received a pay-out on his investment.

In another form, A, the owner of the working interest may assign a portion of the interest to B in return for which B will pay all development, drilling, and equipment costs. B then recovers the grantor's share of the costs from the grantor's share of the revenue. This may be accomplished in three common ways:

1. A may assign a fraction of the working interest to B, B to be given an <u>oil payment</u> covering B's investment to be paid from A's remaining interest.

2. A may assign his entire interest to B until the investment is returned, at which time A regains a portion of the assigned interest.

3. A may mortgage his interest to B as security for a loan for which he is not personally liable.

<u>Working Interest</u>. - The working interest is the total interest minus the royalty. For all practical purposes it is an interest in the oil and gas in place that is liable for the cost of developing and operating a property. It is formed by the granting of a lease by the owner of the mineral rights.

<u>Operating Interest</u>. - The operating interest is that portion of the working interest charged with operational responsibility of the lease. This interest handles all accounting, charging or remitting to each interest its pro-rata share of expenses and profits.

<u>Non-operating Interest</u>. - Non-operating interest is the portion of the working interest not charged with operational responsibility on the lease.

<u>Reversionary Interest</u>. - This is similar to a carried interest, differing only in the sense that the type of interest held by each party changes after a specified set of conditions have been met. A typical example might be found in a farm-out agreement. Lessee A retains a 1/16 of 7/8 overriding royalty until B recovers $300,000 from drilling a successful farmout. After this recovery, the override reverts to a one-half working interest, the override becoming null and void. Therefore, A is said to have a reversionary one-half interest in the property.

<u>Dry Hole Money</u>. - A dry hole contribution is money paid to support a given well by holders of direct offset or nearby acreage. This agreement calls for a well to go to a certain depth or to a certain formation. If, as a result of such drilling, the hole is dry, a specified sum is paid in return for geological and test information. If the well is completed as a producer, no contribution is made.

This is one method of spreading the risk in developing wildcat acreage, since data from a well will serve to help correlate available information on offset acreage.

<u>Bottom Hole Money</u>. - This is paid regardless of whether the well is dry or a producer. In all other respects it is similar to, and contributed for the same purpose as, dry hole money.

<u>Production (Oil) Payment</u>. - A production payment entitles the owner of the payment to a specified portion of the production for a limited time, or until a specified amount of money has

19

been received. It is almost like a royalty, except that it expires when a given amount of money has been received.

It usually arises as a means of paying back a loan from a bank or some other lending institution. It often calls for paying back the principal plus interest from a given percentage of production. This percentage will vary depending on the reserves, the credit rating of the borrower, and his efficiency of operation.

In some extreme cases, where necessary, the bank will take all revenue accruing to the borrower except for necessary operating costs. In some of these cases even operating costs are subject to review by the lending agency. This, however, is not the normal situation and is applied only where the loan is in jeopardy.

Where one assigns royalty for a specific period of time and when such term is not extended by production, the assignment is in reality an oil payment although it may be called by another name.

There are two basic types of production payments - the reserved (retained) type and the carved-out type. A reserved payment is created when the owner of an interest assigns all or part of his interest. In return he obtains an oil payment for a specified amount of money or a specified amount of reserves. A carved-out payment is one in which a party assigns only a payment of a specified sum of money. This money is to accrue to the grantee out of his share of the runs. However, none of his interest has been assigned, merely a specified amount of revenue from the interest. One type of carved-out payment is assigned for materials and services, and/or sold for cash with the stipulation that it be pledged toward development of the subject property. The other type is sold for cash with no restriction on its use.

Proved Property. - A proved property is one probably containing commercial quantities of oil. Because the presence of oil cannot be confirmed positively without drilling, the designation "proved" is an arbitrary one. Although available proof may be sufficient to justify drilling, it may be insufficient for use as collateral. Inasmuch as this is an arbitrary matter, it is necessary to devise some criteria to establish the class of the property. Generally, a property is considered proved if all geological and geophysical information is positive, nearby wells are commercial and apparently on the same structure, and there are no negative values to be considered, except the usual error in the interpretations above. It is likely, therefore, that the amount of evidence necessary for "proof" would probably be much less for a lease broker than for an engineer examining the same property for purchase.

In general, a property must be at least partially developed before it is considered proved. This enables one to establish general formation characteristics and serves as a check on existing geological and geophysical information.

Unproved Property. - Unproved property is possibly productive but, as the name implies, not considered proved. Since there are usually no nearby wells, the only indications of possible oil or gas are taken from regional geological studies and/or general geophysical surveys.

Development Well. - A development well is one drilled on property that is considered proved. Consequently, this designation is arbitrary to the same degree as the property classification.

Wildcat Well. - A wildcat well is one drilled on unproved property. A well is sometimes designated a semi-wildcat for those in-between cases where property classification is uncertain.

Reserves. - Reserves are that quantity of oil or gas that should be produced by the methods outlined in the report under the conditions specified. For oil, the standard unit is either cubic meters or the API barrel (42 U.S. gallons) measured at atmospheric temperature and pressure. For gas the unit is either standard cubic feet or standard cubic meters, as measured at a given base temperature and pressure.

Any reserve figure shown will depend on the evaluator's interpretation of the available data. Because of the inherent leeway in such interpretations the reserve figures for a given reservoir will normally vary at least slightly with the evaluator.

Reserves are generally classified as: (1) proved drilled reserves, (2) proved undrilled reserves, and (3) unproved reserves. They are given in the order of the accuracy with which they may be predicted. Reserves types (1) and (2) are predictable by ordinary reservoir engineering methods, whereas type (3) depends largely on overall regional geological and formation characteristics. General reserve figures of the type published by the API include all reserves other than those:

(1) in unproved portions of partly developed fields

(2) in untested prospects

(3) present in unknown prospects in regions believed to be generally favorable.

(4) that may become available by fluid injection methods from fields where such methods have not yet been applied

(5) that may result from chemical processing of natural gas

(6) that can be made from coal, shale or other substitute sources.

Primary Recovery. - The primary recovery includes the oil and gas recoverable only through the proper utilization of natural reservoir energies. This includes gas cap drive, depletion drive, water drive, or any combination of these.

Secondary Recovery. - Secondary recovery is the recovery of oil and gas made possible by artificially supplying energy to the reservoir. This includes water and miscible flooding techniques, repressuring operations, gas drives, etc.

Rateable Take. - Rateable take refers to the orders sometimes issued by regulatory bodies requiring common carriers (usually pipelines) to take products from all customers in proportion to the number of wells served or the capacity of those wells. It is normally instituted where total well capacity exceeds pipeline capacity to insure that all sellers are able to market their fair share of the oil. It is at best a controversial issue between producers and marketers where used.

Proration. - This refers to the artificial control of producing rate imposed by regulatory bodies in an attempt to enforce good conservation practices and to hold production within the limits of the current market. The methods used vary, none of them being entirely satisfactory to all parties.

Under such a program the allowable production is controlled monthly, either on a per well or a per lease basis. One method is to limit production to a specified amount per day for the full month, whereas another is to let the well produce its tested potential for a given number of days per month.

With the first method; i.e., a per lease basis, it is theoretically possible for one well to produce the entire lease allowable. Where a per well allowable is used, it is difficult, if not impossible, to enforce the order where common tank batteries are used. This fact often makes it difficult to determine individual well production characteristics for evaluation purposes, in the absence of individual well tests.

The allowable is normally established on the basis of market demand, formation depth, the specific producing formation, producing characteristics and any special field rules. For example, when oil production is accompanied by a high gas-oil ratio it may result in lowering the allowable if the ratio is above that established as desirable. The only exception is where such flare gas is properly utilized. Most states make periodic checks for this purpose.

Gas production is normally curtailed only by market availability and the ability of the well to produce as determined by open flow potential tests or the equivalent.

Maximum Efficient Rate (MER). - An MER is a theoretical number that presumably will optimize ultimate recovery and/or economics. It is normally expressed as volume of production per day, per well or per lease (or field) unit.

Such numbers possess some value for general guidance purposes. They are sometimes used (erroneously) as a standard by which the efficiency of a given operation is judged.

Division Order. - A division order is the instructions, signed by all interests, to the oil or gas purchaser showing how the purchase price is to be divided. Normally, all money due the working interest is paid to the operating interest who, in turn, apportions it in accordance with the interests held. All royalty is normally paid directly by the purchaser, although this is not always the case.

Pipeline Proration. - This is a limit on the amount of oil purchased by a line that results in oil production less than that available. It is usually a temporary measure brought on by temporary loss of refining capacity, storage problems or fluctuations in market. This is often a significant factor in past production history or future value of a lease and, as such, should always be investigated.

Turnkey Well. - A turnkey well is one in which the drilling contractor drills, completes, and equips a well under the same contract. This is opposed to the standard drilling contract which usually calls for so much per foot and/or a daily fee, the operator assuming responsibility for completion and equipment.

Payout. - This is a somewhat ambiguous term that is used to denote the length of time necessary for recovery of the original investment. It may refer to the gross payout based only on gross revenue before operating costs and taxes, net revenue before taxes, or net revenue after taxes. When net revenue is used in an agreement it is necessary to specify a means of allocating costs and the bookkeeping methods to be applied.

Lien. - A lien is a legal device used by creditors for the nonpayment of either labor or equipment. It prevents removal of equipment from a lease under court jurisdiction. In preparing a valuation the presence or absence of such liens must always be determined.

Unitization. - Unitization is the process whereby the owners of adjoining properties pool their reserves and form a single unit for the operation of the properties. The revenue from operation of this unit is then divided in accordance with the basis established in the unit agreement.

The purpose of such an agreement is to produce the reserves more efficiently and thus to increase the profit to every participant. It is particularly important where secondary recovery is anticipated.

PROVED RESERVES

Since what is "proved" is somewhat arbitrary, both the Society of Petroleum Engineers and the American Petroleum Institute have been concerned about a more specific definition, particlarly for consistent use of the terms in evaluation.

In June, 1965, the SPE Board of Directors adopted the definition reproduced below published in Journal of Petr. Techn., July, 1965, p. 815). API reserves definitions have been modified to achieve substantial agreement with the SPE version.

Society of Petroleum Engineers of AIME
Definitions of Proved Reserves for Property Evaluation

Proved Reserves – The quantities of crude oil, natural gas and natural gas liquids which geological and engineering data demonstrate with reasonable certainty to be recoverable in the future from known oil and gas reservoirs under existing economic and operating conditions. They represent strictly technical judgments, and are not knowingly influenced by attitudes of conservatism or optimism.

Undrilled Acreage – Both drilled and undrilled acreage of proved reservoirs are considered in the estimates of the proved reserves. The proved reserves of the undrilled acreage are limited to those drilling units immediately adjacent to the developed areas, which are virtually certain of productive development, except where the geological information on the producing formations insures continuity across other undrilled acreage.

Fluid Injection – Additional reserves to be obtained through the application of fluid injection or other improved recovery techniques for supplementing the natural forces and mechanisms of primary recovery are included as "proved" only after testing by a pilot project or after the operation of an installed program has confirmed that increased recovery will be achieved.

When evaluating an individual property in an existing oil or gas field, the proved reserves within the framework of the above definition are those quantities indicated to be recoverable commercially from the subject property at current prices and costs, under existing regulatory practices, and with conventional methods and equipment. Depending on their development or producing status, these proved reserves are further subdivided into:

1. <u>Proved Developed Reserves</u> - Proved reserves to be recovered through existing wells and with existing facilities;

   a. <u>Proved Developed Producing Reserves</u> - Proved developed reserves to be produced from completion interval(s) open to production in existing wells;

   b. <u>Proved Developed Nonproducing Reserves</u> - Proved developed reserves behind the casing of existing wells or at minor depths below the present bottom of such wells which are expected to be produced through these wells in the predictable future. The development cost of such reserves should be relatively small compared to the cost of a new well.

2. <u>Proved Undeveloped Reserves</u> - Proved reserves to be recovered from new wells on undrilled acreage or from existing wells requiring a relatively major expenditure for recompletion or new facilities for fluid injection.

SUMMARY

The question of terminology is an important one. In any evaluation report or summary it is important that the reader place the same meaning on key words the writer intended. This may be particularly critical in international operations.

A phrase like "proved reserves" is a good case in point. It means different things to different people, in different circumstances. One could refer to the API-SPE standards. In some circumstances, the definition intended should be included in a footnote. Where a translation between languages is possible, a Glossary of terms might be desirable as an Addendum.

3

GENERAL ECONOMIC CONSIDERATIONS

An evaluation is made normally on a single unit, or a contiguous set of units, of the overall organization. But ... true value of a property never may be established without some consideration of the overall organization's needs, unique characteristics and the environment in which it operates - now and in the foreseeable future.

As a practical matter the total evaluation of a mineral investment involves three separate functions

1. The geological and engineering calculations on the mineral deposit and the preliminary values associated with these results.

2. The economic assessment of the effects of changes in worldwide supply - demand patterns that alter historical economic values typically used in engineering calculations.

3. The general management considerations which are necessary to select among all possible investments, the combination of investments that satisfy the organizational goals of the company.

Completion of these three functions of the evaluation process requires at least two separate groups, and preferably three, in the organizational structure. Persons making calculations or estimates in each phase seldom possess sufficient facts (or authority) to perform the other functions. In fact, it is seldom desirable that he consider the other functions in explicit detail.

In spite of the need for independent calculations by each function, a general awareness of the techniques, with their strengths and weaknesses, employed in economic and engineering analysis is a necessary requisite for sound mineral investment selection. One purpose of this chapter is to provide an overview of function 2, discussed in more detail in later chapters.

Being aware of the way in which calculations will be used in each phase of the mineral evaluation process is important primarily because functions 1-3 are not totally independent of each other. Geological and engineering calculations are affected by economics, and vice versa. Selecting a mineral investment to meet corporate goals (function 3) likewise depend upon engineering and economic calculations, and, in turn, alter future calculations of engineering and economic parameters.

It is important to remember that any intelligent person can master specific concepts in the geological and engineering, economic, and management functions, given enough time. Time is the important ingredient in selecting the optimal mineral investment. Management oftentimes is well-trained in the technical components of engineering. Many years away from the operating sphere, however, tend to diminish their knowledge of current operating techniques, thereby limiting their understanding of engineering and economic reports from operating divisions.

Awareness of the basis for engineering, economic, and management evaluations is, perhaps, the fundamental ingredient in sound mineral investment planning.

GENERAL STRUCTURE OF MINERAL INDUSTRIES

The basic organizational chart for a mineral company is the same superficially as that for any company of comparable size. Operational and staff functions possess the same general hierarchial characteristics. In reality, though, all mineral extraction, processing and finished product marketers are subject to economic constraints that differ markedly from those organizations primarily concerned with manufacturing.

Today most major companies are organized along functional lines - exploration, production, processing, manufacturing, marketing, etc. Each of these fuctions is then sub-divided by geographical area.

All major companies are international in scope (directly or indirectly). This requires operational functions to meet the unique character of each area plus more centralized support services. Effective communication is a problem. It must span distance as well as the cultural differences that characterize different areas of the world. The financial planner in a central location finds it difficult to comprehend many unique problems of a remote (to him) operation, and vice versa.

The form of the mineral evaluation and the philosophy with which it is written must bridge this gap. One may more often criticize this aspect of evaluation reports than their technical content.[1]

ECONOMIC CONSIDERATIONS

Mineral companies are in the "warehousing" business. They expend money to stock their warehouse (find reserves) which will be withdrawn, finished and sold over a long period of time. The problem is to maintain the warehouse inventory at a proper level - large enough to insure a proper future supply at a cost commensurate with current availability of capital. Setting this level and then finding ways to maintain it is possibly the primary problem facing mineral companies.

---

[1]See "Effective Communications for the Technical Man" - Campbell and Farrar. Part of the Campbell Petroleum Series.

It is necessary to distinguish between "reserves" and "ultimate recovery." Reserves is an economic number reflecting current economics and technology, and the size of warehouse one can afford to develop. It bears no discernible relationship to the amount of any mineral available to be recovered ultimately.

One often hears statements like "we only have 14 years oil reserves so we are in trouble." Even if the 14 years is correct the conclusion is sheer nonsense. As stated in an earlier chapter, this is the maximum amount of warehouse size with current constraints. In fact, the reserve size tends to be rather stable. Of course, the world will run out of oil eventually. The mistake is using a current reserve number as an important factor in ascertaining when and at what rate.

Unless the reserves are replenished at the same rate they are used up, a company is in the process of going out of business - liquidating its reserve assets. A company might show a good profit on successive years and yet be in the process of such liquidation. Maintenance of reserves by successful exploration or by contract is thus an integral part of long-range strategy.

Ownership of reserves is the safest of the alternatives. But, this involves several problems. The first is maintenance of "liquidity" - sufficient cash flow-back into the company account to pay current obligations. This has become more difficult as larger investments are required in projects which have a longer payout (time to recover investment).

As long as readily available, short payout, reserves requiring low investment were available, internally generated funds were often sufficient to provide capital needs without disturbing liquidity. Payout time was short because the time interval was short between investment and substantial cash returns. This circumstance had begun to change by the middle 1950's and virtually ceased to exist by the middle 1960's.

Today, and in the foreseeable future, the mineral industries are faced with even higher investments, longer payouts and the larger risk that time itself promulgates. Not the least of the problems is the fact that the bulk of the large reserves are being found in countries other than where the products are marketed. This will be covered in more detail in later discussions.

Maintenance of liquidity has required the adoption of new strategies. Joint ventures have been instituted to minimize capital outlay. This is seldom better than an economic compromise. Each company in a venture pssesses slightly different economic characteristics. Most joint ventures represent a choice from among less than perfect alternatives.

In addition, external financing is required. Stock issues, debentures, bonds, straight loans, exploration (venture) funds and the like are all being employed. Venture funds are being used by some major companies although exploration funds have primarily been employed by those with a more limited capital structure. These venture funds take many forms but all are an investment in specific ventures, unsecured by the company's total assets. In funding a new exploration program by these means, management is gambling that the assets procured will more than compensate for this dilution.

The need for external financing is not bad. It is merely a change from the historical operating strategy. Assessment of risk and uncertainty - in all its forms - requires a modified decision strategy. This is more difficult to do in practice than in principle. From the evaluation viewpoint it changes the sensitivity of some of the variables involved.

As always, the geological and engineering considerations are a necessity. But .... they are now more nearly the beginning of the decision process rather than the primary part of it. The evaluation establishes a likely asset position and the rate at which the venture might return capital. This is merely the input into the hard decision process. How will the venture be financed? What effect will it have on liquidity and current asset positions (even if successful)? Is the potential return commensurate with the risk involved? What effect would failure of the venture have on company asset position or even solvency (gambler's ruin)? Does the venture "fit in" with long-range company goals? Etc., etc., etc.

As the investment requirement increases, the likelihood of all these questions being favorable decreases. Even if the evaluation results appear favorable, financial considerations may override this fact. Can we compromise the unfavorable aspects? Let us look at farm-outs, reversionary interests, joint ventures, etc.! After all of this, the prospect may be turned down even though it "passes" the conventional evaluation "yardsticks" being used. It is necessary to realize - for a lot of sound reasons - that most of the calculations in this book merely serve to "screen" those ventures worthy of further consideration from those which are not. This is my brief answer to the often heard lament, "I don't understand management turning down my project when it meets or exceeds standards they themselves established."

The above comments apply primarily to the evaluation role in exploration investments. However, those evaluations made to guide production operations, for asset review, sale, and exchange must likewise reflect their economic purpose.

Historical Behavior

Figure 3.1 is a brief summary of several economic yardsticks affecting the U.S. petroleum industry since 1950. The five items shown are of course not the only factors available but they serve our immediate purpose. In reviewing this figure remember that the vertical position of any one factor compared to another has no particular significance. It is the change in relative value with time that is important.

In preparing Figure 3.1 the year 1967 was arbitrarily taken as the reference year (index = 100). A value on the index scale lower or higher than 100 simply indicates a price or cost in that year when compared to the year 1967. The slope of any one line is a measure of the inflation or deflation in that time period.

Even upon casual inspection it is apparent that the lines are essentially parallel from 1950-1970. Possibly the two lines of most general interest are oil price and the consumer price index of the U.S. Bureau of Labor Statistics. These two curves "track" each other rather well until 1974. This is not too surprising since they are not independent variables in a free marketplace. The cost of energy has always been a factor in both agriculture and manufacturing. Furthermore, what the energy industries pay for goods and services must affect the selling price of its products.

From 1970-73 there was a noticeable increase in oil prices which once again tracks the general trend in prices. Thereafter, U.S. oil prices showed a sharp increase as a result of OPEC price actions. The wholesale price index also increased but at a slower rate. Was oil price a factor? Certainly, but it is erroneous to suppose that it was the sole cause. During this same

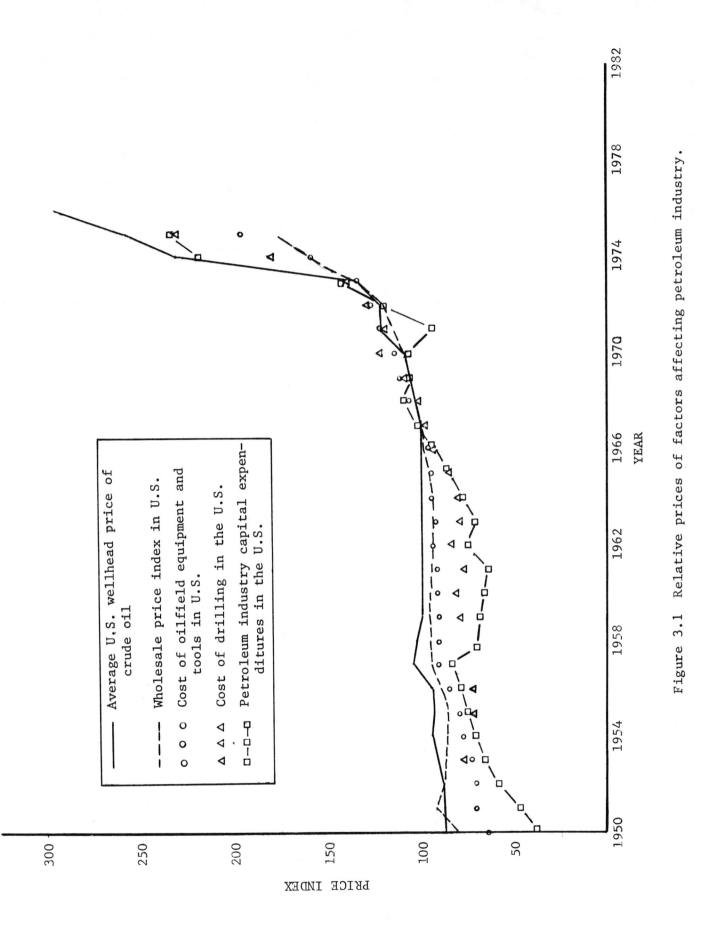

Figure 3.1 Relative prices of factors affecting petroleum industry.

PRICE INDEX

YEAR

Average U.S. wellhead price of crude oil

Wholesale price index in U.S.

o o o Cost of oilfield equipment and tools in U.S.

△ △ △ Cost of drilling in the U.S.

□–□–□ Petroleum industry capital expenditures in the U.S.

29

period the U.S. government also pursued policies that were inflationary in nature independent of oil prices. It is important not to draw over-simplistic conclusions from presentations like this. It is also important to note that the OPEC control of energy price imposes a largely uncontroll- able impact on economics. A cartel like OPEC cannot ignore supply-demand-economic considerations if it is to survive but within fairly broad limits it can take unilateral actions that adversely affect consumer nations. Although the data shown in Figure 3.1 are for the U.S. comparable curves could be developed for other countries or regions.

A most interesting curve is the trend in U.S. capital expenditures. There was a notice- able decline in the period 1957-1961. It was not until 1965 that the level of U.S. spending re- attained the 1957 level again. However, during this 1957-61 period worldwide expenditures (not shown) remained essentially constant. Why? Stated simply, the U.S. price was not attractive compared to price relative to foreign sources of oil. Available capital was diverted to the latter arena. Whereas prior to this time about one-third of the expenditures were outside the U.S., it jumped to about 44% in 1957 and about 50% in the following year. In 1971 foreign ex- penditures reached a peak, when they constituted two-thirds of total available capital outlays.

These data reconfirm a fairly obvious fact that the ability of a financial entity to in- vest is dependent on available capital. The manager of any asset has an obligation to use it for the maximum benefit of the owners. This includes picking the best investment from among the alternatives available. Thus the diversion of capital outside the U.S. was not necessarily arbitrary, it was the result of a U.S. economic climate that made investment there less attractive among the alternatives then available.

Since 1974 the combination of increased foreign taxes and royalties and substantial nationalization have curtailed the attractiveness of non-U.S. investments. In some areas profits on drilling and production have declined to as low as 10-15 U.S. cents per barrel. This is un- attractive for investment, basically. The major reasons for a company to stay is protection of source of supply.

There is one other conclusion in Figure 3.1 worthy of a brief mention. Notice that since 1974 the cost of drilling and equipment has outstripped the rise in the consumer price index. This probably confirms the old rule-of-thumb that the cost of doing business rises to meet your income. It also confirms one of the reasons why mineral company rate of return on investment has not risen with increasing gross income.

What will the curves look like in the future? That involves forecasts using techniques like those outlined in Chapter 9. Any forecast made will depend on the political scenario used because political uncertainty is now probably the primary factor affecting the economic yardsticks shown.

The OPEC countries have used the U.S. dollar as a basis for pricing. But, as this cur- rency fluctuates in value compared to the cost of purchased goods, the true value received for the oil varies. The logical step (seemingly) would be to raise oil price to maintain purchasing power. But this has an adverse effect on consuming country balance-of-payments which in turn causes internal economic probelms. Thus, the eoncomics of producing and consuming countries is on a collision course. The net result will become some sort of political compromise difficult to predict.

The U.S. is a good example of the problem. In 1953 Americans paid about $565 million for imported oil. By 1977 this had risen to about $45 thousand million (billion), an increase of about 8000%. The 1977 value represented almost half of U.S. oil consumption and was the largest component of a large negative balance-of-payment.

Complicating the process somewhat are large reserves in areas like the North Sea and Alaska which are rapidly entering the marketplace. This will depress demand for OPEC oil assuming some conservation of usage is involved. Sooner or later the general populace will begin to practice conservation. It could happen rapidly. They are slow to react but ultimately a kind of mass psychology takes over. Oil consumption could be curtailed about 20-25% in the U.S. for example without any noticeable infringement on life style.

Add to these factors like changes in tax structure, regulation costs and like factors, and extrapolation of past trends is not routine.

<u>Natural Gas</u>. The cost of natural gas has been controlled by regulation since 1954. Consequently there are no basic data on even a semi-free market basis in the U.S. One can, however, draw some conclusions based on recent trends. The most important of these is that the price of natural gas will (and should) reach the price of alternative fuel sources regardless of political pressures. This is bound to happen in spite of the long time, gross ineptness of U.S. government policy.

Figure 3.2 shows the equivalent cost of alternate fuels based on their gross heating value (3.1). All volume and value units are in standard U.S. values; MMBTU is read "million BTU." This is equivalent to 10 therms.

If you compare prices one quickly finds that natural gas historically has been woefully underpriced. At 1978 liquid prices the natural gas price on an equivalent basis should have been somewhere between $2.50-3.00/MMBTU, at least 25% above the average price at the wellhead.

This is important in forecasting the future because bargain prices tend to stimulate demand. This has been particularly true of natural gas for it was not only cheap, it was also clean and convenient.

What will be the effect of rapidly increasing gas prices? It is too early to tell exactly. Commercial users obviously possess the capability to switch to alternative fuels which are available on a non-interruptable basis and may be cheaper. The homeowner will tend to stay with gas and electricity. We have some indications that rising gas prices tend to curtail demand. According to the American Gas Association gas consumption generally averaged 20,000 Btu/customer/degree day before 1973. It dropped to about 18,000 in the winter of 1973-75 and then further to 17,000 in the abnormally cold winter of 1976. This undoubtedly results from construction standards, insulation and usage patterns. Another 15-20% in demand probably could be achieved without any real disruption of current comfort standards.

A degree day is the temperature relative to $65^{\circ}F$. If on a given day the average temperature is $60^{\circ}F$, that is 5 degree days. This is a fairly acceptable device to compensate for the effect of weather conditions on gas deliverability. It is used as such by gas companies.

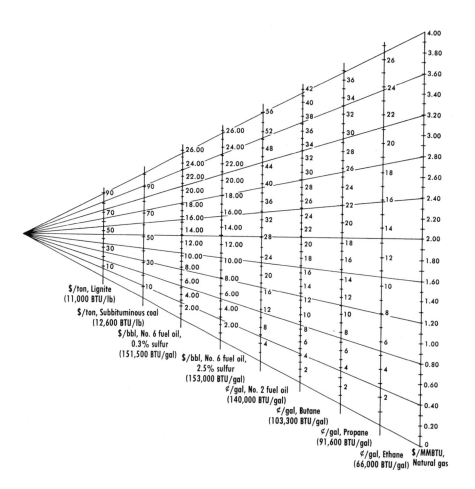

Figure 3.2 Comparative costs of alternate fuels based on
gross heating value. (Ref. 3.1)

Predicting future trends depends to a large degree on regulation policies. Among knowledgable persons there is little doubt that deregulating natural gas prices will increase reserves which will tend to stabilize prices. This would be a very reasonable assumption particularly if tax incentives were provided for re-investment in new exploration and tax penalties were
imposed on excess profits.

Coal. The historical price and demand for coal probably possesses little significance.
It has been a depressed industry for many reasons. One has been the price and convenience of
natural gas as a fuel. This industry is now in a state of rapid change.

Reserves are very large and appear to be well over 100 years at current rates of consumption. But there are shifts in source. Over half of the proven U.S. reserves are west of the
Mississippi River. Also, much of these western reserves can be recovered by strip mining. With
proper equipment one man can strip mine 90-135 metric tons/day compared to less than 9 metric
tons/day in deep shaft mines of the east. Western coal also contains low sulfur which minimizes
the need for environmental control equipment at the power plant site. For all of these reasons,
western coal can be shipped 1000-1500 miles into the U.S. midwest and remain competitive with
locally produced coal.

32

The growth in western coal production has been fantastic. In 1970 about 40 million metric tons were produced which represented about 4% of the U.S. total. One forecast predicts western coal production will rise to about 300 million metric tons by 1980, about 40% of total forecast production.

There are a number of factors affecting total coal production and the distribution of that production. One is the basic unsecurity of eastern coal reserves. Eastern coal miners are a part of an inbred sub-culture that has become increasingly militant, and has shown little compassion for total society needs. From recent experiences in England as well as the U.S. relying on their production for critical needs introduces a serious risk. On the other hand, strip mining in the west is primarily an equipment operating problem and to date the personnel have been anti-union and anti-strike.

Other factors include the availability of water, environmental actions, the rate of development of fluid hydrocarbons from coal, and many like factors.

Oil Shale. This possesses no real commercial history and is not likely to in the near future. Production of oil or gas from oil shale requires a capital cost at least 1.5 times that for conventional oil and gas exploration. So long as supplies of these materials are available at more attractive rates, oil shale will develop slowly. Plants will be built but the primary motivation will be to develop technology, gain experience and provide a general basis for the large need which could conceivably develop in the 1985-1990 period, depending on the world level of prices.

Nuclear power is a controversial alternate to the use of fossil fuels for power generations. By virtue of its political liabilities and the higher cost of the power, development rate will be by default - justified only by the lack of availability of other sources. There will certainly be a continued growth but it will be slower than many predict.

One basic technological problem is water availability. Large amounts of cooling are needed. Furthermore, the effluent water is radioactive and must be impounded for a short length of time until radioactivity drops to acceptable levels. If the water released is for human consumption there are both emotional and rational problems. In seaboard areas we deem it likely that nuclear plants will be located on artifical islands offshore for both political and water availability reasons.

Geothermal energy will provide a large local impact in areas where it is commercially available. Chapter 12 is an overview of this source.

Capital Requirements

The development of new energy resources will require enormous amounts of capital. Is that capital available? The answer to this question is an important one. Different studies of capital needs provide different detailed results but the overall conclusions are rather consistent.

Figure 3.3 is a summary of demand and supply of primary energy for the years 1974-1990 prepared by Bankers Trust Co. and summarized in Reference 3.2. The main assumptions in this study were:

1. Gross national product (GNP) growth of 3½% per year.

2. 3% per year growth in energy consumption.

3. No new and unforeseen environmental constraints.

4. Availability of petroleum imports.

5. Essentially free market conditions for energy development.

Each of these assumptions affects capital availability. For one thing if affects how energy development will be financed, internally or externally. It is estimated that the capital needs will be about $800 billion U.S. doallars (based on the purchasing power of a 1974 dollar) for the period 1975-1990, based on Figure 3.3. Thus, no inflation effects are included. It is predicted that the percentage needed for each segment of the industry are as follows: coal - 6%, oil and gas - 38%, synthetic fuels - 4%, electric utilities - 53%.

It is estimated that <u>external</u> financing of about $149 billion dollars will be needed for the fossil fuel industry between 1975-1990 which is about 40% of the total projected capital outlay. The electric utilities will have <u>external</u> financing needs of about $217 billion dollars for the same which is about 65% of the total capital outlay. Total external financing is estimated to be $366 billion dollars in the 15 year period. Is this going to be available?

From 1966-1975 the energy industries captured about 20-21% of the capital available in the capital marketplace. It can be presumed this share would be available in the future. If the share had to increase substantially above this percentage availability of funds could become questionable. This leads to the next pertinent question. What will be the size of the future capital market? This is estimated by Bankers Trust in Table. 3.1.

TABLE 3.1

**PROJECTION OF GROSS NATIONAL PRODUCT
AND RELATED CAPITAL MARKETS**
(Billions of Dollars)

		1974 Actual	1975 Est.	1980	1985	1990
1	**GROSS NATIONAL PRODUCT**	1,397	1,337	1,671	2,053	2,462
	FUNDS RAISED FOR —					
2	U.S. Government	13.0	70.9	8.4	0.0	0.0
3	State & Local Govts.	16.3	16.0	16.7	20.5	24.6
4	Consumer Credit	9.6	4.0	21.7	26.7	32.0
5	Home Mortgages	34.0	28.1	36.8	34.9	36.9
6	Foreign	2.0	2.7	1.7	2.1	2.5
	Business Sector					
7	Corporate Equity	3.5	5.3	13.4	16.4	19.7
8	Corporate Bonds	19.7	24.1	25.1	30.8	36.9
9	Mortgages	22.5	21.4	23.4	28.7	34.5
10	Business Loans	56.9	11.0	35.1	43.1	51.7
11	**Total Business Sector**	102.6	39.8	97.0	119.0	142.8
12	**TOTAL FUNDS RAISED**	177.5	161.5	182.3	203.2	238.8

SUPPLY OF PRIMARY ENERGY

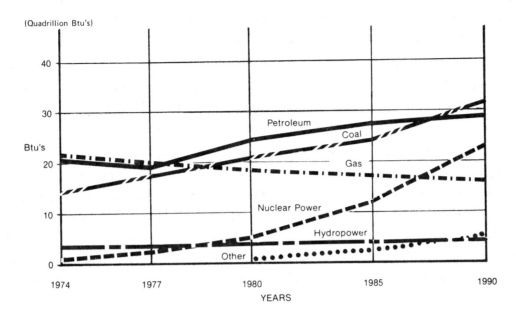

(Quadrillion Btu's)

Btu's

Petroleum

Coal

Gas

Nuclear Power

Hydropower

Other

YEARS

DEMAND FOR PRIMARY ENERGY

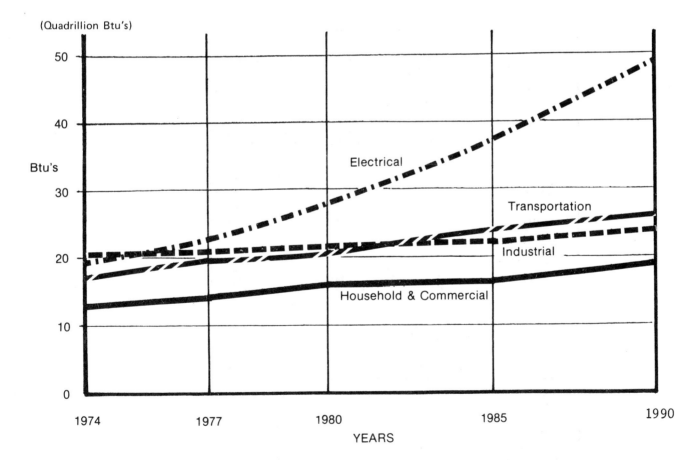

(Quadrillion Btu's)

Btu's

Electrical

Transportation

Industrial

Household & Commercial

YEARS

Figure 3.3 Forecast of supply and demand for energy. (Ref. 3.2)

35

convince a nation's leadership to accept higher prices now in order to develop competition for other nation's minerals that will lower prices in the future. Political wisdom dictates keeping prices as low as possible now and letting the future take care of itself, since most politicians are most conscious of perpetuation of their office. This concentrates thinking in a narrow time frame.

How does one predict future political occurrences and then incorporate them into economic planning? The answer: One cannot. One may analyze the reservoir by the most sophisticated methods possible - using the biggest computer - and still not have the most important answers for considering a prospect.

As supply-demand figures show, an ever burgeoning world demand for minerals requires aggressive search for new reserves. For a publicly held company, for which liquidation is not a goal, this replenishment of reserves is essential. Recent events in the petroleum industry exemplify the problems facing other mineral industries.

The bulk of the overseas reserves discovered prior to the early 1970's were contained in relatively few giant oil fields in the Middle East and North Africa, under long term concessions, at a time when nationalism was a minimal factor. Due to revision and re-negotiation forced by producing countries, the net value of these reserves to the oil companies has diminished. The basis for the large continuing investments has changed (for the worse). More pressure is being exerted for reinvestment of profits from these older (cheaper) reserves into development of new ones (in the same country) under terms which may be economically marginal. The cash flow-back into the corporate treasury from which the investment was made is thus curtailed (and less than prior planning would indicate).

What then can one expect in the future? A partial answer seems feasible even though I do not claim to be clairvoyant. Certain trends have developed which seem likely to continue.

- "Operating agreements" will continue to replace concessions or long-term leases.

- The bulk of new exploration will be directed toward politically stable countries (when feasible).

- Transportation and marketing patterns will alter as supply-demand changes on a regional basis.

- The internal pricing structure will change so that each segment of the industry - from production through marketing - will be more self-sufficient.

- Petroleum decisions will be guided, if not restricted, by national energy policies. Government mineral treaties are even likely.

There are more, but one could write an entire book on this subject alone.

Operating Agreements

An operating agreement, as I use the term, is an arrangement whereby ownership of a substantial protion of the mineral rests with the producing country. The formal contract is likely with a nationalized company which serves the country as an agency. The contractor (outside company) might or might not have an equity in the petroleum found. However, he secures all rights for the purchase of some fraction of the mineral.

The contractor, as operator, is entitled to some protion of the proceeds - in the money or mineral - for performing his function. In the final analysis, the contractor's net revenue must reflect his risk as well as his capital investment. Ideally, both parties will likely share the capital investment. It is merely a guess on my part, but I visualize a common arrangement whereby the contractor absorbs all exploration costs but the parties share in all capital costs incurred thereafter.

Some form of agreement like that above may be a reasonable compromise. It tends to nulli-fy those emotional concerns that foreign interests are "exploiting our country." The parties are partners. Each is satisfying a need of the other. From the contractor's viewpoint legal owner-ship is important but less so than a reliable continued source of supply.

This trend, predicted in predecessors to this book, has continued as forecast. The form of the agreements has fluctuated with an "accordion" effect. The host country has adopted mili-tant attitudes which stifle outside capital and reduce exploration. Then the host country "backs-off" and liberalizes the arrangements to attract such capital. But, the rules available during capital influx may not exist during the life of the property. A contract between a government and a private company is an expression of intent at that time but it provides limited long term security. As political and economic circumstances change there inevitably will be changes forced on the private company. This is true of all countries regardless of the form of government.

A country offers concessions to private capital which finds commercial reserves. Said concessions involves payment of certain royalties and taxes but the equity remains with the in-vestor. At some later date the host government announces it wishes to buy a controlling per-centage. In effect this is not a request, it is in reality an order. The only real questions remaining are how much will be paid, how it will be paid and when. Often the payment is in miner-als the company already thought it owned. It is equivalent to "buying" an apartment house by paying the owner out of the rents he would have obtained anyway.

There is no reason to anguish too much over this. It is a fact of life and the risk built-in to the investment in the first place. The company has two choices - acquiesce or lose a critical source of supply.

Government Regulation
The political and military importance of energy requires some form of government regula-tion. The primary political concern of any large consuming nation like the U.S. is preservation of a secured supply at minimum cost to the consumer.

The avalanche of rhetoric in the U.S. tends to obscure the basic underlying problem. The consumer has enjoyed a plentiful supply of cheap energy. All of a sudden the petroleum industry, whose public image has never been very good anyway, raises price-supply issues. For the unin-formed, there is a credibility gap. Into this gas step the self-styled consumerists, ecologists, and general "crusaders." The motive of the individual - good or bad - yields the same result ... confusion. Even the informed public servant in the process of seeking a compromise between

39

cost of new supply and maintenance of a traditional price structure contributes to this confusion. Unfortunately, a minimum cost policy endangers long-term secure supply. At best, it buys some time by treating symptoms rather than the basic problem.

On a purely rational basis regulation should confine itself to those areas affecting the health and welfare of the general citizenry. It should not attempt to control day-to-day decisions except as they affect safety and the environment. Unfortunately the line between control and interference is a very indistinct one. There is always the question about whether the benefit is worth the cost. The cost of regulation is seldom discussed. But on a financial level only, it is good business to spend a dollar for regulation when the consumer benefit is less than this. The consumer ultimately pays all costs.

Consider the following. In 1978 the budget of the U.S. Department of Energy was about $10.6 billion. To place this amount of money in perspective consider that it represents

- $2.59 for each barrel of crude oil produced in the U.S. in 1976.

- a hidden 10 cents per U.S. gallon surcharge for each gallon of gasoline consumed in 1976.

- an amount of money that exceeds 1975 capital and exploration expenditures to explore for and produce U.S. crude oil, natural gas and natural gas liquids (estimated at $9.4 billion by Chase Manhattan Bank).

Is the benefit of the DOE worth the cost? We will leave that decision to the reader.

The cost of regulation goes beyond government budgets. It is estimated that 20-30% of mineral company professional overhead is directly primarily to regulatory matters and is not available for the finding and production of new mineral resources. Another hidden and somewhat obscure factor is the effect improper regulation has on the investment climate. Whether rational or not, most professional managers believe that it has, and does, inhibit investment.

By virtue of the pressures present, all political entities tend to over-regulate. The philosophy is "protect ourselves from criticism by assuming the role of a tenacious watchdog." This action causes several reactions. One is a series of confrontations that serves no interest very well. A second is that more time is spent protecting a "position" than trying to solve a problem. Proffesionals concerned with energy regulation are engaged more nearly in preparing reports and summaries resulting from confrontation than in solving the basic problems. The time cost and lost productivity resulting from merely talking about the problems would have contributed to their solution.

We point out these things because in any evaluation future regulatory and statutory actions are likely to be inconsistent. The emulsion of emotion and ration is so "tight" that it forestalls a completely cohesive, logical, long-range policy.

One thing is certain, the confusion that has been prevalent in recent years must subside to some degree. At worst this will occur only when a large enough segment of the population is inconvenienced by an energy shortage.

Also ironically, the role of government will more likely be governed by actions (or lack of same) instituted by the energy industries and OPEC than by government itself. Most regulation fills a void or controls irresponsible actions. The petroleum industry has not been irresponsible. It has simply concentrated on efficient production and marketing of its products; too little attention has been paid to its public position.

The answer involves several steps:

1. Creating public awarness of the true short - and long - range situation in a manner that is not merely self-serving.

2. Positive recognition that future U.S. energy development will undoubtedly involve some kind of joint venture arrangement between industry and government.

The disadvantages of a joint venture arrangement are obvious and well-known. But, most undeveloped U.S. energy sources exist in lands owned and/or controlled by some governmental unit. The only alternative is expensive lease sales and extension of controls by people without direct responsibility for operations. Regulation has a way of differing (favorably) when the one regulating has an equity interest. A partnership also muffles to some extent those charlatans who continually wail that "private companies are exploiting the people." The situation in the U.S. is not that much different from other countries with government owned minerals.

A carefully planned joint venture relationship could minimize a current deterrent to greater reserves development - the initial high cost of government leases. Under the traditional approach, the capital requirements are burdensome. Because of the inherent time lag between lease payment and cash flow-back (if any), the debt service charges become significant. Both the company and consumer lose. Source of supply is curtailed and the consumer (as always) pays for the interest overhead.

We visualize some form of arrangement where the government uses a "carrying interest" approach with "reversionary or back-in" provisions. Companies would bid on a lease competitively. The winner would pay a certain percentage down. The remainder would be a type of oil payment only from production from that lease. The government would have an equity interest in the venture. In essence, what one would have - on a larger scale - is the venture fund approach used by some companies to finance individual ventures. Some forms of this have been used in other countries.

For all of its faults, some form of the above might be superior to the real alternatives. Tax incentives and subsidies are subject to political whim. They always supply fuel for charges of "favoritism" and talk of tax "loop holes."

Attempts to promote awareness of the problem among the general public are laudable and must always continue. But, this is too slow for the purpose of change. The "lead time" for development of any large new energy sources is at least five years. Public reaction will be slow until it is actually inconvenienced.

Some sort of miracle is needed to promote a greater degree of mutual trust between the private and public sectors. The basic lack of trust is the source of most improper regulation.

One example is the belief held by some that companies are withholding large amounts of minerals from the market to wait for higher prices. This is tantamount to saying that the mineral industries managers are economically illiterate. Charge them with what you will but that particular charge is utter nonsense. It is contrary to basic value of money concepts and liquidity problems. Attempts to control this phantom problem lead to policies that arbitrarily set norms like maximum effective rates (MER), production delay times, etc, etc. These kinds of things are "Mickey Mouse" regulation at its worst.

Lest we appear too harsh, we reiterate that intelligent regulation is a political necessity. The bulk of this should consist of simple monitoring activities. There are several very basic ingredients necessary - a greater degree of trust, a clearer understanding of what regulation can and cannot accomplish and restriction of regulation to those areas where a positive financial benefit accrues to the consumer. Can these things ever happen? Let us pray that they can.

ACCOUNTING PRACTICES

Accounting practices is an important economic consideration. A change in practice can effect drastically the apparent profit from a given venture. The major problems are how

1. One charges off intangible investments.

2. The method of amortization (depreciation) of capital items.

3. The time lag between the initial investment in reserves and the cash returns from that investment assuming that the amount produced must be replaced by investment to find new reserves.

An intangible investment is an investment the result of which produces no tangible value, even as salvage. This is of particular importance to minerals industries because unsuccessful exploration is an obvious intangible. Whether it be exploration for coal, oil shale or petroleum a hole in the ground has no real value. Exploration for petroleum emphasizes the exploration accounting problem for the entire minerals industry.

In the U.S. there is a Federal Accounting Standards Board (FASB) charged with developing accounting rules. Equivalent agencies exist in other countries. A good example of the arbitrary nature of said rules is afforded by the handling of exploration expenses. There are two basic philosophies.

1. Successful Efforts Method - unsuccessful efforts are written as expenses immediately since a dry hole is not an asset.

2. Full Cost Method - capitalize through amortization all exploration cost as a deductible against income from successful wells which amounts to treating a dry hole as an asset.

There are strong supporters for both divergent viewpoints. Most large diversified companies tend to favor successful efforts accounting. In addition to size there is a difference in attitude between stable and growing companies and the relative amount of capital invested in intangibles.

Proponents of successful effort accounting point out that a dry hole is not a true asset so that including it as such can be misleading. They point out further that a very unsuccessful company could appear successful upon casual scrutiny of a balance sheet with full cost accounting. Thus the claim is made that successful efforts accounting is basically more honest.

Most of the proponents of full cost accounting are from small, growing firms who spend a relatively large amount of their income on exploration. Successful efforts accounting can depress the reported earnings of such companies drastically, as much as 300% in some cases. This effects their ability to attract capital.

Pipeline and utility companies also resist successful efforts accounting since it would sharply increase their debt-to-equity ratios. This is important to them because of their reliance on external sources of capital.

We have a simple solution to this dilemma. Allow both under disclosure provisions which make it clear which is being used. However, for the purposes of this book the above is one example of how accounting affects economic conclusions.

Amortization or depreciation is another arbitrary area. This is the device for recovering money invested on which tax has already been paid. In the U.S. one can use an approved accelerated depreciation to recover the investment faster or a straight-line method based on units of production or per unit of time. Regardless of the method, inflation raises a problem. The replacement cost of the capital item being depreciated increases with time.

For example, suppose an item costing $100,000 is to be depreciated at the rate of $10,000/ year for 10 years. At the end of 10 years the item has to be replaced but its cost is now $200,000. The amount put aside does not represent the true cost of that item to the company during the intervening years. The result is that _true_ profit may be overstated. The question - should depreciation be based on original or replacement cost? Once again viewpoints differ.

Then there is the inventory delay problem. At any one point in time the warehouse inventory is composed of reserves of different age and of different cost to find. Some may be petroleum reserves found 20 years ago at a cost of 50 cents per barrel; some may be new and cost $8.50 per barrel to find. As these reserves are produced they are co-mingled and lose their identity. Profit is based on current cash flow. In effect old, cheaper to find minerals (regardless of the type) are subsidizing that profit. But these are being depleted faster than new minerals.

Can accounting accurately represent this cost of inventory situation? Only to a degree. There are two basic systems. One assumes that the first into the inventory is the first out (FIFO). The other assumes last in first out (LIFO). These two approaches do not give the same net result and can affect economic appraisal.

There are other factors affecting accounting results. But these three alone should be sufficient to make the point that accounting prices can significantly alter the apparent economic value of mineral properties. It is therefore necessary to be cognizant of this and compensate as necessary in the decision process.

For several reasons the value yardsticks shown in Chapter 5 will not check accounting rates of return. This is not bad and there is no reason why they should check. The important thing is to recognize the effect of accounting practices on apparent value.

Transportation Patterns

The large markets for mineral products are the industrialized nations, primarily the U.S., Europe, and Japan, who import the minerals as raw materials, convert them to manufactured goods, and either consume them at home or export them. Other markets are also growing at a rapid pace but those like Australia are served presently by internal production.

Of course, actual reserve percentages fluctuate with new discoveries, but the basic pattern illustrates the dependence of industrialized nations on other countries for their raw materials.

Since the primary markets are located some distance away from the "warehouse", transporation considerations must play a key role in the economic considerations. Mere cost is not the only concern. One major concern must be preservation of supply lines in the face of local political problems and wars between countries.

Even a casual student of history should be aware of the impact caused by closing of the Suez Canal and periodic interruptions of service in pipelines by countries they cross. In a major war involving naval action, the tanker is a major target. History shows clearly that wars are won by that combatant who can maintain (or develop) energy supply lines and comparable manufacturing capability. The long supply lines between supply and market introduce a consideration over and above simple economics. It dictates some form of political control that might not be conducive (in at least the short range) to sound economics.

There are plenty of purely economic considerations. If one does not develop a large tanker fleet - by outright ownership or long-term lease - he is at the mercy of the "spot market." In 1970, for example, the demand for tankers far exceeded availability. Libyan production was curtailed, the Suez Canal was closed, and fuel shortages loomed in some of the markets. This spurred new construction. Spot charter rates soared. By the middle of 1971, many tankers were idle as production increased again from North and West Africa. In the tight period some had let contracts for tankers or negotiated long-term leases for capacity that was not currently needed.

These kinds of fluctuations occur with an unknown frequency. The very instability of the costs complicates future economic planning. The very uncertainty and the possible prohibitive cost of reducing that uncertainty adds even greater incentive to the procurement of reserves closer to the market.

The growing concern for ecology, worldwide, even affects the picture. The bulk of the Middle East oil and gas contains sulfur compounds. These will have to be removed at some point in the system. Suppose large quantities of "sweet" petroleum are found closer to the market? There is a double saving - sweetening and transportation. The companies with large Middle East investments are faced with a dilemma. To produce such reserves to protect a large investment will cost more money for transportation and treating. As more petroleum is produced in Southeast Asia, for example, the substantial Japanese market for Middle East oil could be compromised, depending on supply-demand considerations then in effect.

Both of these forces affect the demand for other minerals also, since low sulfur coal and, natural gas, and other non-polluting minerals will increase in demand.

44

Whatever the ultimate shift in marketing channels, the net value to the producer may decrease in comparison to the other alternatives available. In the late nineteen sixties and early seventies there was no real alternative to Middle East oil. Necessary supply lines were fixed by this fact. This basic need dictated a strategy that could best be characterized as a "holding action."

This will change only to the degree that the major consumers can develop reserves in their own areas. What will result? The price structure must alter to reflect the true value of a reserve in relation to others available.

At some point, the source of supply will alter drastically. Coal, oil shale, tar sands, etc. will become competitive as crude oil prices rise. North America, rich in such resources, will become more self-sufficient. Other supplies will have to compete with "local" sources.

As one examines the situation there is one inescapable conclusion ... source of supply, and the resultant transportation patterns, is not a static thing. Long-term planning based on present circumstances is ill-advised. For a number of reasons - political, national security and economic - long-range capital commitments involving long distance transportation are becoming less desirable as time passes.

INTERNAL PRICING STRUCTURE

Mineral pricing structure has "grown like Topsy." Although it is difficult to change, such change is overdue. In the past twenty years the change in producer country attitudes, consumer country tax policy and marketing are several of the factors dictating such consideration.

The average major, integrated company will make about 10% annual return on its total investment. This is less than many other basic, consumer-oriented industries. It is too low for the risk involved and the cost of using money for the relatively long time period between investment and cash flow-back. This is the basic culprit for any fossil fuel energy shortages that may occur in the foreseeable future.

Internal pricing strategies have become even more complicated as the nature of integration has changed. Mineral companies were historically vertically integrated in that they handled all phases from exploration to retail sales of one mineral. In the sixties, movement began toward horizontal integration where companies began to branch out and acquire other minerals. The objective in expanding company investment horizons is to increase return on investment, while simultaneously reducing the risk involved.

Petroleum companies are investing in coal, nuclear, and other energy forms. Steel and manufacturing firms are involved in mineral investments. In each instance the companies are trying to protect their present investments and hedge against future economic changes which may make their present major activity less attractive.

How did the current pricing situation develop? To obtain any kind of an answer one must look to the U.S. petroleum industry. It was the first large producing and marketing entity. As such, it has set the pattern for energy industries worldwide and for many other, newer mineral industries.

45

It is convenient to divide the industry into segments - producing, transportation and marketing. Marketing covers refining, petrochemicals, fertilizer and all activities involved in the conversion and sale of finished products made from raw material of fossil origin.

Historically, about 50-60% of the total capital expenditure of the U.S. petroleum industry has been spent on the exploration for and exploitation of petroleum reserves. Since this was the largest single expenditure and involved the greatest risk, it is not surprising that pricing policies were geared to ensure profit in this area. In fact, it was not too uncommon to find minimum profit in transportation and refining, with marketing being little better than a "break-even" operation. Philosophically, at least, these latter functions were simply strategic investments necessary to generate profit from reserves development.

It must be remembered that the bulk of the early managers and investors were "oil-finders." The manufacture and marketing of finished products was secondary in their thoughts. After all, the problem was routine (in their minds); demand continued to keep pace with the finding rate. The problems were simply logistical in nature - saturate the marketing area with service stations to maintain a competitive position, provide a suitable transportation and refining network and promote an "image" that would support the desired competitive position. Under the existing circumstances this position was both logical and successful.

At that point in time the "cut-rate," independent marketer was merely a nuisance. His stations were few in number, generally inferior in appearance and service, and served only a local area on a cash basis. In fact, he served one useful purpose for major refiners - as a market for surplus refined products during normal supply-demand fluctuations. Such excess products could be "dumped" on the open market at lower price than normal, using the differential cost concepts in vogue. With this concept, products beyond a basic amount cost less to produce if fixed overhead is borne largely by those products marketed internally.

This approach remained unchanged until the early 1970's even though the conditions upon which it was based started changing noticeably in the late 1950's.

The net result has been a continuing series of periodic gasoline "price wars." One station in an area lowers prices, often based on a cheap purchase of gasoline. This is a useful tatic to hopefully capture a greater share of the market temporarily in the hope that some of this advantage will remain beyond the "war." A neighboring station will retaliate by lower its prices and the thing spreads. Price cutting continues until no profit is made by anyone. At this point everyone is motivated to call a halt to the thing.

During this war refiners have supported their marketers by absorbing some of the price cut through "allowances." Both thus share the burden imposed.

The reasons and methods causing this situation are understandable. The logic is not. Anyone possessing even minimal economic intelligence has recognized the need for change. Price wars are economically disruptive and do not improve gasoline consumption. They even contribute to a long-range energy deficiency by limiting liquidity.

A corollary problem is the saturation marketing technique. In its grossest form you add stations in an area that already has too many. You capture some of the market and up your total refined gallonage, which reduces your cost per gallon. A good tactic? Maybe, in the short run.

46

But, the individual station suffers from inadequate cash flow. Quality of personnel and quality of service suffers.

Now, someone decides to use gimmicks. Give away glasses, plastic bowls, steak knives and other such trivia; or, play games. Also push credit card sales and try to make "your card" more attractive by selling radios, appliances, tool kits and the like to card holders at attractive prices. Seemingly forgotten are two facts - almost everyone does it (which nullifies advantage) and total gasoline usage does not increase. I have not yet known of anyone who took a trip because of a price war to get some "free" glasses. Such customers must at least be rare.

This traditional model had to change and is. First of all profits on production are too small in many cases to subsidize marketing. This means that each element in the vertical chain must be more autonomous from a profit picture. Secondly the companies now are all competing internally somewhat. The oil shale division is competing for capital against tradiational exploration. A marketing manager will ultimately have to purchase oil from both, maybe at a different price.

The growth of the independent marketer has been dramatic in recent years. Many factors have undoubtedly affected this. One is the closing of many major company stations that were uneconomical. They have been replaced in increasing numbers by "cut rate" stations as part of a convenience store where overhead may be split among several functions. A portion of this has been dictated by rising sales of TBA items (tires, batteries and accessories) by shopping center chain stores.

A factor affecting "cut rate" sales has been the U.S. entitlements program. In effect, the purpose is to balance out costs for small independent refineries so they can compete with major marketers. The concept possesses some merit but the actual program serves as a horrible example of bad regulation in practice. It typifies the result when one tampers with basic economic parameters. Nevertheless, the net result is the ability of some to sell at cut rate prices.

Commercial fuel sales. The marketing problems confronting the petroleum industry are not confined to retail gasoline sales. There is a large commercial market which includes

> Aircraft fuel sales
>
> Fuel oil
>
> Petrochemical products (usually for use by others in the manufacture of finished consumer products)
>
> Gas and gas liquids for fuel and blending purposes

All of these complicate the planning. The raw material for all products comes from the same warehouse (reservoir). It is simply a matter of sub-dividing or converting the raw material into those products which possess the greatest market value. In a real sense the planner is caught in a "bind." There are several incompatible constraints at work which prevent "easy" decisions.

- The unique position of energy in a country's strength and in the personal welfare of its citizens.

- Unpredictable political changes which stem from the public's needs and concerns.

- Continual changes in the quantity and quality of product needs.

- The internal competition of fuels - from the same warehouse - for the market.

One could list more but these suffice to illustrate the character of the economic model.

In order to view these properly one must recognize that the plants and facilities necessary to produce finished petroleum products require very large capital outlays. These capital facilities are reasonably inflexible from the standpoint of what products they may produce most efficiently. By any economic yardstick - with the usual practical constraints - the facility must be used for 5-10 years, at close to full capacity, to be a sound investment. Any factor which compromises the time and cost factors involved, in turn compromises the strength of the industry.

No country has ever attained, and remained, in a sound economic position without a sufficient, secured supply of energy and mineral resources. A high productivity of manufactured consumer products - salable in a competitive market - has always been the cornerstone of prosperity. Security of raw materials implies control. Such control is always more positive within a country's own boundaries. Those nations without internal energy and mineral resources have undergone dramatic cycles of prosperity and depression due to uncontrolled supplies and markets. Many major wars have had as their root cause such considerations.

The minerals industry - in all of its facets - is critical to the welfare of a country like the U.S. This imposes obligations and constraints beyond purely economic models.

Political constraints exist to an unusual degree. Any intelligent politician will try to compromise the problem; insure maximum development of resources at minimum cost to the nation's consumers.

Any value system must be cognizant of these factors. A value system based on money alone is neither suitable nor practical. The economic model must include as pertinent variables the environment, national security, individual service and like factors. In essence, the political factors must be guided to some degree rather than merely reacting to and trying to live with such factors. The degree of certainty achieved probably would be less than desired but certainly better than historical experience.

Continuing market change. The continuing change in market is an ever-present uncertainty. Some elements may be predicted to a degree; some may not. Demand for jet fuel, natural gas, motor fuel, etc. follows a trend basis that may be predicted within a given period. The amount of product storage that is justified economically must be rather small compared to the market.

There is a limit to the way a crude petroleum feedstock may be split efficiently. During the winter months fuel oil demand normally is higher than in the summer, with motor fuel demand decreasing. But, if the changes in demand are not compatible with the character of the feedstock, there is something left over. Or, suppose a very mild winter season causes a fuel oil surplus, storage is full, and other markets do not take up the slack. Refining rate must decrease. The capital cost per unit volume produced must rise. The flexibility needed is only achieved through increased capital cost for equipment that will operate at full capacity only part of the time. Short term shortages may only be prevented through higher investment in relation to total sales income.

In addition to this form of capital investment, there is the regular change in market demand as engine characteristics change. One example of this was the "no-lead" regulation made in the U.S.

Product competition. In the commercial fossil fuel market natural gas, fuel oils, and coal are internally competitive. The first two offer a special concern. They both come from the same production operation. Commercial users feasibly may interchange their use; individual home consumers cannot. To some extent, the relative availability of each is fixed for much of the natural gas is produced in association with crude oil.

If fuel oil is in surplus, it may be sold by lowering price at the expense of natural gas sales. This might be attractive to the refining department but not to natural gas sales, providing both are in surplus. Internal pricing policy has sometimes thus been inconsistent with the cost of the common raw material source.

The product competition for the same market is disappearing as the demand for energy taxes the supply. A good case in point is natural gas. In order to keep transmission lines at capacity during periods of low domestic fuel sales, gas has been sold at a lower price in the U.S. to industrial consumers on an interruptible basis. This means simply ... "We will sell to you only at this price (or at all) when surplus capacity is available." Once again this policy was dictated by the small differential cost to move additional gas as opposed to no sale at all.

By 1970 U.S. interruptible sales had begun to diminish. The reason - domestic sales utilized more of the capacity. This condition will continue. The use of fuel oil, coal, and other forms of energy will continue to grow. Coal will be gasified - crushed, transported and burned as a slurry - or used as a solid fuel. The former two methods may become predominant because of lower cost and/or greater convenience.

The preceding brief discussion outlines many of the factors which complicate petroleum economics. Certain of the historical patterns will be changed. It is not a matter of if, only when and how. Some concern for the timing and magnitude of such changes is an integral part of petroleum property evaluation. The historically standard assumption that current prices and market practices will prevail throughout the property life introduces potentially serious error.

Later sections will explore political and future economic factors. Other mineral industries, while often not integrated to the degree occuring historically in the petroleum industry, face similar problems. Few other mineral industries sell their mineral to consumers in original form. Most minerals are used as inputs to produce a saleable item that differs significantly from the original mineral. Copper, lead, zinc, phosphate, mercury, tin, iron ore, etc. are seldom bought off the shelf in a retail store. During periods of excess supply, producers have an incentive to acquire companies that process their mineral to insure an outlet. Users of such minerals, conversely, have an economic motive to purchase suppliers when supply shortages exist.

SUMMARY

The preceding pages overview briefly a number of factors affecting mineral property economics. Many of these will be explored in greater detail in future chapters.

An important philosophical point we are attempting to make is that value involves a whole host of dependent factors. Many of these are not technological in nature but involve politics, accounting, economics and other associated disciplines. While the mineral specialist cannot hope to be an expert in all of these areas, he or she must be cognizant of their influence.

REFERENCES

3.1 Leffler, W.L., Oil Gas J., Aug. 15 (1977), p. 75.

3.2 Krupp, H.W., SPE Paper No. 6113, Fall Meeting (1976).

4

VALUE CONCEPTS — VALUE OF MONEY

One cannot make an evaluation without using a value system. An integral part of evaluation in a world of commerce must be value of money concepts. However, using a money value system to the exclusion of all other values is illogical as well as irresponsible.

Most of the emphasis herein focuses around money, particularly the time value of money. Money is not only important basically but is a finite parameter that may be handled arithmetically. Other value elements must be considered in the money parameter.

What are some of these other values? Environmental, cultural, political and human factors are probably the primary ones. In most cases they must be considered subjectively.

It is necessary to consider questions like:

Will this project, if undertaken as planned, be compatible - and not
harmful - to its physical environment?
Is this project compatible with both company and human goals?
Will the cultural and personal goals of the people affected be
advanced by the subject project?
Is the project in the national self-interest of the country(ies)
involved?
If the answers to any of the above type of questions are negative,
can the project be modified to meet the needs of the total
value system (including monetary)?

Concern for a total value system has become too idealistic. What is proposed is merely that people - as individuals and as organizations - cannot ignore their responsibility as members of the human race. This does not mean that profit motivation is in opposition to these things. In fact, history shows that the quality of life is usually superior inherently in those areas where the capitalistic system prevails. Concern for profit need not be in conflict with other values.

It is lamentable that advocacy has replaced cooperation in the attempt to meet common goals. There are effective compromises available but it is difficult to achieve. We continue to subscribe to the total value system but deplore most of the positions assumed by advocacy groups.

51

Concern for those things covered in the questions above almost always requires higher investment costs than if they were ignored. Much of this may be truly regarded as strategic in nature. The word strategic as used here covers any investment cost that may not be assessed arithmetically using conventional economic yardsticks.

Concern for the sociological affairs of man are strategic in nature. Which is less expensive, investing in positive approaches to fulfill a society's needs or having to absorb an overhead dictated by constraints (often unrealistic) imposed from without? Either way, the cost must be passed on to the consumer. This introduces problems, but the former approach provides an easier basis than the latter.

Money is the only standard capable of including all of these values in one system. The real problem is transforming each of the value elements into one common denominator - money. That is the only reasonable way to compare costs and benefits.

ABOUT MONEY

Money is a convenient value system. It is readily transported and exchanged. The value of money involves several facets - time and the amount of goods or services it may purchase. Time value is a finite part of the evaluation procedure and is discussed throughout this book. The primary purpose of this chapter is to look at real purchasing power in terms of non-calculable time factors. Some concern about the history of money provides a measure of understanding about the current aspect of many of these factors.

The establishment of a value system was a necessary part of any organized society, regardless of how primitive. Based on anthropological studies, it is reasonably apparent that man only survived as an animal species because he organized into bands wherein each individual served a unique function. There were hunters, tillers of the soil, weapon and tool makers, food preparers and the like. The maker of weapons had to obtain some of the "kill" in order to survive. This involved some degree of bartering. Some value had to be placed on each function so that each party could exist at a satisfactory level. As functions became more specialized, this process became more complex but the primary system simply still involved the maintenance of life.

As man became more agrarian a three way bartering system had to develop between the hunter, farmer, and merchant. There was a concept of value but not of wealth, as we know it.

Simultaneously, the concept of the family as the basic unit developed - for many practical and social reasons. This, in turn, centered a man's attention on that unit. As a sign of "love" and his natural aggressiveness there was a desire to provide better for his group than did his neighbor. As new opportunities were visualized a kind of commerce developed, but it was limited in geographical scope. The bartering was a continual thing between people in intimate contact.

Groups of families banded together into tribes for mutual protection and to satisfy an apparent need for orderly activities. This led to leaders and followers by a natural process. Even in this primitive society barter was oftentimes difficult and inefficient. Money (in the most general sense) was developed to facilitate bartering. This was merely something of recognized value that could be easily exchanged and was of a size easily carried and stored.

Money was not necessarily a sign of a sophisticated civilization. Rather it was a matter of convenience and practical need. A century before Christ, the Indians of Great Britain stamped

coins. Yet, a thousand years later, in the same area, the monasteries and manors operated almost exclusively by means of direct bartering. With some exceptions, direct bartering was the
primary method of commerce up to the 13th Century A.D. The household or tribal economy was the
paramount governing factor.

Money first became used by traders and "bankers." In the beginning most people never
owned (or even saw) the coinage used. The growing use of money as a medium of exchange necessarily required the development of several concepts:

> Conception of private property
> A more formalized system of relative values
> Differentiation between money and real wealth
> The changing value of money as a function of supply and
> > demand for the goods it represents

The advent of money also brought on individual freedom. A man could freely use the monetary
rewards for his efforts. The degree of reliance on his leader was diminished.

The first recorded use of a coin-like money was in Lydia around 600 B.C. The Greeks under
the leadership of Athens were the first to make money available to the masses on a controlled
basis. Silver coins were issued by the priests of the Parthenon. They retained control over
both the quantity issued and the value assigned. They had a basic understanding that the value
of a coin was only representative of what it would buy. One early coin was called an "ox" because that weight of coin was worth exactly one ox. As the relative value of an ox varied in
the economy, the coin likewise varied.

The Greeks also recognized that money was not wealth per se, but was merely a mechanism
by which wealth could be exchanged. This rather subtle differentiation is important in truly
understanding our monetary system.

Such a system was reasonably successful so long as money exchangers were within the
control of the issuers of the money. As world commerce developed, such control became less positive. One now had the problem of assessing the value of one currency in terms of another. This
led to the first crude international monetary organizations - the prelude to the monetary groups
we know today.

Token versus Intrinsic Value

In the absence of formalized governmental units and monetary unions, most early coins
were made of a metal with recognized value. Even though the face value of the metal was not
necessarily equal to the good the coin would purchase, this recognized value promoted acceptance
of the coin. Copper, silver, and gold coins have thus dominated coinage.

A token coin is one whose content has little value in relation to its purchasing power.
As copper became plentiful it came to be used primarily in token coins of low value. It wears
well with repeated use and is easily formed. Silver has been used because of its desirable
metallic properties and relative scarcity. Historically, the value of the silver content has
matched the expressed value of refined silver, although this is seldom the case currently.

Man has always had a special fascination for gold. It has been relatively scarce, works
beautifully and is universally accepted. Its use in coinage is limited by its softness. As long

as there was little routine handling of coins this was not critical. Wealth (as represented by gold) was in the hands of very few and the coins were not continually exchanged. Paper money was instituted for convenience when exchange became common. Even though paper possessed no real intrinsic value (from its content), it was not true token money for it could always be exchanged for equivalent gold or silver.

One problem with use of rare materials, with intrinsic value, is that the supply must always be in some balance with demand so that the market value of the metal is in balance with the value of the coin in commerce. This is unlikely unless the source of metal may be controlled. (Diamonds are an example of a precious material whose supply is carefully controlled in an attempt to stabilize price.) Consequently, the price of gold ultimately had to be pegged in terms of some arbitrary value. Even this is self-defeating if the supply becomes inadequate to meet the needs of commerce. This is the case of gold currently.

So long as token money may be exchanged for desired needs and services, it offers no problem (except emotional). Private company scrip, coupons, and the like have always been a part of the total monetary system.

Token money is less valuable only to the extent that the lack of backing reduces faith. When the U.S. changed from true silver coinage to a "sandwich" type, no change in real purchasing power of a given coin changed. Yet, there was a noticeable emotional reaction.

From a purely rational viewpoint, money is not wealth per se; it is a conveninet vehicle merely for the exchange of wealth. It is essential that this be understood clearly when appraising economics.

INTERNATIONAL MONETARY SYSTEM

So long as only one system of currency was involved, monetary value was relatively easy to establish; a natural kind of economic equilibrium developed. At this point in time though, commerce is truly worldwide and involves different nations with different currencies. This means that value is not established so easily. Throughout much of man's economic history this inherent problem has been compromised by pegging all currencies to the one which was predominant.

Roman coinage was the basis for many years. The British pound sterling was the basic world standard during that period when the British Empire dominated the economic scene. Each of the nations within the empire's control might issue its own coinage but it was tied to sterling for the value purposes. Vestiges of this remain to this day.

In the period just preceding - and during - World War II, the dominance of the British Empire had waned and the United States was becoming the dominant economic nation. That war completed the change for all practical purposes.

By 1944, when the Bretton Woods conference was held, the U.S. dollar was truly the only currency possessing a stable value worldwide. The dollar and sterling were then made the world standards for exchange of all other currencies. For all practical purposes the dollar was the basis. One could use sterling as an exchange basis, but it in turn was "supported" by the U.S. dollar.

This was the basic nature of things until 1971. During this period other nations developed as significant contributors to world commerce - particularly Japan and West Germany. The formation of the European Economic Community (EEC or "common market") affected the economic balance. All of these things, plus changes in internal U.S. economics, made it ever more awkward to use the U.S. dollar as a fairly inflexible standard. What was happening, very simply, was that productivity of the U.S. economy was not rising commensurately with the rest of the world. As a result, true relative money value was changing to a degree different from that allowed by existing monetary agreements. The relative value of trade between "currencies" and the price of comparable goods was different from the official exchange rate. This was reflected in the U.S. balance-of-payments.

By 1977-78 the U.S. dollar had fallen in value. This caused a net decrease in real purchasing power for OPEC nations and others who pegged prices on the U.S. dollar. This caused serious economic ramifications. If the price was raised 15% to counteract a dollar decline of 15% (and keep purchasing power constant) there could be serious recessionary pressures which would affect sales volume. One alternative was to base prices on a mix of currencies.

Having one's currency used as a world standard may be flattering but it also imposes some significant economic burdens. Some economies have collapsed for this primary reason.

The simple truth is that using a country's currency as a world monetary standard is contradictory to that country's self-interest. An ideal international payments and exchange system must facilitate international commerce by making it easy. Relative conversion values should remain "steady" to permit reasonable prediction of the value of future cash flows for a given investment. This desire inserts a time inflexibility.

Now, during a given time period, the domestic situation changes in the country whose currency is the standard. A deficit balance-of-payments results. The imbalance can continue only so long as the deficit country has sufficient reserves or available credit. Neither of these is ever unlimited.

This is what happened to the U.S. in the 1970's. Some payments adjustment was necessary to preserve U.S. internal fiscal goals. Several devices were used to make the adjustment:

1. A 10% surtax on most imported goods was temporarily imposed.

2. Pressure was exerted on principal trading partners to revalue their currency relative to the U.S. dollar.

3. Dollars could no longer be redeemed in gold.

The purpose - obtain a favorable balance-of-payments. If a major trading partner of the U.S. revalues their currency upward, such as Germany did in 1970, U.S. products become cheaper to he German citizen, thus encouraging importation of U.S. goods; while German goods become more expensive in the U.S., which discourages U.S. citizens from importing German goods. The result, hopefully, is a balanced balance-of-payments for both countries.

This, however, is a somewhat simplistic conclusion. Market reallocation does not occur instantaneously. Involved are capital investments, closing and opening plants, developing new market channels, etc. No reasonable degree of economic equilibrium may occur.

In addition, it complicates longer term investments. "Expensive" investments may return "cheaper" money in terms of purchasing power.

The name of the game is compromise - maintaining a necessary, workable international system that is compatible with the selfish, internal interest of each participating country. The extent to which any one country can afford to compromise depends on how sensitive its internal fiscal strength is to a favorable balance-of-payments. Since the U.S. exports only 4-5% of its gross national product, it inherently would be more sensitive to internal fiscal requirements than say Japan.

Mechanisms for Payments Adjustment

Some understanding of the mechanisms used to preserve a degree of equilibrium in world monetary affairs is necessary to appreciate the probelm.

Devaluation. - One is devaluation of your currency. Great Britain used this mechanism when they devaluated the pound from $2.80 to $2.40 (U.S.). The purpose was to encourage exports and discourage imports - hopefully accompanied by an increase in productive efficiency. To some degree, the British accomplished this. By changing the ratio, the pound that used to buy $2.80 of goods in America would now purchase only $2.40 worth. Unless the British consumers had an extremely inelastic demand, foreign purchases should probably slow down. On the other hand, the American importer could then buy 14 per cent more goods with the same outlay of money. Of course, this presumes that the citizens of the devalued currency country will be content with this situation and don't demand an increase in wages to offset the devaluation.

Changing the exchange rate is a good method for temporarily achieving the desired goals except when nations, unwilling to allow each other a trade advantage, follow suit - leaving the original country in the same relative position as before. Following the British devaluation, Denmark and many other countries reduced the value of their currency by the same 14%, resulting in shifting the burden to the supposedly stronger nations, such as the U.S. If the U.S. had reduced the value of the dollar at the same time the British devalued, nothing would have been accomplished from Britain's standpoint. In essence, the changing of exchange rates is only effective when other nations do not reciprocate. Failure to reciprocate places a burden on these nations' potential balance-of-payments (places pressure on their currency).

Price changes. - The exchange rate is a price, but a very special price, because changes in it alter the relationship between the prices of goods and services at home and the cost of foreign goods and services. It is fairly obvious that the same goals could be accomplished by holding exchange rates constant and changing prices at home, making your goods cheaper to aliens and theirs more expensive. For instance, a 14% change in home prices would serve the same function as devaluation, if politically feasible. Although price changes are mainly caused by market forces, discretionary use of government powers will also enable reductions to be made. It is not a favorable ploy of most Western governments.

Income changes. - The next means of change revolves around the ability-to-buy mechanism. Obviously, an increase in income induces a willingness to import more from abroad. To look at it the other way, as Europe's incomes grow, the more they will be willing to import more from the U.S. or vice versa. If, then, the U.S. has a balance-of-payments deficit, one way to help remedy the problem is to reduce personal incomes by heavier taxation (like a 10% surcharge) or letting

56

the natural operations of the market take care of it. This presumes the government does not increase expenditures as a result of the taxation.

Supposedly, the natural market mechanisms work on the principle that if exports exceed imports, there is a net inflow of goods making the exporting country richer. With enlarged income, the exporting country begins to increase imports, until theoretically the two begin to balance. However, for the market mechanism to work it must be able to stop at the place where imports equal exports (unlikely). Even with help from governmental controls, which can either aid or deter the market mechanism, there is no assurance the income changes will bring about the desired results, due to inadequacy of the market and the problem of determining just how much to change incomes.

Deflationary monetary and fiscal policies tend to reduce balance-of-payment deficits, while expansionary policies increase deficits; but these methods are not very popular.

Controls. - While all the previous methods have attempted to work through the market, adjustment through controls seeks to suppress the market forces. Quotas and tariffs are the best examples of this system.

The critical question in discussing any such idea is its effectiveness, ease of operation, and cost. Since controls do not change desires of the citizen to import and export, there is always the tendency to circumvent the rules. Even if the regulations are adhered to, there is no assurance they will impose the desired effects. A good case would be the restriction of imports, causing the demand to shift to domestic goods, usually resulting in higher prices; or initiating a shift of the limited goods from export industries into the domestic industries, which will not reduce the disequilibrium at all. Although it is a long, complicated process with unlimited wants and scarce resources, controls that do not change desires cannot be completely effective. In some situations this may be the only possible mechanism of adjustment. In all fairness to the use of controls, situations arise when market factors cannot alone produce the desired results.

Basic Economic System

There are several basic (pure) systems depending on the criteria used. The principal criteria are (a) the degree of stability of exchange rates and (b) the extent to which market forces are allowed to operate. Depending on (a) and (b), a third criterion is the balance-of-payments adjustment mechanism. Using these criteria, there are four pure systems.

Pure gold standard system. - The pure gold standard system consists of stable exchange rages, with market operations free from direct controls and with primary adjustment through automatic price and income changes. The exchange rates of each country are based on the gold content of the monetary unit, set by the government, so countries are really comparing gold contents rather than just money. A major characteristic of this system requires each government to buy and sell gold at some base price in order to maintain the value of the currency. In the 1930's when the U.S. and Britain were on the gold standard, the pound sterling was defined as 113 grains of fine gold, while the U.S. dollar contained 23.22 grains of fine gold. The determination of the rate of exchange involved simple arithmetic: 113 divided by 23.22, or 1 pound = $4.86+.

The critical point for the pure gold standard system is the point where the actual rate of exchange varies more than the base rate of exchange by more than the transportation and insurance costs. When it does, it allows dealers to earn profits by buying gold in one country, selling it in another and then exchanging the profits for the original currency. Suppose the market rate of exchange changed to 1 pound = $2.42, with the base rate being maintained at $2.40 to the pound. Dealers would buy gold in the U.S. for $2.40, ship it to England for pounds, and then trade the pounds for dollars on the open market. These dealers have bought 1 pound for $2.40 and sold it for $2.42. The net result is a gold outflow from the U.S.

Gold is not the only means used to dispense with the disequilibrium problem because short term capital movements will, if the dealers feel the rate of exchange will never exceed $2.42, send money into the deficit country, restraining the movement of the exchange rate.

Adjustment problems can be very serious under the gold standard because of the limited quantity of gold each country can hold and the amount of foreign short-term assets available to each country. Naturally, as the deficit continues, foreigners will be less and less willing to accumulate claims for protection against possible devaluation of currency. The changing of exchange standards and initiation of quantity controls are anathema to the gold standard, thus only allowing the use of price and income changes.

The stability maintained by the constant exchange rates served as an impetus to world trade, especially during the last quarter of the 19th Century to World War I, when it was used extensively. These characteristics have been the basis for the large number of people wanting to return to the pure gold standard.

Unfortunately, conditions changed so much that most countries were not willing to accept the basic tenet of the pure gold standard system, namely the subordination of national economics to a world order of things. To maintain stable exchange rates, price and income within your country must be manipulated to work on disequilibrium. If price and income deflation is fought, the probable result would be the abandonment of fixed values of currencies - free trade. This is exactly what happened in the early 1930's because almost all countries began looking to national problems, not the world's.

In essence, the failure of the pure gold standard today is the unwillingness of countries dedicated to the maintenance of full employment and price stability to accept internal disorder for the sake of external equilibrium. The problem is caused by the conflicting policies needed to reach both goals at the same time.

Freely fluctuating exchange rates. - With the collapse of the gold standard system in the '30's, many economists diagnosed the cause as the inflexible exchange rates and proposed the logical cure, flexible exchange rates. These would allow each country the opportunity to carry on a domestic policy of full employment and price stability (and balance-of-payments equilibrium). The flexible exchange rate means that market forces determine what the exchange rate will be rather than the gold backing of each currency. Instead of using price, income, or quantity controls, the entire process of stabilization is brought about by changing the exchange rates.

Although the freely fluctuating exchange rate system advances internal or domestic programs, it is directly opposed to creating an atmosphere conducive to international trade. Since stable exchange rates are the only consistent means of communication between international traders, the use of flexible rate makes international cost comparisons from day to day extremely difficult and fosters an element of risk, giving rise to unproductive speculation.

If the nation is reasonably stable, speculation can aid in maintaining stability of the exchange rates, but should serious disequilibrium occur, speculation only hastens the breakup of the currency. When the exchange rate begins to rise and speculators feel it will continue to do so, domestic importers will accelerate their foreign orders before they become too expensive, while foreigners will want to put off buying goods on the hunch they will be cheaper in a couple of days, causing the disequilibrium to increase instead of becoming stable.

When evaluation of this system is rendered (pro or con), it must be done with extreme caution because of the lack of factual data as to its effectiveness. Flexible exchange rates were used to some degree in the early 1920's and '30's; but, due to lack of participation by the major powers, no conclusive judgments were reached as to the workability of flexible exchange rates, even though more faults than virtues were noted. Canada used a similar system from 1950 to 1961 and achieved surprising stability. Unfortunately, this cannot be considered a reliable test because of the closeness of the Canadian dollar to the American dollar.

Exchange controls. - As quantity controls are used to influence an economic disequilibrium, there are international systems based on exchange controls. Once the exchange rate is set, the exchange controls attempt to solve balance-of-payment problems by non-economic methods. In this type of system all economic and financial transactions are subject to the approval of a control board. In addition, all foreign exchange, bought or sold, has to be channeled through control board offices. Usually the exchange rates are at fixed levels except for transactions in a free market area. Also, the home currency is usually overvalued - that is, the rate is held below the market equilibrium level.

Various devices are used to keep the balance of payment system under control, with the heart of the system being power over international transaction. Probably the most commonly used method is "multiple" exchange rates, meaning the setting of different rates for different goods. By setting a high rate of exchange on luxury goods and low rate for raw materials, they, in effect, are discouraging the imports of luxury goods and encouraging the buying of the goods considered important.

Exchange controls are usually thought to be a mere continuation of a centrally controlled economy, but all countries have used these powers during an economic or monetary crisis, deep recession or inflation, persistent balance-of-payment problems, or as a part of a national program for development.

With the advent of the Great Depression over the world, exchange controls came into popular use and lasted in Western Europe until 1958 when the countries felt they were strong enough to compete with the rest of the world. Many of the Latin American countries and less developed nations still maintain rigid controls (as do the U.S. and Western Europe to a limited extent).

In spite of the inadequacies of exchange controls, conditions may arise where they are merely the lesser of many evils. If a severe domestic inflation occurs, a freely fluctuating system will only encourage the growth rate, while with a gold standard with stable rates there may be enough gold in reserve to meet all the demands. A non-economic factor necessitating controls would be a political instability where the only way to keep capital in the country is to prevent its exit.

Generally, the only logical basis for the use of exchange controls rests solely on their ability to do the job with less pain than any other mechanism. Exchange controls, price, and income effects may not be strong enough to handle the job. Even if they can, the consequences may be worse than the original problem, leaving only exchange controls to work on the deficit.

Flexible exchange. - In an attempt to combine the advantages of the gold standard and flexible exchange rates, theoreticians developed the flexible exchange system. Here the exchange rates are neither kept at fixed, predetermined levels, nor allowed to freely fluctuate with the private market forces. There is controlled freedom of market transaction. This function is the responsibility of some government agency, usually called a Stabilization Fund, which indirectly controls movements by selling and buying foreign exchange, buying to retard a fall in exchange rates and selling to prevent a rise. In most cases the Stabilization Fund is operated under, and supplied by, the national bank or treasury.

The Bank of England serves such a function. Pounds may be bought on the open market, when the pound drops in value, and then sold when the rate starts to rise above a desired level.

The attractive part about this system is the ability to have either a stable exchange rate, as under the gold standard, or a freely fluctuating rate, depending on how the men running the system react to the problem. If they remain passive, the freely fluctuating section will become active, while active participation to keep the same rate is reminiscent of the gold standard. In hopes of counter-acting the disadvantages of one, the directors merely bring into play the advantages of the other; by temporary fluctuations in rates through market operations, the problems of free fluctuating rates are alleviated. Allowing rate change insures against the rigidity of the gold standard.

Unfortunately for the world, this system is no panacea because the system requires the almost instantaneous appraisal by fallible men of the nature of the problems, their causes, and the best means of curing them (resulting often in either over-reaction or under-reaction).

In an attempt to help determine the right tactics to use, purchasing power parity was developed. This holds that the equilibrium rate is one equalizing the purchasing power at home and abroad, found by multiplying the past period's equilibrium rate by the changes that have taken place in the purchasing power of each country. Unfortunately, this theoretical device is almost as imperfect as the original idea because the theory assumes only relative prices affect balance-of- payment and ignores cylical or structural changes.

The major weakness of the plan is the need for close cooperation among the countries of the world. Because there is no international organization with the power or capability to prevent national policy boards from running on a collision course, problems are inevitable. Such conflicts were not uncommon in the 1930's until cooperation of the national monetary authorities was instituted under the 1936 Tripartite Agreement between the U.S., Great Britain and France. With this as a basis, the post-war monetary system was developed from the flexible exchange rates with a few revisions.

Post World War II

The contemporary money system, as envisioned by the originators at the Bretton Woods Conference in 1944, closely approximated the flexible exchange rate system, with several major changes, hopefully adding to its effectiveness. The dream of the creators was to avoid

the rigid exchange rates and deflationary adjustment mechanism of the
gold standard, the instability of freely fluctuating rates, the conflicts of national policies and competitive exchange depreciation under
the flexible rate plan, and the distortions caused by exchange controls,
while capturing the stability of the gold standard, the easy adjustment
and market freedom of freely fluctuating rates, discretionary control
over rates under the flexible plan, and selective use of quantitative
controls when needed.

In addition, in trying to combine the best of all systems, these men realized the desper-
ate need for an international organization to aid and control international commerce. This
prompted the creation of the International Monetary Fund (IMF) at the same conference.

It was fairly obvious that any new system made from the older types would involve compromises on all sides. But, with the inadequacies of the other four possible choices, all the
delegates were willing. From their work sprang the adjustable peg exchange system. Essentially,
each country has its money pegged at a certain par value, based on gold, but any object would do
(such as elephants or oil wells). These rates were subject to repegging if the situation called
for it. The most important objectives of the system were freedom from quantitative controls
and domestic economic stability with full employment. The planners were trying to prevent the
clash between domestic policy and international transaction, which was one of the major problems
of the previous systems characterized by stable exchange rates. Probably as important as pre-
venting international clashes was their realization that changing the value of a currency is not
the exclusive prerogative of each country, but should be under the jurisdiction of an international agency, such as the International Monetary Fund.

The IMF has become the hub of the monetary world with a membership of over 100 countries,
including all the major trading powers outside the Soviet Union and mainland China. The IMF is
guided by a Board of Governors selected from member countries with the responsibility for providing assistance to the individual nations and carrying out the obligations or rules set down
by the Articles of Agreement.

Maintenance of stability and freedom from quantitative controls are the goals of the IMF.
Due to the limited quantity of reserves in the possession of each nation, a larger cache of reserves had to be made available in times of emergencies. This fostered the creation of a revolv-
ing pool of credit called drawing rights in the IMF. The source of this extra reserve is made
available by member quotas determined by the Board of Governors, based on each country's national
income and amount of foreign trade - payable in gold, 25%, and currency of country, 75%. When a
member is running a balance-of-payments deficit and has insufficient reserves to meet this outflow, the country simply borrows the kind of foreign exchange needed from the IMF. The deficit
country exchanges part of their reserves in the IMF for the desired currency. Hypothetically, if
France owed the U.S. larger sums of money than it had reserves, say a 4 billion dollar deficit,
France could exchange 20 billion francs (at 5 francs = $1) and receive the necessary 4 billion
dollars to pay the U.S.

Although the example indicates a purchasing operation, it is really a credit operation,
because France now has to repurchase all of its currency over the quota set by the IMF and pay

interest on the outstanding drawings (in gold or acceptable foreign exchange). However, the drawing rights system is designed and intended only to meet temporary and occasional distrubances, thus limiting its effectiveness.

In addition, an added burden is placed on the performance of the system by the regulation against allowing countries the right to borrow any amount that would increase IMF's holding of their currency by 200% (or 25% in any 12 month period). Because the IMF considered short-run fluctuations inevitable, they only considered helping member nations for that length of time. The IMF has always felt that the responsibility for long-run equilibrium rests with each country maintaining a sound economic policy. Making provisions for dealing with long-term deficits was not the IMF's responsibility.

Unfortunately, economics run by men are not perfect, and a few carefully placed mistakes at the right time, or shifts in demand and technological changes, can lead to events no one has any control over. Both necessitate the use of more drastic measures, such as devaluation or repegging. To safeguard against flagrant use of this power, IMF rules originally required any devaluation of over 10% to be approved by the IMF member countries. Nations will not go along with devaluation unless the disequilibrium is not amenable to correction at the present rates. The threat of exchange controls by the deficit country, or of serious unemployment or inflation in the deficit country without devaluation must be very positive.

As good as this system sounded on paper, in the actual operation of the system (versus the theoretical operation), drastic modifications (especially from the U.S. point of view) changed the entire outlook from an adjustable peg mechanism to that of the gold standard of the 1920's. The IMF adopted reserve currencies which were considered "as good as gold" and actually used as payment instead of gold, such as the U.S. dollar. All IMF currencies were considered pegged to gold, with the U.S. Dollar being regarded as being "more pegged" than the others. This meant, in effect, that the U.S. was still on the gold standard, while the rest of the world followed the essentials of the adjustable peg system.

The burden placed on the U.S. was a heavy one, because as the volume of international trade grew, more reserves were needed by each country to pay for their international trade. But..
... the gold supply remained fairly constant, leaving only the "good as gold" dollar to increase countries' reserves.

Simple arithmetic indicates that for the reserves of all the other countries to increase, the supplying country must dispense more than is coming in, resulting in a deficit. Since the dollar has been considered a substitute for gold, many billions have been held as reserves by nations and individuals. Devaluation of the dollar from $35 per 1 oz. of gold to something higher would involve capital losses for these holders of the dollar. The U.S. was therefore "damned" if they changed the dollar value of gold and equally "damned" if they didn't.

The IMF has developed special arrangements to meet the basic confidence problem. The IMF gives the U.S. preferential access to reserves in which the six largest members contribute additional resources from their own private reserves. The other measure is made outside the IMF between national banks to exchange currency in cases of emergencies.

The use of these two methods came to the public's attention after the devaluation of the British pound in November 1967, and the rise in speculation over the future of the U.S. dollar placed added pressure on U.S. gold supply. At this time the "gold pool" composed of the U.S., Switzerland, and other nations met regularly to lend support to the attempts of the U.S. to stave off the run. Probably the key to the limited success of the U.S. in fighting off the loss of confidence was the aid received from these other countries rather than having them join the "run."

The confidence problem is essentially a short-run liquidity problem. The U.S. has enough total assets to pay off the claims against it, but does not have sufficient international liquid assets, such as gold or foreign exchange. In the long run there simply are not enough international reserves to finance a growing volume of trade throughout the world. Additions to reserves depend on the gold supply, which is not produced according to need, leaving only the reserve currency to supply the increased demand. This leads to a dilemma: As the proportion of key currency reserves to gold increases, the liquidity of the key currency declines. The vulnerability of key currency countries is increased, causing the further growth of reserves to be stunted by the lessening acceptability of key currency balances and pressure on key currency countries to eliminate their payments deficits. This has been the dilemma of the U.S.

When the reserve country reaches this stage, aid is needed from the IMF, but IMF funds can only serve to finance temporary payment deficits and cannot cure these problems. Also, the key currency countries cannot change the exchange rate for fear of destroying the entire system. It is also unlikely that the reserve country will resort to price or income deflation, except in the most critical of situations. Only the exchange controls are left and they are not permitted to be used by the constitution of the IMF. In any event, there has been no satisfactory means of adjustment developed for the reserve currency country. The only real recourse is a sound fiscal policy and high productivity.

Table 4.1 represents the model balance of payment sheet in a very simplified form used by the U.S. and most countries in the world. It is extremely useful to a nation running a deficit, because only by determining in what area the deficit occurs can a country decide the appropriate adjustment mechanism to tackle the problem. Random choice of cures, such as a ban on tourist travel to foreign countries, ignores the realities involved and in some radical cases can do more harm than good, resulting in an increased deficit instead of curing the problem.

This simplified form is divided into four sections, each with many different subdivisions based on the characteristics of the transaction. This accounting procedure uses the double-entry form of bookkeeping. At the end of each fiscal year the budget will balance, but this is of no analytical interest in evaluation of the system.

The U.S., in the course of a fiscal year, might show an extremely large profit in the current account. Obviously, the source of any trouble must come from the deficits of capital and unilateral transfer accounts exceeding the profits obtained in the current account. The simplest way of stating the cause of the deficit is the failure of the current account to adjust fully to changes in the capital and unilateral transfer accounts, known as the transfer mechanism. This is the adjustment process by which monetary transfers (capital movements and unilateral payments) are converted into real transfers in the form of net movement of goods and services.

The best way to understand the transfer mechanism is to think of it as the large outflow of capital (including unilateral transfers) creating a large deficit. By purchasing goods and services the net credit balance on the current account should just equal the net debit balance on the capital account. If, for some reason, the two don't balance, the logical approach is to change the exchange rate since it is an indication that the dollar does not have the correct par value. But, the U.S. position in the current monetary system precludes any use of this devaluation mechanism.

The price and income effects could be brought into play. The domestic policy of full employment begun by the Kennedy administration and continued thereafter - allowing the market forces to work on prices and income - only make the goals of full employment and balance-of-payments equilibrium harder to achieve. The real problem is not the lack of an adequate transfer mechanism, but the result of trying to achieve two goals in direct opposition to each other.

One very important concept has been indicated, but not dealt with in the detail it deserves; namely, just who has the power to aid in curing the causes of a deficit? The main burden has been placed on the country incurring the deficit, but in most cases the nations earning a profit could do more than the deficit country, because when a nation such as the U.S. has its currency used as a reserve currency, it cannot afford to depreciate or devalue the dollar very much. If the other major trading nations would appreciate their money, making them worth more, the same result would be achieved.

One should avoid making overly simple conclusions from commonly published balance-of-payments numbers. What they are often quoting is simply item I(A) in the current account shown in Table 4.1, the difference between import and export of merchandise. For example, in 1977 the U.S. trade deficit was $26.7 billion. The oil import bill for that year was $42.1 billion. From this one can conclude that except for oil the U.S. exported $15.4 billion more goods than it imported. Such numbers are important but they are not the only elements affecting total balance-of-payments.

Recent Events

A series of events starting in 1970-72 illustrate the U.S. problem. The U.S. raised the price of gold which constituted tacit dollar devaluation. Dollars were no longer redeemable in gold. Countries like West Germany and Japan with a surplus balance-of-payments increased the value of their currency. The IMF broadened the support range somewhat to provide greater flexibility. Everyone knew this was a stop-gap measure to "buy time" until some more permanent solutions could be found, hopefully.

In 1972 the British pound was allowed to "float" to counteract speculation in that currency, causing a crack in the temporary arrangements. In the free market the pound sterling attained a value of $2.45-$2.50 (from $2.60). The need to float (a unilateral decision) was made in the face of a current balance-of-payments surplus. This indicates the sensitivity of any international monetary system to an individual country's fluctuating circumstance.

By 1977 the pound had dropped to about $1.55 and then rebounded to about $1.95 in 1978. These kinds of fluctuations had many adverse effects. Nevertheless, there has been a pronounced trend for countries to let their currency "float," with some intervention when the occasion dictated.

Summary

As this book goes to press we can look forward to continuing upsets. Every industrialized country faces

- Increasing demands for higher salaries without commensurate increases in productivity.

- Increased costs of government services in the face of growing taxpayer rebellion against higher taxes to support said services.

- The almost impossible political demand of full employment while curbing inflation - two incompatible problems.

- The realization that internal fiscal policies must be compatible to a reasonable degree with that of other countries.

The net result is a situation that is most sensitive and can lead to fluctuating prosperity and recession during short time periods.

The question of productivity is a most perturbing one for it cannot be solved positively by government. Decreasing productivity results from some combination of two circumstances:

1. Inadequate capital investment in machinery and plants to allow efficient use of labor.

2. Worker attitudes which promote the idea of more pay for less work.

Simply replacing workers by machines causes a serious secondary problem - unemployment. Unemployment is an overhead against productivity through taxes. When demand is elastic the forces of overhead (capital cost, taxes, wages, etc.) are on a collision course with salable quantity of production.

Since minerals are a basic international product, it is affected by all this. The worth of any mineral property will fluctuate in some random manner as the economic system fluctuates. Even though the calculations in this book do not formally incorporate internal monetary matters, they are nebertheless a factor affecting value.

SUGGESTED READING

1. Snider, D. A., Introduction to International Economics, R. D. Irwin, Inc.
2. National Planning Assoc., Pamphlet No. 130; Washington, D. C. (1971).

Table 4.1 General Accounting Method for Balance of
Payments Calculations

	Debit	Credit
I. Current account:		
A. Merchandise trade:		
1. Merchandise imports	X	
2. Merchandise exports		X
B. Service transactions:		
1. Transportation:	X	
a) Rendered by foreign vessels, airlines, etc. . .		X
b) Rendered by domestic vessels, airlines, etc. . .		
2. Travel expenditures:		
a) In foreign countries	X	
b) By foreigners in home country		X
3. Interest and dividends:		
a) Paid to foreigners	X	
b) Received from abroad		X
4. Banking and insurance service:		
a) Rendered by foreign institutions	X	
b) Rendered to foreigners by domestic institutions		X
5. Government expenditures:		
a) By home government abroad	X	
b) By foreign government in home country		X
II. Capital account:		
A. Long-term:		
1. Purchase of securities from foreigners*	X	
2. Sale of securities to foreigners		X
B. Short-term:**		
1. Increase of bank and brokerage balances abroad. . .	X	
2. Decrease of foreign-held bank and brokerage balances in home country	X	
3. Increase of foreign-held bank and brokerage balances in home country		X
4. Decrease of bank and brokerage balances abroad . .		X
III. Unilateral transfers:		
A. Private:		
1. Personal and institutional remittances to non-residents	X	
2. Remittances received from abroad		X

	Debit	Credit

B. Governmental

    1. Grants, indemnities, and reparations made to other countries   X

    2. Grants, indemnities, and reparations received from other countries   X

IV. Gold account:

    A. Import of gold and increase of earmarked gold abroad***   X

    B. Export of gold and increase of earmarked gold for foreign account***   X

*Includes new issues, transactions in outstanding issues, and transfers resulting from redemption and sinking-fund operations.

**Also includes currency holdings, acceptances, and other short-term claims not listed.

***"Earmarked" gold is gold physically held in one country for the account of another.

5

TIME VALUE OF MONEY

The chief justification for money is that it is a convenient medium of exchange and serves as a measure of purchasing power. Its value is a function of its purchasing power and the demand for it. The physical value of money is small, as it consists only of a small amount of paper and/or metal. In this respect it is no different from a piece of steel machinery whose value is determined not by the pounds of steel present but by what it can do or the profits it can render. Money is, therefore, a commodity of convenience.

Governments issuing money, of course, have assets to lend support to its value and to instill confidence; i.e., increase its demand. This demand will logically fluctuate with the economic status of its issuer. Any demand must be determined by how much purchasing power a unit of money possesses.

This may be illustrated by common practice in the money exchange markets of the world. There are official exchange rates between different currencies, but often the actual rates do not coincide with them. This is invariably the case when the relative purchasing powers of two currencies differ from the established rate between them.

For example, suppose the official exchange rate between Mexico and the U.S.A. was 12 pesos to one dollar. Further suppose that heavy machinery costing $10,000 in the U.S.A. could be purchased in Mexico for 100,000 pesos. Based on purchasing power, the equivalent exchange would be 10 pesos per dollar. If this same ratio were expended over enough commodities, the exchange rate would undoubtedly reach this level in the money market.

Time Value

Another example, among many, is afforded by money issued by the Confederate States in the 1860's. It has no value except to collectors, for it possesses no purchasing power. These facts establish the premise that money is a commodity representative of purchasing power, and as such its value is dependent on: (1) time, (2) relative supply and demand, (3) the cost of using it, and (4) its tax position.

This chapter is concerned with the effect of time, and the cost of using money, on the value of money.

COST OF USING MONEY

When a person uses money belonging to another person, he must pay rent for it, just as he would for the use of another's equipment or facilities. The rent for money is called interest, the amount borrowed being the principal.

The rate of interest that the borrower should expect to pay and the lender expect to receive will depend on several primary factors:

1. What is the current rate of interest for the term and type of financing being used?

2. What is the rating of the borrower in the financial world? More importantly, what is the risk of the borrower becoming insolvent during the term of the loan?

The lender is presumably not a party to any risk associated with the borrower's use of the money. He is merely renting money. As a practical matter though, the interest demanded depends on the reputation and capital structure of the borrower. The lender wants reasonable assurance that the money will indeed be repaid.

A corporation is in no different position than an individual in this regard. If you can offer security (by reputation and personal worth), you would normally borrow from a bank rather than a loan company which specializes in high risk loans at higher interest rates. Comparably, an established company with a sound record and competent management can borrow money at less cost than one without these attributes.

The size of the loan compared to the capital assets is a factor. The smaller the ratio, the greater the "security." The lender must likewise exhibit a cash flow which appears large enough to cover principal and interest payments as well as other obligations.

When raising money using venture funds, convertible bonds or issuance of more stock, the interest rate is not a firm number. The convertible bond carries a promise to pay so much at a certain rate for a given period of time, but the holder may convert it into stock which nullifies that rate. He is now an owner of the general assets rather than merely holding a mortgage against those assets. Holders of stock expect some combination of dividends and asset appreciation to preserve stock value, but there is no specified interest rate per se.

For time value of money purposes in the evaluation, it is common to use an interest rate that is compatible with bank loan rates, if any form of outside financing is used.

If internal funds are to be utilized in an investment, an interest charge is still made. In this case the proper charge is the recent historical rate of return of the organization on its capital investments. Often this figure is determined by dividing net cash flow in a given year by book value of assets. The interest rate thus developed is usually 9-12%; 10% is used commonly for evaluation purposes. This is inherently a "no-risk" return on the basis that it includes both successes and failures when employing an investment strategy that is expected to continue.

This charge is made to keep project money on a consistent basis. The given project is renting money from the general company treasury. Presumably, if said money was not invested in the project, it could be invested elsewhere at this average opportunity rate.

Before discussing the value of money from the investment viewpoint it is necessary to understand the basic interest equations from which the models are derived. These equations simply outline the effect of time on money value for a given repayment schedule, using one or more interest rates as a measure of time value.

BASIC INTEREST EQUATIONS

Many formulas have been developed for the calculation of these quantities, and these have been widely shown in the literature.

Simple interest. Simple interest rates are usually quoted as the percentage of the principal for a period of one year. With simple interest, the amount to be paid when the loan is repaid in full is dependent on the time the loan has been in effect. This type of loan may be for any period, and is usually the type obtained by an individual from a bank. If one borrows $1000 for one year at 6 per cent simple interest, he must repay $1060.

Compound interest. With compound interest the amount of interest must be paid when due at the end of the interest period, and, if not, is added to the principal for calculation of interest during the next period. In cases where the money has been borrowed for a fixed length of time, the principal plus interest being due at that time, the following sample calculation will illustrate the principal involved. In this example, it is assumed that we are borrowing $2000 with no payments until the end of year 4. Since we must return both principal and interest we could find that payment due is $2524.95 as:

Year	Principal at beginning of Year	Interest, 6 Per Cent	Amount Owed at End of Year
1	$2000.00	$120.00	$2120.00
2	$2120.00	$127.20	$2247.20
3	$2247.20	$134.83	$2382.03
4	$2382.03	$142.92	$2524.95

(Final payment)

Although the interest rate is usually shown per annum, it may be compounded for periods less than one year. It may compound daily, monthly, quarterly, semi-annually, or annually.

The following nomenclature will be used in the interest formulas developed:

n = number of interest periods (normally in years)

i = effective annual interest rate, a fraction

j = Nominal annual interest rate, a fraction

m = number of interest periods per year

C = principal sum subject to interest (loaned or invested at time zero)

S = a future sum, n years away from principal investment C, equivalent to C as related by an appropriate interest payment

i_m = effective interest rate per interest period, $i_m = j/m$

I_n = income (or payment) made at the end of each interest period for equal-payment series

Using this nomenclature, the amount of the principal plus interest at the end of the first interest period would be $C(1 + i)$, at the end of the second period $C(1 + i)^2$, etc. If these are summed up for n interest periods, one would obtain the expression

$$C(1 + i)^n = S \qquad (5.1)$$

or

$$C = S \left[\frac{1}{(1 + i)^n} \right] \qquad (5.2)$$

Equation (5.1) says that if a sum C invested at the present is compounded at i per cent interest, it would have a value S at the end of n interest periods. Equation (5.2), which is simply an algebraic rearrangement of (5.1), says that C is the present value of a sum S received n interest periods away.

The values C and S are equivalent. In a previous example, $2000 compounded at six per cent annually was worth $2525 at the end of four years. These two sums are equivalent, for any-one regarding six per cent interest as fair would be willing to pay $2000 to receive $2525 four years hence, or vice versa, disregarding any inflationary trends.

The term in the brackets is known as the _deferment_ or _discount_ factor. The value S received at some future time is then known as the _undeferred_ value. The discounted value of S is then the _deferred_ value. This latter value, when corrected for all costs of obtaining the revenue, is known as the _net deferred_ value. This value represents all future net profit from the investment converted to present value. It has been common for purchasers of producing oil properties to pay from 2/3 to 3/4 of this value where a realistic bank interest rate has been used.

Nominal and Effective Interest Rates

The interest rates i and j are equal if the principal sum earns interest once per year. If interest is credited to the investment for periods less than one year

$$i = \left(1 + \frac{j}{m}\right)^m - 1 \qquad (5.3)$$

The _nominal rate_ is the one usually used by banks and other lending agencies in outlining rates of payment. The _effective rate_ is the real annual rate, payed or earned, reflecting the interest period used for compounding purposes.

Example 5.1: A bank advertises an interest rate of 4% on savings accounts, compounded quarterly. What is the effective rate?

$$i = \left(1 + \frac{0.04}{4}\right)^4 - 1 = 0.041$$

This means that the effective annual rate is 4.1% for the nominal rate quoted.

71

Continual compounding. It is convenient in many calculations to assume that interest is compounded continuously. This means that there are an infinite number of interest periods (m) per year.

When m equals infinity in Equation 5.3

$$i = e^j - 1, \qquad j = \ln(1 + i) \tag{5.4}$$

and

$$e^{jn} = (1 + i)^n \tag{5.4a}$$

where: ln = logarithm to the base e

From Equation 5.4a it is possible to substitute the expression e^{jn} for $(1 + i)^n$ in Equations 5.1 and 5.2 when interest is compounded continuously.

Continuous compounding is seldom used in banking circles but is convenient for use in evaluations. This will be summarized in detail in Chapter 19.

Equations 5.1 and 5.2 represent the basic compound interest equations. It is convenient to develop specific equations for given circumstances using the same basic principles. The nomenclature used by Thuesen (1) will be employed for this purpose.

Equal-payment-series, compound-amount factor. A case may arise where a series of equal payments is made at the end of each interest period. What is the value of the payments plus accrued interest at the time the final payment is made?

For this purpose we will define I_n as the income (or payment) made at the end of each succeeding interest period.

$$S = I_n \left[\frac{(1 + i)^n - 1}{i} \right] \tag{5.5}$$

Example 5.2: An amount $1000 is paid at the end of each year for 5 years. The effective annual interest rate is 6%. What is the total compound amount of principal plus interest accumulated in five years?

$$S = 1000 \left[\frac{(1 + .06)^5 - 1}{0.06} \right] = (1000) \ (5.637) = \underline{\$5637}$$

Under the circumstances specified, the account would contain $5637.

Equal-payment-series, sinking fund factor. Suppose it is desired to accumulate a fixed amount of money at the end of a specified period by making a series of equal payments at a fixed interest rate. This merely involves solving Equation 5.5 for I, given a value of S.

$$I_n = S \left[\frac{i}{(1 + i)^n - 1} \right] \tag{5.6}$$

Example 5.3:   It is desired to accumulate $10,000 by making a series of six equal, annual payments. If the effective annual rate is 8%, what is the amount of each payment?

$$I_n = 10,000 \left[\frac{0.08}{(1 + 0.08)^6 - 1} \right] = (10,000)\ (0.1363) = \underline{\$1363}$$

Six payments of $1363 plus accumulated interest would yield $10,000 at the end of six years.

Equal-payment-series, capital-recovery factor. This merely provides the amount of a series of equal payments when a principal amount C is invested at a given rate.

$$I_n = C \left[\frac{i(1 + i)^n}{(1 + i)^n - 1} \right] \tag{5.7}$$

Example 5.4: $10,000 invested at 6% will provide for ten equal payments of what amounts?

$$I_n = 10,000 \left[\frac{0.06(1 + .06)^{10}}{(1 + .06)^{10} - 1} \right] = (10,000)\ (0.1359) = \underline{\$1359}$$

Equal-payment-series, present-worth factor. This is merely Equation 5.7 solved for C, for specified values of I_n and i.

$$C = I_n \left[\frac{(1 + i)^n - 1}{i(1 + i)^n} \right] \tag{5.8}$$

Example 5.5:   What is the present value of ten annual payments of $1359 each, discounted at 6%, if the first payment is made at the end of the first year.

$$C = (1359) \left[\frac{(1 + .06)^{10} - 1}{0.06(1 + .06)^{10}} \right] = (1359)\ (7.36) = \underline{\$10,000}$$

Use of tables. Equations 5.1, 5.2 and 5.5-5.8 each have a bracket quantity containing i and n. It is convenient to represent these brackets in tabular form so that each calculation is reduced to simple arithmetic. Tables 5.1-5.6 are provided for this purpose. The values in each of the above examples were taken from these tables.

Note (in using tables): In some books using interest tables like these, i is defined as the interest rate per interest period (our i_m); n is defined as the number of interest periods; and I_n is defined as one of a series of equal payments at the end of each interest period. These definitions may be used with these tables as well as those in this book. Our definitions herein have been made for evaluation convenience.

More complete tables are available from handbooks and Reference 1.

73

TABLE 5.2

Present Worth Factors - $1/(1+i)^n$

(Find C for a given S)

Interest Rates - i

n	5	6	7	8	10
1	0.952	0.943	0.935	0.926	0.909
2	0.907	0.890	0.873	0.857	0.826
3	0.864	0.840	0.816	0.794	0.751
4	0.823	0.792	0.763	0.735	0.683
5	0.784	0.747	0.713	0.681	0.621
6	0.746	0.705	0.666	0.630	0.565
7	0.711	0.665	0.623	0.584	0.513
8	0.677	0.627	0.582	0.540	0.467
9	0.645	0.592	0.544	0.500	0.424
10	0.614	0.558	0.508	0.463	0.386
11	0.585	0.527	0.475	0.429	0.351
12	0.557	0.497	0.444	0.397	0.319
13	0.530	0.469	0.415	0.368	0.290
14	0.505	0.442	0.388	0.341	0.263
15	0.481	0.417	0.362	0.315	0.239
16	0.458	0.394	0.339	0.292	0.218
17	0.436	0.371	0.317	0.270	0.198
18	0.416	0.350	0.296	0.250	0.180
19	0.396	0.351	0.277	0.232	0.164
20	0.377	0.312	0.258	0.215	0.149

TABLE 5.1

Compound Amount Factors - $(1+i)^n$

(Find S for a given C)

Interest Rates - i

n	5	6	7	8	10
1	1.050	1.060	1.070	1.080	1.100
2	1.103	1.124	1.145	1.166	1.210
3	1.158	1.191	1.225	1.260	1.331
4	1.216	1.262	1.311	1.360	1.464
5	1.276	1.338	1.403	1.469	1.611
6	1.340	1.419	1.501	1.587	1.772
7	1.407	1.504	1.606	1.714	1.949
8	1.477	1.594	1.718	1.851	2.144
9	1.551	1.689	1.838	1.999	2.358
10	1.629	1.791	1.967	2.159	2.594
11	1.710	1.898	2.105	2.332	2.853
12	1.796	2.012	2.252	2.518	3.138
13	1.886	2.133	2.410	2.720	3.452
14	1.980	2.261	2.579	2.937	3.797
15	2.079	2.397	2.759	3.172	4.177
16	2.183	2.540	2.952	3.426	4.595
17	2.292	2.693	3.159	3.700	5.054
18	2.407	2.854	3.380	3.996	5.560
19	2.527	3.026	3.617	4.316	6.116
20	2.653	3.207	3.870	4.661	6.727

TABLE 5.4

Sinking Fund Factors - $i/[(1+i)^n - 1]$
(Find I for a given S)

Interest Rates - i

n	5	6	7	8	10
1	1.000	1.000	1.000	1.000	1.000
2	0.488	0.485	0.483	0.481	0.476
3	0.317	0.314	0.311	0.308	0.302
4	0.232	0.229	0.225	0.222	0.215
5	0.181	0.177	0.174	0.170	0.164
6	0.147	0.143	0.140	0.136	0.130
7	0.123	0.119	0.116	0.112	0.105
8	0.105	0.101	0.097	0.094	0.087
9	0.091	0.087	0.083	0.080	0.074
10	0.080	0.076	0.072	0.069	0.063
11	0.070	0.067	0.063	0.060	0.054
12	0.063	0.059	0.056	0.053	0.047
13	0.056	0.053	0.050	0.047	0.041
14	0.051	0.048	0.044	0.041	0.036
15	0.046	0.043	0.040	0.037	0.031
16	0.042	0.039	0.036	0.033	0.028
17	0.039	0.035	0.032	0.030	0.025
18	0.036	0.032	0.029	0.027	0.022
19	0.033	0.030	0.027	0.024	0.020
20	0.030	0.027	0.024	0.022	0.017

TABLE 5.3

Compound Amount Factors - $[(1+i)^n - 1]/i$
(Find S for a given I)

Interest Rates - i

n	5	6	7	8	10
1	1.000	1.000	1.000	1.000	1.000
2	2.050	2.060	2.070	2.080	2.100
3	3.153	3.184	3.215	3.246	3.310
4	4.310	4.375	4.440	4.506	4.641
5	5.526	5.637	5.751	5.867	6.105
6	6.802	6.975	7.153	7.336	7.716
7	8.142	8.394	8.654	8.923	9.487
8	9.549	9.897	10.260	10.637	11.436
9	11.027	11.491	11.978	12.488	13.579
10	12.570	13.181	13.816	14.487	15.937
11	14.207	14.972	15.784	16.645	18.531
12	15.917	16.870	17.888	18.977	21.384
13	17.713	18.882	20.141	21.495	24.523
14	19.599	21.015	22.550	24.215	27.975
15	21.579	23.276	25.129	27.152	31.772
16	23.657	25.673	27.888	30.324	35.950
17	25.840	28.213	30.840	33.750	40.545
18	28.132	30.906	33.999	37.450	45.599
19	30.539	33.760	37.379	41.446	51.159
20	33.066	36.786	40.995	45.762	57.275

TABLE 5.6

Present Worth Factors — $[(1+i)^n - 1]/i(1+i)^n$
(Find C for a given I)

Interest Rates — i

n	5	6	7	8	10
1	0.952	0.943	0.935	0.926	0.909
2	1.859	1.833	1.808	1.783	1.736
3	2.723	2.673	2.624	2.577	2.487
4	3.546	3.465	3.387	3.312	3.170
5	4.329	4.212	4.100	3.993	3.791
6	5.076	4.917	4.767	4.623	4.355
7	5.786	5.582	5.389	5.206	4.868
8	6.463	6.210	5.971	5.747	5.335
9	7.108	6.802	6.515	6.247	5.759
10	7.722	7.360	7.024	6.710	6.144
11	8.306	7.887	7.499	7.139	6.495
12	8.863	8.384	7.943	7.536	6.814
13	9.394	8.853	8.358	7.904	7.103
14	9.899	9.295	8.745	8.244	7.367
15	10.380	9.712	9.108	8.559	7.606
16	10.838	10.106	9.447	8.851	7.824
17	11.274	10.477	9.763	9.122	8.022
18	11.690	10.828	10.059	9.372	8.201
19	12.085	11.158	10.336	9.604	8.365
20	12.462	11.470	10.594	9.818	8.514

TABLE 5.5

Capital Recovery Factors — $i(1+i)^n / [(1+i)^n - 1]$
(Find I for a given C)

Interest Rates — i

n	5	6	7	8	10
1	1.050	1.060	1.070	1.080	1.100
2	0.538	0.545	0.553	0.561	0.576
3	0.367	0.374	0.381	0.388	0.402
4	0.282	0.289	0.295	0.302	0.315
5	0.231	0.237	0.244	0.250	0.264
6	0.197	0.203	0.210	0.216	0.230
7	0.173	0.179	0.186	0.192	0.205
8	0.155	0.161	0.167	0.174	0.187
9	0.141	0.147	0.153	0.160	0.174
10	0.130	0.136	0.142	0.149	0.163
11	0.120	0.127	0.133	0.140	0.154
12	0.113	0.119	0.126	0.133	0.147
13	0.106	0.113	0.120	0.127	0.141
14	0.101	0.108	0.114	0.121	0.136
15	0.096	0.103	0.110	0.117	0.131
16	0.092	0.099	0.106	0.113	0.128
17	0.089	0.095	0.102	0.110	0.125
18	0.086	0.092	0.099	0.107	0.122
19	0.083	0.090	0.097	0.104	0.120
20	0.080	0.087	0.094	0.102	0.117

BASIC MONEY MODEL

The above equations summarize the basic relationships reflecting the time value of money. All working equations used to reflect time value of money are derived in some manner from these basic interest equations - particularly Equations 5.1 and 5.2.

The basic model for any profit oriented organization is shown below.

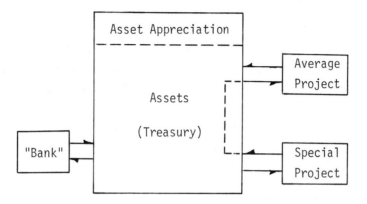

The inflows to the treasury projects are considered to be net values after operating expenses and taxes. The "bank" includes any source of outside financing - loans, bonds, stockholders, etc. In order to remain financially sound, the organization must provide money for payment of principal plus interest, as well as dividends to stockholders.

The goal of any "healthy" organization is to provide continuing liquidity to meet all current obligations while appreciating its asset value (equity position). Some compromise between liquidity and long range appreciation is necessary.

The special project is the investment being evaluated. Capital needs are provided from the treasury. The net income from this project is returned to the treasury for re-investment and to pay current obligations. We might note that our model assumes that the treasury is "loaning" the capital required by the special project and that the interest required in return is equal to the opportunity rate.

The average project represents the average re-investment opportunity of the organization. In the basic model it is assumed that money from the special project will be re-invested at the average opportunity rate for the life of the special project. The asset appreciation shown is that resulting from the special project investment being made and profits from re-investing in the average project.

Unless any outside funds from the "bank" are secured only by the special project, said funds are presumed to flow into the treasury and represent a general obligation of that treasury. In most cases debt obligations are secured by the total assets and not merely those of the special project.

There are several assumptions inherent in models like these which are realistic - from a practical viewpoint:

1. Money entering the treasury from a given project loses its identity with that project. It is used to pay general treasury obligations and for re-investment at the average, no-risk opportunity rate. This rate varies, but 10% per year is often used in the absence of more specific information.

2. The cost of using the money required for investment in the special project is fixed by the average opportunity rate, unless outside funds are secured by that specific investment only.

3. When computing the time value of money by some interest formula, the time at which a cash flow is received or disbursed does not need to coincide with the timing of the interest periods. Continuous compounding might be used (for computational convenience) even though the cash flows are considered as daily, monthly, or "lumped" at the midpoint of the year. So long as a system consistent with company fiscal policy is employed, the timing factors chosen will have no significant effect on the final investment decision.

It must be remembered that many factors are involved when making a given decision. One non-economic factor discussed earlier is the total value system. In addition to sociological factors, the following are always involved in a decision:

1. Risk factors
   a. Prediction of future events and economic climate
   b. Probability of success or failure

2. Rate of return on investment

3. Effect that failure(s) would have on organization's economic future

4. Tax ramifications

5. Current investment needs and opportunities

6. Cash generation needs in future years to remain in a sound and dynamic position. (Might involve deferral of revenue for economic reasons.)

7. Romance factors

8. Organization's financial structure

Our discussion will be limited to Items 1 and 2. These are the ones usually formally considered by the practicing engineer while the remainder are usually management prerogatives. Some understanding though of Items 3-8 is essential for intelligent engineering appraisal. Space does not permit a complete discussion of those latter factors, but a few comments are essential.

In theory there are always an infinite number of investment opportunities available for the investment dollar. In practice this is never really true. There are always limitations imposed by organizational policy, personnel capabilities and governmental interference in economic affairs.

A decentralized region, area or division, for example, has certain geographic limitations that restrict investment potential. An oil company management is not likely to seriously consider a project to manufacture television sets unless they have an insufficient number of attractive oil investments. If they do consider it, they must include in their cost considerations the acquisition of new qualified personnel. Most managements inherently limit the largest portion of their investments to areas where they have experience and are in a position to make qualified judgment decisions. This is one reason why diversification is necessarily slow.

No comment should be necessary about the limitation imposed by governmental regulation.

We must also recognize that not all investments are profit motivated. There are some that are made for strategic reasons. Strategic, as I use the word, means that said investment is necessary to achieve long range company goals. This class of investment must necessarily strengthen the organization so that it enhances the probability of success or profit motivated investments. One example of this is funds expended for laboratory research and field tests (pilot floods, special tests, etc.). These cannot be compared with profit motivated investments by means of a single yardstick such as a return that is impossible to measure in dollars and cents. Can anyone cite the cash flowback resulting from research, public relations, college aid programs and similar efforts? No! Yet, any enlightened executive recognizes their value.

No definite discussion is possible on such things as the romance factors. However, they exist. Investment decisions are made by human beings who in turn are emotional creatures. How else can we explain the irrational decisions every one of us has made at one time or another. Let us also not forget that there can be a lot of romance in the numbers that we plug into time value of money equations.

It is certainly true that the type of organization is of some consideration. A publicly held corporation that theoretically has an infinite life is inherently different from a private company when it comes to very long range planning. Tax structure plays a major role in this difference.

Differences in corporate financial structure also play a role. In many instances investment trusts, insurance companies and the like have paid more money (accepted lower rate of return) than oil interests for proven reserves in the ground. This does not reflect error on the part of either group but rather points up the difference in their financial structure; i.e., the effective alternatives to purchase of this property that were open to them. The insurance company by law and practice engages in non-speculative ventures exclusively and therefore compares this purchase with bond and stock dividends, bank interest rates and the like. The oil company compares it to the average return that they can expect on all their ventures - both good and bad. This difference in outlook can lead to differences in investment attractiveness that is sometimes significant.

Many of the factors involved in the investment decision involve one facet or another of the need for financial continuity from year to year. Stable income must be provided to meet those routine expenses needed to operate the organization. This includes fixed direct and indirect overhead, reasonable return on the investment to the owners (stockholders), maintenance of a sufficient operating fund and provision for funds to meet any financial obligations previously incurred (interest on loans, preferred stock, debentures, etc.). Consequently, when approving a budget for drilling and development, these items first must be subtracted from estimated income. Planning also must involve making such decisions so there will be enough income in future years to meet all obligations plus investment needs.

It would be sheer coincidence if anticipated income and expenditures properly balanced during a given year without continuing adjustments. During some years cash outlay for dividends or investments can be increased above planned levels; while in others short term borrowing is necessary to achieve proper balance. Long term borrowing usually falls outside this immediate area for it serves to increase the size of the treasury in order that it may engage in investment activities of a size and magnitude beyond the current income picture. The resulting investments would also normally be of the nature that would not only yield a satisfactory rate of return but also would have long range income potential. The issuance of 20-30 year debentures is a common method of accomplishing this.

Capital planning and budgeting may sometimes involve the deliberate deferral of revenue to a later date. This is not a typical petroleum problem, but it has arisen in cases where current income is tending to generate more investment capital than an investor can wisely invest within the scope of his operations. One relief for this is diversification. Such deferral can take many forms, some of which are designed to take full benefit of existing tax statutes.

All of the above points up that you don't always pick the project yielding the highest rate of return. Said rate must be high enough to meet your investment criteria. No company, for example, would be on a sound basis if all of their assets were tied up in short term investments. This might involve the future necessity of making a larger number of re-investments in a given period of time than the organization could intelligently cope with; or, that would exceed the number available. Drilling deals have a way of following a feast or famine cycle and you can't hold one for long periods of time. In other words, capital planning must provide a sufficient degree of financial stability so that you can take advantage of the most attractive opportunities that arise. To be able to do so markedly reduces the cost of capital and thus increases profits.

No decision is ever complete before one considers the effect that failure in the venture may have on your financial well being. If you can't afford to fail, you really can't afford to invest, providing there is some strategic value in maintaining the organization. It should be obvious that the large corporation is in a somewhat different position than the individual in this regard. This is one of the areas where the application of the laws of probability and statistics play an important role in formal decision making.

Payout, Leverage

The general pattern of petroleum investments is shown in the following sketch.

The curve shown is for the simplest case. An initial investment (C) is made at time zero. In many cases some income is generated from that investment without much time delay. At payout the net income (after taxes and project expenses) equals the amount of the investment. Beyond payout new money is generated for re-investment by the treasury and payment of general treasury obligations. At any time following payout, the cumulative amount of plus money shown represents undiscounted profit - profit generated without regard for the time value of money for the time interval involved. The curve terminates at that time when the investment no longer generates any net income.

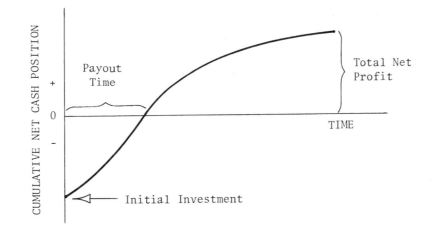

The detailed method for determining net income in any time period varies somewhat in each company. The general practice is to deduct operating expenses from gross revenue to obtain net revenue. The net income is found by subtracting taxes from net revenue. The person preparing the evaluation will usually employ some arbitrary company standard to estimate taxes since detailed tax ramifications are outside his expertise (and responsibility).

Payout is that point in time where the undiscounted net income equals the capital investment. Payout is one convenient "yardstick" for measuring the attractiveness of an investment. It is a measure of the rate at which revenue is developed early in the project. It is furthermore a good barometer of the time risk involved. The less the payout time, the less the risk imposed by uncertain future events.

Until payout, the project represents an asset liability to the treasury. At payout the amount of risk capital becomes zero. From that time on the project becomes a positive asset because it generates treasury appreciation. In those areas where political uncertainty is a major concern, payout may be a most critical consideration.

Payout for ventures in hostile areas where marketing or environmental problems may cause lengthy delays, causes the curve presented above to have a much different look. Obviously, a delay, for any reason, would extend the payout time by the time required to satisfy any such problem. In these cases we must be very careful in accepting a given payout value. Does the value being utilized represent payout from the date production commences or from the date of the initial investment? Acceptance of a payout value without clarifying the ground rules may lead to faulty decisions.

In spite of its obvious advantages, payout alone is not a complete measure of value for the following reasons:

1. It does not indicate the rate of earnings following payout.

2. It does not measure total profit.

3. It does not measure in any way the asset appreciation potential for the investor - presumably a primary goal of most investments.

4. Time value of money is not formally included (although some indirect correlation may be possible).

Some of these disadvantages are tempered by circumstance. A basic strategy of any major, integrated company involves developing reserves to protect future needs as well as providing a balance between fluids produced and marketed. If the reserves are very large, and payout may be achieved by the production of only a small percentage of said reserves, other considerations may be considered secondary if the project may be financed properly.

Payout is not the only factor calculated from the cumulative cash flow curve.

Leverage is defined as that ratio found by dividing undiscounted total net income by investment. Total net income on the curve would be total net profit plus initial investment.

Profit-to-investment (P-I) ratio is defined as the undiscounted total profit divided by the investment.

Example 5.6: A project is estimated to require an investment of $2,850,000. From geological data a total net income of $6,000,000 is estimated. Estimated payout time is four years.

$$\begin{array}{ll} \text{Net income} & \$6,000,000 \\ \text{Investment} & \underline{\$2,850,000} \\ \text{Net profit} & \$3,150,000 \end{array}$$

$$\text{Leverage ratio} = \frac{6,000,000}{2,850,000} = 2.11$$

$$\text{P-I ratio} = \frac{3,150,000}{2,850,000} = 1.11$$

The numbers show that the treasury should accumulate over three million dollars beyond year four. The decision maker using the ratios shown will compare them to his past experiences (as a judgment factor) to determine minimum acceptable values.

Such ratios are simple to compute and do not require detailed cash projections (which might be inaccurate very early in a project anyway). Although they serve a purpose, they do not replace the input information attained by time value of money considerations.

Income and Investment Delays

The cash flow curve we had been discussing assumes that any investments after time zero are smaller than current revenue so that the curve always has a positive slope and is continuous. Although this is commonly encountered, it is not always the case. If the timing and amount of large subsequent investments can be anticipated, the curve should reflect this and would have a shape as shown.

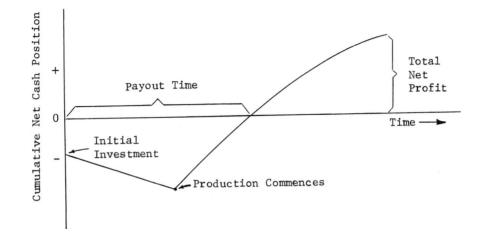

It is common to have a perplexing time lag between the start of major investments and the beginning of income production. Early developments offshore and remote areas without current transportation are examples of these. The first part of the cash flow curve may be horizontal or have a negative slope. In this case the initial investment may include the capital expended plus the debt service on that capital before income begins.

BASIC TIME VALUE OF MONEY CONCEPTS

Some time value of money concept must be included in any quantitative appraisal regardless of what other yardsticks may be used. The basis for all equations is the compound interest concept outlined previously in this chapter. Even though the basis is simple, it is possible to express it in an almost infinite number of ways.

Summarized below is a brief review of the most common approaches employed. With each I have pointed up the plus and minus characteristics. My purpose is neither to praise no condemn. Nor, have I attempted to "rank" them in order of preference.

The choice of a method is somewhat arbitrary - rather like the choice of religious or political preferences. Fortunately, it probably is true that understanding a given method's characteristics and its intelligent use in a given economic circumstance is more important than the method itself. Intelligent use does at least imply though that the method employed is reasonably compatible with the economics governing the investment. For this reason, the more enlightened practitioners do not use one "magic formula" for all time value concerns, but rather, use all the tools which may be available.

Before discussing the application of time value of money equations, let us review briefly some of their basic forms. At present we will consider only the form wherein annual income is received only at the mid-point of the year with annual compounding of interest.

At some point in a mineral reservoir, production declines, Chapter 19 discusses these patterns. The latter part of that chapter incorporates the ideas which follow - with decline curves - to develop simple appraisal procedures.

83

TIME VALUE OF MONEY EQUATIONS

In summarizing the equations below the following terminology will be used:

C = amount of any capital investment

I = net income from operations for any time period indicated by a subscript;
I without a subscript refers to total net income for the project.
$$I = I_1 + I_2 + I_3 + \ldots + I_n$$

i = effective interest rate of "bank rate" of interest, a decimal fraction

i_o = average opportunity rate of return (time value of money) for treasury funds, a decimal fraction

i_r = internal rate of return in the equation being used which makes the present value of all incomes equal to the present worth of all investments, a decimal fraction

n = subscript or exponent to denote occurrence in nth year

t = general time term; any year in a sequence when used as subscript

t' = time of delay between investment and start of income

t_p = payout time

Subscript "n" is used with a term to denote that value for any general time period.

It should be noted that the exact definition of each term is fixed by the equation in which it is used. A portion of the confusion that tends to exist between methods results from failure to recognize this.

What constitutes income I is a good example. Income is not merely new money generated by an investment. It may result from any savings, expense avoidance or expense reduction that is expected to result from the investment. The equation used, and the subsequent interpretation of results, must properly reflect these differences.

The term i_r is called by many names. In the financial world it may be called "effective yield to maturity," "marginal productivity of capital," "marginal efficiency of capital" and "internal rate of return." In the petroleum business phrases are used like "discounted rate of return," "profitability index" and "return on investment (ROI)," to name a few. I will simply refer to it as rate of return. The semantics is confusing but not important if you take the trouble to understand the system.

One must distinguish between investments designed to generate new returns and those that merely accelerate the rate at which returns are received from a given project. One might acidize or fracture treat some wells with a major purpose of accelerating production of reserves, not increasing reserves.

The equations summarized are often used to rank alternative investments. In doing so, remember that investments may be divided into two broad categories - generative and sustaining. A generative investment is one which leads (hopefully) to asset appreciation. A sustaining investment is one necessary to maintain the organization. Investments to develop necessary reserves, repair existing facilities so they can operate properly, etc., are examples of the latter. These two types cannot be judged by the same "yardsticks."

Rate of Return - Single Interest Rate

There are two simple ways to write a compound interest type equation to satisfy the definition of i_r (equate present value of net receipts to present value of investments).

One form commonly used is to find the rate of return (i_r) which makes present value profit equal to zero.

$$\frac{I_1 - (C_1 + C_i)}{(1 + i_r)^{0.5}} + \frac{I_2 - C_2}{(1 + i_r)^{1.5}} + \frac{I_3 - C_3}{(1 + i_r)^{2.5}} + \frac{I_4 - C_4}{(1 + i_r)^{3.5}} + \ldots + \frac{I_n - C_n}{(1 + i_r)^{n-0.5}} = 0$$

(5.9)

Equ. 5.9 assumes that all income for a given year occurs at the mid-point of that year. The exponents reflect this. The term i_r is the effective annual returned as defined by Equ. 5.3.

In Equ. 5.9, the initial investment is assumed to occur at the mid-point of the first year, at the same time as the first year's receipts. Should a given situation require the initial investment to be at the beginning of year 1 we would simply modify this equation by taking the C_i term out of the numerator.

A similar form may be used alternatively.

$$C = \frac{I_1}{(1 + i_r)^{0.5}} + \frac{I_2}{(1 + i_r)^{1.5}} + \frac{I_3}{(1 + i_r)^{2.5}} + \frac{I_4}{(1 + i_r)^{3.5}} + \ldots + \frac{I_n}{(1 + i_r)^{n-0.5}}$$

(5.10)

At first glance, Equ. 5.10 appears to differ from 5.9 only in the method of handling C. The initial investment (C_i) is assumed to occur at time zero. Any subsequent investments in the project will be discounted back to the present at the opportunity rate rather than the rate of return of the project and added to the initial investment. Equation 5.11 shows a common way of doing this.

$$C = C_i + \sum_{t=1.0}^{t=n} \frac{C_t}{(1 + i_o)^{t-0.5}}$$

(5.11)

If the numerator of each term in Equ. 5.10 is the same (series of equal payments), the equation may be written in the form of Equ. 5.8. For payments received at mid-year the exponent in the denominator only will be the total number of years minus 0.5(n - 0.5).

With both Equ. 5.9 and 5.10 one must estimate values of C and I_n for each year. The equations are then solved for i_r - by trial-and-error. The number of trials required for

85

solution of either of these approaches may be minimized with graphical solution. A plot of PVP versus i_r is shown as Figure 5.1.

Comments on Equ. 5.9 and 5.10. These equations possess certain implicit characteristics that govern their applicability.

1. Provide a profit indicator independent of the absolute size of the investment.

2. The rate of return is particularly sensitive to errors in estimating investment requirements and cash returns in the early years.

3. Equ. 5.9 may not be solved meaningfully for i_r if cash flows are all negative (dry hole), all positive, or the total undiscounted revenue is less than the investment (project does not pay out).

4. The amount of revenue early in the life of the project has the greatest influence. Revenues beyond 10 years have an almost trivial effect on rate of return.

5. No direct measure is supplied regarding the absolute size of the profit generated, particularly on long life projects.

6. All money generated from a project may be re-invested at the rate of return calculated (or that re-investment potential of said funds is immaterial). (2,3,5,6).

7. Multiple rates of return are possible when using Equ. 5.9. As many solutions are available as there are minus cash flows. If two terms are negative in the series, two values of i_r satisfy the equation (2,3,5,6,7,8,9).

8. In acceleration investments, Equ. 5.9 may give meaningless results.

When applying these two equations to any project, one must be aware of the above characteristics.

Example 5.7: Two investments D and E are being considered. The table below summarizes the calculation using Equ. 5.9.

Year	Investment		Net Revenue		Net Cash Flow	
	D	E	D	E	D	E
1	$71,785	$71,785	$20,000	$40,000	$-51,785	$-31,785
2			20,000	25,000	20,000	25,000
3			20,000	15,765	20,000	15,765
4			20,000	0	20,000	0
5			20,000	0	20,000	0

	D	E
Rate of Return	20 per cent	20 per cent
Life of Investment:	5 years	3 years
Undeferred Net Profit:	$ 28,215	$ 8,980
Total Treasury at End of Year 5, if Net Revenue Re-invested at 20 per cent:	$162,980	$162,980
Total Treasury at end of Year 5, if Net Revenue Re-invested at 10 per cent:	$128,036	$116,306

Which would you choose if only one could be accepted? They both require the same investment and have the same rate of return. The choice would depend on strategy, but most would probably pick investment D. It has a substantially higher undeffered (undiscounted) net profit and greater treasury appreciation at the assumed average opportunity rate of 10%. Notice that if this opportunity rate is the same as the rate of return, treasury appreciation is the same for both at the end of the longest life project (5 years).

Net Present Value (NPV)

The NPV is the present value of all future returns discounted back to time zero at the average opportunity rate of the company. This average opportunity rate (i_o) is normally 9-12%; 10% is commonly used in planning. The equation possesses exactly the same form as Equ. 5.10.

$$NPV = \frac{I_1}{(1 + i_o)^{0.5}} + \frac{I_2}{(1 + i_o)^{1.5}} + \frac{I_3}{(1 + i_o)^{2.5}} + \frac{I_4}{(1 + i_o)^{3.5}} + \ldots + \frac{I_n}{(1 + i_o)^{n-0.5}} \quad (5.12)$$

Note on Equ. 5.12: This equation is oftentimes written with -C on the right hand side and the result called NPV. I denote NPV as equal to PVP+C. Using my nomenclature, if -C is placed on the right hand side, the result is PVP not NPV. This designation is arbitrary, but I deem it convenient to separate investment, incomes and profit.

If NPV > C, the rate of return (i_r) is greater than i_o. If NPV = C, the rate of return equals i_o. When NPV exceeds C, it means that the investment will earn a rate of return i_o plus an additional amount of profit.

When using an equation like 5.9 or 5.10, a company might not consider a venture based on return only if said rate of return is less than 18-20%. This is about twice the average opportunity rate. The difference is an informal method of introducing risk into the computation.

NPV-C may be designated as present value profit (PVP). PVP is oftentimes divided by C to obtain a present value profit-to-investment ratio. This introduces size of investment needed into the comparison of alternative projects. The larger the investment needed for a given PVP, the less attractive the investment (on a relative basis).

87

Present value profit profile. One method for presenting profitabilities is shown in fig. 5.1.

The point at which each curve crosses the zero axis represents the value of i_r for investments A, B and C (graphical solution to Equ. 5.9). The value on the ordinate represents present value profit at each corresponding rate of return value. At 0%, the undiscounted profit is shown. The value at i_o, expressed as a percentage, is the PVP discussed above.

This type of profile is very informative. Project A has the highest profit but the least rate of return. Project C possesses the reverse characteristics. Project B is in-between. PVP and rate of return would rank these projects in reverse order, if used as the sole criterion.

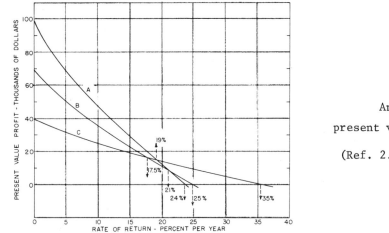

Fig. 5.1

An example of a present value profile plot

(Ref. 2. Courtesy SPE)

The reason for this dichotomy is probably due to the fact that Project C produced most of its returns early in its life. This produced a high rate of return independent of profit.

The moral: Different approaches necessarily do not always give the same ranking to alternative investments.

Example 5.8: Investments D and E of the previous example are examined for NPV using Equ. 5.12 and an average opportunity rate of 10%.

| | Net Revenue | | Discount | Net Present Value | |
Year	D	E	Factor	D	E
1	$20,000	$40,000	0.953	$19,060	$38,120
2	20,000	25,000	.867	17,340	21,675
3	20,000	15,765	.788	15,760	12,423
4	20,000	0	.716	14,320	---
5	20,000	0	.651	13,020	---
				$79,500	$72,218

The present value profit (PVP) is as follows:

D: $79,500 - 71,785 = $7715

E: $72,218 - 71,785 = $433

88

The discount rate shown above in the fourth column represents the inverse denominator of each term in Equ. 5.12 for the year in question. The values shown were taken from Table A.1 in Appendix A.

The figure below shows the character of the PVP plot for Prospects D ane E. Even though the rate of return calculated by Equ. 5.9 is the same, Project D offers far more profit (both undiscounted and discounted). For most, D would be the clear choice of these two alternatives.

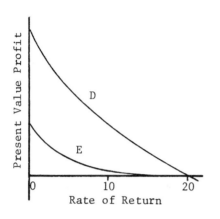

For reasons like those pointed out in this simple example, most companies use some form of present worth or profit ratio as a primary value of money yardstick in the decision process.

Appreciation of Equity

An alternative to the above methods is one where treasury growth is the primary yardstick (2,9,16). In this model each return from a project is assumed to grow at the average opportunity rate of the treasury for a length of time equal to the life of the longest lived project being compared.

The basic, general equation is

$$C(1 + i_r)^n = I_1(1 + i_o)^{n-1} + I_2(1 + i_o)^{n-2} + \ldots + I_n \qquad (5.13)$$

The right hand side represents the sum of each annual return plus accumulated interest compounded from the time it is received until project conclusion. The term i_r is then the effective rate of return earned by investment C. Once again C is the present value of all capital investments. Each side of Equ. 5.13 is equal to the asset appreciation of the treasury.

The rate of return in Equ. 5.13 still equates investment with returns at the end of the project; an investment C compounded at rate i_r for n years would yield the same amount of money as the project returns compounded at rate i_o.

Example 5.9: For prospects D and E of the previous two examples, Equ. 5.13 reduces to

(D) $(71,785)(1 + i_r)^5 = 128,036$, $i_r = 0.123$

(E) $(71,785)(1 + i_r)^5 = 116,306$, $i_r = 0.101$

Utilization of this approach is consistent with our time-value theory and also respects the rate-of-return concept, which says that rate of return is that value which equates the future value of all cash flows to the initial investment C. We also have eliminated the most unrealistic characteristic of conventional rate-of-return calculations (assuming $i_o = i_r$) by compounding at the historical opportunity rate of the organization. Since any project usually requires a relatively small percentage of the capital assets of an organization, it is reasonable to assume that few, if any, successful ventures will alter the historical rate of return significantly. The effective rate of return, i_r, is equivalent to Capen, et al's (22) growth rate of return (GRR).

Appreciation of equity will give the same ranking as present worth. The measure of worth is just slightly different. From Equ. 5.13, when $i_r > i_o$, PVP is a plus number. On Project E the PVP was only \$433 which is comparable to i_r of only 0.101 (essentially i_o). When $i_r = i_o$, PVP = 0.

Appreciation of equity possesses several advantages over net present value:

1. It provides a positive number for treasury growth when that is a primary part of the investment strategy.

2. It considers the impact of project length on a consistent basis when comparing projects with a different life.

3. On long life projects it probably gives a more realistic weight to cash flows many years in the future. NPV discounts cash flows beyond 10 years more severely than the real value of long range, future receipts - for a corporation presumed "infinite" life. (See section entitled Long Life Projects.)

4. The value of rate of return, although usually lower than that derived from the discounted cash flow rate of return, is usually more realistic since it accounts for re-investment of future cash flows at i_o instead of the rate of return of any individual project.

5. The one major disadvantage of this approach is that it is very difficult to compare projects of different lengths.

Where either NPV or appreciation of equity are reasonable "yardsticks," most seem to prefer NPV. Presumably, it is easier in principle and therefore easier to comprehend. However, appreciation of equity is usually preferred by those who like to rank projects by their rate of return.

Other General Two-Interest Methods

Appreciation of equity is one of several methods available using two interest rates (21). These other methods possess one characteristic in common; they use one interest rate to establish the time value of money invested and the second to evaluate the time value of returns from the investment. Each, however, defines the economic system somewhat differently.

One of the earliest of these was by Hoskold (13,15) in 1877. This is sometimes referred to as the sinking fund approach. It is used with municipal bonds and some elements of the mining industry.

This approach calls for setting aside a fixed sum I_n each year during project life. In its pure form, this amount will yield an interest i_r on investment C plus a sinking fund deposit which at the same interest rate will amount to C in n years.

In its modified form

$$I_n = Ci_r + C \left[\frac{i}{(1 + i_o)^n - 1} \right] \tag{5.14}$$

The interest rate i_0 is a rate at which each return may be re-invested to return the investment. You will note that the term in brackets in Equ. 5.14 is the equal-payment-series, sinking fund factor shown in Equ. 5.6.

Kaitz (13) has proposed a similar approach which he designates as percentage gain on investment (PGI). The equation for his version of i_r is

$$i_r = \left[\frac{(1 + i_0)^n}{C} \right] \left[\frac{i_0}{(1 + i_0)^n - 1} \right] \quad \text{(PVP)} \quad (5.15)$$

If one compares Equ. 5.6, 5.12, 5.14 and 5.15, internal similarities are apparent. In principle, an amount of money $(i_r)(C)$ is being subtracted from each annual cash flow and termed a profit in excess of i_0. The remainder of the cash flow when re-invested at rate i_0 will return the initial investment C plus interest at rate i_0.

This is similar to Hoskold. Kaitz wishes to recover the investment plus interest where Hoskold wishes to recover only the investment, or, $i(\text{Hoskold}) = i(\text{PGI}) + i_0$.

If one rearranges Equ. 5.15, it may be shown that the PGI approach is equivalent to dividing PVP into a series of equal annual payments to be received over the life of the project.

Arps (14) has proposed a method in an attempt to reconcile accounting methods with those above. His average annual rate of return is used to equate the present value of net income minus depreciation to present value of the average investment. In other words, the rate of return is based on some average investment rather than the initial investment.

Accounting Rate of Return

The on-going measure of a project or company's success is easily obtained through conventional accounting. Commonly used in accounting for return on investment is the ratio of net book income to net book value of assets. This is sometimes called book yield. The normal accounting process leads to its rather widespread use.

Is this accounting rate of return compatible with the type outlined above? Not necessarily! Policies with regard to amortization, and capitalization or expensing of investments, affect the difference. A major variable is the timing of cash flows into and out of the treasury. Stated generally, book yield tends to overestimate the true rate of return for typical petroleum projects where the primary investment occurs early in the project life.

Which is the more realistic? Solomon (3,5) documents the answer quite clearly. He concludes that book yield is not a reliable measure of return but will continue to be used since accounting is the only mechanism available to an investment unit to measure on-going performance.

If this is realistic, this simply means that one must temper the use of book yield "by a far greater degree of judgment and adjustment than we have employed in the past." As a practical matter, initial decisions will be made by one or more of the methods outlined herein and will be checked by a book yield. Any inconsistency will be reconciled only by the good judgment of those responsible for decisions (which is their basic responsibility anyway).

APPLICATION OF TIME VALUE METHODS

To anyone who has considered seriously the concepts of value summarized in this book to this point, some element of confusion probably exists. As if future risk was not enough, in addition, one can develop a number of reasonable sounding equations from the same compound interest principles. What is the "magic" answer? Truthfully, there is none. Those who profess to have found one are practicing a form of "alchemistry," have simply picked one they think they understand, or have (or been told to) accepted the arbitrary decision of another.

Each approach possesses its own inherent strengths and weaknesses. With the variety of strategies and circumstances encountered in the petroleum industry, no one approach offers total reliability. The practical answer is to establish several time value criteria, use them consistently and continually check back on performance.

This array should logically include payout, one or more rate of return criteria, and at least one method of characterizing present worth or asset appreciation. As we shall illustrate in the next section, these will not necessarily give a consistent ranking of the projects being considered. The final choice must depend on the compatibility of a method with the primary goal of the investment.

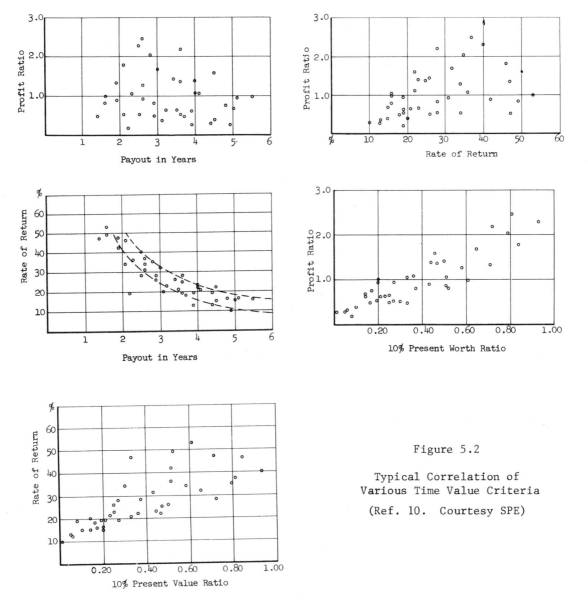

Figure 5.2

Typical Correlation of
Various Time Value Criteria

(Ref. 10. Courtesy SPE)

In Fig. 5.2 the following nomenclature is used:

Profit ratio = undiscounted profit divided by amount of investment

Rate of return = rate of return by an equation like 5.9 or 5.10

Payout = time for undiscounted net earnings to equal the investment

10% present value ratio = ratio of profit to investment with both discounted to

time zero at annual rate of 10%

Consider, for example, a proposed investment in a country considered politically unstable. Payout time might be the primary consideration if the value of the reserves (discounted or undiscounted) are large in relation to the investment. Rate of return per se might be only a secondary or even trivial concern. One only invests in politically unstable environments if it is needed to sustain the organization's activities.

Fig. 5.2 is an example by Northern (10) showing how different criteria correlate. The points shown are for 40 exploration investments with the primary amount coming in the beginning of project life.

It is apparent that any relative correlation is less than desirable. The best correlation is between payout and rate of return. This is expected because both emphasize the effect of returns very early in project life.

In this plot I have superimposed some dashed lines. The lines outline the range of values obtained when correlating these two criteria for a large number of projects. Although no exact correlation exists, an average line drawn midway between the dashed lines would provide an approximately relationship between payout and rate of return.

Concern about these inter-relationships have been explored by others (9,11,12). The information by Brons and Silbergh (11) is summarized in Chapter 19 since it involves the use of decline curves.

LONG LIFE PROJECTS

The energy industries historically have maintained a profit position with the use of conventional economic parameters. However, with rapidly increasing costs and smaller reserves in new discoveries, the investment criteria utilized in decision making has become, and will continue to be, more critical than ever. The base income of many large corporations today is in long-term projects that were initiated in the 1950's or early 1960's. In an attempt to replace production, companies operating in the U.S. are generally choosing high initial potential, short-term projects, and are foregoing participation in the long-term projects that would provide the base income of the future. Many companies are unknowingly pursuing this approach and liquidating their assets because the conventional investment criteria greatly favor short-term projects.

This section will be concerned with presenting new parameters that will add insight into the effect of long-term projects on achieving financial goals. These relatively new parameters are necessary since the political climate, costs, and size of discoveries in the oil industry are very much different than even 5 years ago. Although the fundamentals are the same, the evaluation procedure presented herein will only restructure the conventional mechanics to allow for these dynamic changes. More importantly, these new parameters re-emphasize that all parameters are just models of the corporate treasury, and that models must continually be modified to reflect changing political and economic climates.

Discounted Payout

Equations reflecting the time value of money tend to understate the true value of cash flow to the company. For example, the discount factor for Year 25 at 10 per cent interest is 0.0968. Thus, a dollar received 25 years from now is worth only $0.0968 today. According to present value rate-of-return calculations, this is theoretically valid, but does it reflect the value of the revenue in Year 25 to an organization or investor who plans to be in business 25 years from now? Not really!

When a bank lends money for a new car, the borrower repays the principal plus interest-no more and no less. The bank does not change its position at the end of the finance period arguing that the car has appreciated in value so you owe more than the contractual agreement. This same transaction is true when money is "loaned" to a project - be it from an external source or the treasury. It makes absolutely no difference to the project since we have established that the bank makes the loan to the treasury - not to the project. In turn, the treasury "lends" capital to the project. Having made the loan, the treasury (now the bank) expects nothing more than return of the principal plus the required interest (the average opportunity rate of the treasury). Hence, any money generated to the treasury over and above this amount is new money - free from any charge against the original investment since all risk capital and the required interest has been returned. Thus, discounted payout is a measure of the time required for a project to return this principal and interest.

Discounted payout (DPO) is a much more critical parameter than is undiscounted payout since it truly represents that point in time when a project is no longer a liability to the treasury. If DPO is represented by the symbol t_p and the average interest rate to be charged by the treasury is represented by i_o, then DPO is reached when the summation of the right-hand side of Eq. 5.16 equals the present value of investments (C).

$$C = \frac{I_1}{(1 + i_o)^1} + \frac{I_2}{(1 + i_o)^2} + \ldots + \frac{I_{t_p}}{(1 + i_o)^{t_p}} \qquad (5.16)$$

As with interest equations, the exponents can and must be altered to be consistent with the compounding (discounting) timing such as continuous, midyear, etc. This equation assumes annual end-of-year compounding (discounting).

Example 5.10: Investments D and E from the previous example, discounted payout based on mid-year discounting is:

Year	NPV D	Cum NPV D	NPV E	Cum NPV E
1	$19,060	$19,060	$38,120	$38,120
2	17,340	36,400	21,675	59,795
3	15,760	52,160	12,423	72,218
4	14,320	66,480	---	---
5	13,020	79,500	---	---
	$79,500		$72,218	

DPO for Project D would obviously be somewhere between year 4 and 5 since cost is $71,785. Consequently, the discounted payout can be approximated by finding the proportionate part of year 5 income:

For D
$$DPO = 4 \text{ years} + \left(\frac{71,785 - 66,480}{13,020} \right) = \underline{4.41 \text{ years}}$$

Following the same procedure for E would result in a discounted payout of 2.97 years.

Treasury Worth Index

Acceptance of discounted payout as a paramter is a large step toward revamping the evaluation system. However, it does not solve other inherent problems with the discounted value approach to long-life projects. The first critical question we must ask is: Will a dollar generated 25 years from now buy 9.68¢ worth of goods and services in Year 25, or will it purchase $1.00 worth of goods and services? Will you send a check to a contractor 25 years from now for $10,330.58 ($\frac{1}{.0968}$ x $1000) for a $1,000.00 invoice just because the $10,330.58 was generated by a project which began 25 years ago? Of course not!

We must remember that money is not wealth per se, but is merely a convenient means by which we exchange wealth (things of value). Once all "seed money" plus its cost (interest) have been retunred, new money at some future time may be related to its value at present only by the time change in its purchasing power. The decay of money value (inflation) normally is different from the real or accounting cost of using money. The real value of project net income is quite different after DPO than before. Comparing units of money, using some interest rental rate, is not comparing relative true wealth (purchasing power).

Acceptance of the philosophy that the value of money is quite different after DPO leads one to conclude that a proper value system would consider a dollar generated after DPO as a dollar by which wealth can be exchanged. We have no argument with the idea that one must allow for inflation. Inflation could be handled one of two ways. We can discount at the average rate of inflation (i_i) after PDO, or we can assume that the average rate of inflation will be matched (and quite possibly exceeded) by the price increases of products.

Should you elect to discount at the average inflation rate, please note that according to U.S. governmental cost indexes, the average inflation rate from 1838 to the early 1970's was about 3 per cent per year. We feel that an inflation rate of this magnitude (or possibly up to

95

5 per cent), even upon considering today's inflation rate, must be utilized since large fluctuations have occurred historically on both sides of 3 per cent. Inspection of the historical data reveals that most periods of high inflation are followed by "recessions" that maintain the over-all slope.

Election of the second method of handling inflation not only simplifies the calculations, but is also believed to be fairly responsible. The discussion in this section assumes product prices will increase at a faster rate than the average future rate of inflation. It sounds somewhat ridiculous to state that oil prices will be $20 per barrel by 1985, but how many evaluations conducted in 1970 utilized a $14 oil price? Very few. Nevertheless, we have no argument with either approach so long as sound judgment, based on the facts, prevails.

The concept of treasury worth that follows assumes that product prices will match or exceed the rate of inflation. Accordingly, we have ignored all discounting after DPO, if for no other reason than to simplify the equations. Treasury worth is defined as the value of the discounted cash flow prior to reaching DPO plus the undiscounted value of the cash flow after DPO.

$$TW = C + I_{(t_p+1)} + I_{(t_p+2)} + \cdots I_n \qquad (5.17)$$

where: t_p = discounted payout time

Treasury worth can be related to investment by the treasury worth index (TWI). TWI is the ratio of treasury worth to the investment (C). Note that treasury worth index is somewhat equivalent to leverage. TWI may be expressed as

$$TWI = \frac{TW}{C} \qquad (5.17a)$$

TW and TWI account for the true difference in cash flow before and after DPO. As a parameter it is easily seen that it can yield a more realistic comparison between short- and long-term projects.

Example 5.11: Consider the case of investment F, which has the same cost and cash flows as E, but also has cash flows in year 4 of $10,000 and in year 5 of $5,000. Assume that DPO is at the end of year 3. Consequently, treasury growth and the treasury worth index for Project F are:

$$F: \quad TW = \frac{40,000}{(1.1)^{0.5}} + \frac{25,000}{(1.1)^{1.5}} + \frac{15,765}{(1.1)^{2.5}} + 10,000 + 5,000 = \$87,218$$

$$TWI = \frac{87,218}{71,785} - 1.21$$

The actual return to the treasury of these cash flows, adjusted for interest payments to the treasure is $87,218 or 1.21 times the initial cost.

96

<u>Accrued Treasury Assets</u>

Like the appreciation of equity calculation which compounds to a future date, accrued treasury assets (ATA) is a measure of total treasury growth resulting from participation in a given successful project. By definition, ATA is the summation of discounted cash flow until discounted payout (DPO) and the accrued value of all cash flows after discounted payout. In essence, we are assuming in this model that we cannot earn interest (compounding) at the same time that we are paying interest (discounting).

We may express ATA in equation form if we represent the discounted payout time with the symbol t_p and total life of the project with the symbol n.

$$ATA = \frac{I_1}{(1+i_o)^1} + \frac{I_2}{(1+i_o)^2} + \ldots + \frac{I_{t_p}}{(1+i_o)^{t_p}} + I_{n-(t_p+1)} \, (1+i_o)^{n-(t_p+1)}$$

$$+ I_{n-(t_p+2)} \, (1+i_o)^{n-(t_p+2)} + \ldots + I_n \tag{5.18}$$

Notice that the sum of the terms prior to DPO is equal to C. Thus we could replace all of the terms being discounted by the value of the initial cost (C). Accrued treasury assets differs from treasury worth in that values are compounded to a future time period instead of discounted.

Example 5.12: For investment of the previous example the accrued treasury assets would be

$$F: \quad ATA = \frac{40,000}{(1.1)^{0.5}} + \frac{25,000}{(1.1)^{1.5}} + \frac{15,765}{(1.1)^{2.5}} + 10,000(1.1)^{5-(2.5+1)} + 5,000(1.1)^{5-(2.5+2)}$$

$$= 38,120 + 21,675 + 12,423 + 11,537 + 5,244 = \$88,999$$

Accrued treasury asset exceeds treasury worth by amount of the money earned from reinvesting the cash flows after DPO. This process takes into consideration the discounting of funds to pay back the treasury and the profit from reinvesting at the opportunity rate of the organization. We may now use the value to determine the amount of interest required on the initial investment so that it would accumulate an equivalent amount of money as this project (ROR).

<u>Treasury Growth Rate</u>

The treasury growth rate (TGR) is another measure of return on investment utilizing the concept of ATA. This rate-of-return approach is similar to that of appreciation of equity in that we compound all cash flows after DPO to the end of the project, discount all cash flows prior to DPO and relate this value (ATA) to investment.

In equation form we may now solve for the return on investment.

$$C(1 + TGR)^n = ATA \tag{5.19}$$

Treasury growth rate is similar to Equation 5.13, with the exception of the way ATA is calculated. TGR must be modified in much the same manner as the appreciation of equity rate of return was modified by the concept of GRR by Capen et al. (22). This modification really deals with replacing ATA with a variable that has a common time reference point. We still need to allow for the difference in value before and after DPO. Thus, ATA can be modified by choosing a reference point (preferably 7 to 10 years), compounding cash flows after DPO to the reference point and discounting all cash flows after the reference point time at the rate of inflation (i_i). If you wish to maintain the convention that product prices will exceed or match the inflation rate, simply use the undiscounted value of the cash flow after the reference point time.

Example 5.13: Treasury growth rate for investment F is

$$F: \qquad 71,785(1 + TGR)^5 = \$88,999$$

$$1 + TGR = 1.044$$

$$TGR = 4.4\%$$

As we can see from Example F, these parameters most assuredly result in a much different value for return on investment. This value for return is correct for the model which was utilized. However, if this model is correct and we had not performed these calculations, we might have made a hasty decision. Remember, we always have the alternative of choosing none of the present projects. These parameters provide additional insight into the decision process as discussed in the next sub-section.

Investment Alternative Examples

Most people view this type of theory with skepticism. They ask, and rightly so: This theory is very nice but how can I use it? Table 5.7 contains six projects that represent typical investment alternatives and their respectable cash flows.

Projects A and B represent very typical cash-flow projections for infill wells or "cornershots." Project C is a typical cash-flow projection for a well drilled in frontier areas - a long time delay with high potential. Project D is a typical medium-life project that has a negative cash flow in Year 3 due to a major workover, installation of pumping equipment, drilling of a water injection well, or some other operational cost that will increase productivity. Project E represents a frontier area (such as the North Slope) with constant investment to develop a market outlet at the expense of the operator. It is definitely a long-life project with high potential. Project F represents a domestic, onshore project that does not have high potential but does possess a steady, low declining cash-flow profile that will require an investment in Year 8 for conversion to secondary recovery. Project F is realistic since an operator can expect (by rule of thumb) to withdraw as much oil under secondary recovery as by primary.

If we were using DCF rate of return as our main parameter, we would rank these projects in order, with Project D being the most desirable and Project F the least desirable (Tables 5.8 and 5.9). Likewise, if we were using PW/C, Project D would still be the most desirable and Project B would be the least desirable.

Using the conventional parameters and assuming only three projects can be financed from the treasury, the task of the investment analyst is to select the three best projects. If the

Table 5.7

Six Sample Projects

Year	A	B	C	D	E	F
0	-100	-200	-300	-200	-200	-200
1	60	180	0	180	- 50	60
2	40	100	0	120	- 50	50
3	30	40	-100	- 20	200	45
4	25		200	160	150	40
5	20		500	120	100	40
6	15		300	90	95	35
7	10		250	60	90	35
8	5		200	40	85	5
9			150	30	80	60
10			100		75	50
11			60		70	45
12			35		65	40
13			25		60	40
14			20		55	35
15			15		50	35

Table 5.8

Various Parameter Values for Six Sample Projects

	A	B	C	D	E	F
Payout *	2.0 Yrs.	1.2 Yrs.	4.4 Yrs.	1.2 Yrs.	3.6 Yrs.	4.1 Yrs.
DCF ROR	42%	63%	33%	90%	29%	24.3%
PW/C	1.63	1.45	2.76	2.86	2.2	1.68
GRR_7	19.4%	17.8%	27.6%	29.1%	23.5%	18.1%
DPO	2.4 Yrs.	1.4 Yrs.	4.8 Yrs.	1.3 Yrs.	4.63 Yrs.	5.9 Yrs.
TW	$195.40	$310.10	$1630.30	$771.60	$1051.40	$559.60
TWI	1.95	1.55	4.34	3.86	3.63	2.80
ATA	$246.50	$325.40	$3432.00	$1163.00	$1842.50	$801.40
TGR	12%	17.6%	15.9%	21.6%	13.1%	9.7%
ATA_7	$231.00	$397.30	$1696.00	$973.00	$1081.90	$531.30
TGR_7	12.7%	10.3%	24%	25.4%	20.7%	15%

Table 5.9

Ranking of Projects by Major Parameters

DCF ROR	PW/C	GRR_7	TWI	TGR_7	DPO
D	D	D	C	D	D
B	C	C	D	C	B
A	E	E	E	E	A
C	F	A	F	F	E
E	A	F	A	A	C
F	B	B	B	B	F

only investment criteria is to select the three best investments, Investments D, C, E are indicated as the most desirable according to PW/C, GRR_7, TWI, and TGR_7 in Table 5.9. Likewise, the DPO and DCF ROR suggest the selection of D, B, A.

The ranking of projects by major parameters illustrates the difficulties associated with selecting investments when different parameters are involved. Except for D, which is favored by all parameters because of its immediate payback followed by 6 years of revenues, a choice must be made between B, A, C, and E. Projects B and A would be selected over C and E by DCF ROR and DPO primarily because of the shorter payout - indicating again that these investment parameters are biased in favor of short-term investments. Project C, with a TW of $1,630, and Project E, which contributes $1,051 to the company's treasury, clearly would add more to the corporate treasury than either Project A or Project B and might be selected if no liquidity problems are anticipated.

The apparent agreement among parameters such as PW/C, GRR_7, TWI, and TGR_7 falls apart if our budget were restricted to one project or extended to four projects. Project C would probably be selected if only one project is undertaken, according to the TWI parameter, whereas Project D is the choice of all other parameters. Is a project that returns more than 4.34 times its initial investment after repaying the treasury interest on the borrowed money of less value than a project that only returns 3.86 times its original investment? Not necessarily, but neither is it necessarily a better investment.

In situations such as this Projects C and D should both be given closer inspection to determine whether D is preferred because of its shorter time horizon (9 years vs. 15 years), smaller initial investment ($200 vs. $300), or a combination of both. Biases introduced in the investment selection process by lower growth rates associated with longer time horizons may be checked by employing parameters such as TWI, ATA, and ATA_7 which look at just the increase in the corporate treasury from the project alone.

A corresponding project selection problem occurs when we have to select four projects. Project A is the candidate for the fourth project proposed by GRR_7. Parameters PW/C, TWI, and TGR_7, however, would select Project F. In this case GRR_7 appears to favor the project with the smallest investment cost and shorter time horizon. Project F, which returns approximately 2.84 times its initial investment plus interest, certainly exceeds the value to the treasury of a project returning only 1.95 times its initial investment.

ACCELERATION PROJECTS

An acceleration project is defined normally as one wherein the primary purpose is to increase the rate at which revenue is returned from a given asset. Can we justify an investment of a given size to accelerate cash flow from this venture? This could involve in-fill drilling, well treatment, reducing overhead in some manner, replacement of equipment, etc.

Example 5.14: The following table summarizes a projection on a given project.

	With Acceleration		Without Acceleration		
	Annual		Annual		
	Prod. Rate	Net	Prod. Rate	Net	Incremental
Year	(bbl)	Income	(bbl)	Income	Income
1	90,000	$185,000	50,000	$85,000	$+100,000
2	55,000	97,500	41,500	63,750	+ 33,750
3	33,000	42,500	34,500	46,250	- 3,750
4	20,000	10,000	29,000	32,500	- 22,500
5	0	0	24,000	20,000	- 20,000
6	0	0	20,000	10,000	- 10,000
	198,000	$335,000	199,000	$257,500	$ 77,500

If one used incremental income as I_n in Equ. 5.9, the following results are obtained for various investments.

Investment	Rate of Return
$77,500	0
79,000	0.085
85,000	.049
87,500	.069
95,500	.153
100,000	.21
110,000	.59 and .87

The results (in this case) are ludicrous on two counts: (1) rate of return increases with increasing investment and (2) multiple rates of return are obtained for some investments.

NPV, equity appreciation or TGR provide a reasonable answer. If the incremental income is re-invested at 10%, the result is shown below.

Year	Incremental Cash Flow
1	$($100,000)(1.1)^5 = $161,000$
2	$(33,750)(1.1)^4 = 49,400$
3	$(- 3,750)(1.1)^3 = - 4,990$
4	$(-22,500)(1.1)^2 = -27,225$
5	$(-20,000)(1.1)^1 = -22,000$
6	$(-10,000)(1.1)^0 = -10,000$
Total	$146,285

A given investment will increase the treasury by $146,285 by the end of year 6, compared to doing nothing, at an average opportunity rate of 10%. For an investment of $75,000 the rate of return from Equ. 5.13 is 11.8%. The NPV from Equ. 5.12 is $86,583.

The above example points up the potential problems when using Equ. 5.9 and 5.10 for acceleration projects on an incremental income basis. One must use any rate of return with caution. It tends to make all acceleration projects look good. Common sense dictates this is not always the case.

Time Delays

The delay between investment and the first returns is a routine time value problem. This must occur before payout and a charge against the cash flow is absolutely necessary. Each of the cash flow equations summarized in this chapter automatically shows the influence of time delay.

There is only one word of caution. In original planning do not be too idealistic about time delays. Offshore and in areas remote from manufacturing facilities, a several year delay is rather normal. If some type of governmental approval is needed for production or transportation, adding at least one more year for a major project may be reasonable.

FAIR MARKET VALUE

One last subject which must be mentioned is the very common practice of borrowing against mineral assets and the establishing of fair market value. In many cases, these two factors go hand-in-hand. Often however, the amount which can be borrowed from a lending institution will be somewhat lower than fair market value.

Determination of these values, even for large corporations, is often done by unbiased, outside parties. In the case of a bank loan, the appraisal is used to satisfy two basic concerns - total reserves and the annual cash flow capabilities. The property must have sufficient value to repay the loan if it were sold on the open market. At the same time the bank will require that the property have sufficient revenue producing ability to pay operating costs and taxes, as well as provide reduction of principal and interest.

Determination of property value always begins with determination of the fair market value (FMV). The most commonly accepted definition of fair market value is the price that a knowledgeable and willing buyer is willing to pay, and the price that a knowledgeable seller is willing to accept. There are as many methods for determining fair market value as there are evaluators. However, most determinations are based on one of the four following basic methods with every other approach being a variation and/or combination.

Payback Method. The FMV equals the cumulative future undiscounted net value for the first few years after the property is purchased or developed. The maximum length considered in this value is usually no more than one-third of the remaining life.

Going-Unit-Price Method. With this system FMV would be based on a percentage of the future undiscounted revenue (FGR). Properties in the early stages of depletion may have a fair market value equivalent to 40 per cent of the future gross undiscounted net value while an older property would usually have a value closer to 25 per cent.

Two-Thirds Present Worth Future Net Income Method. This method is derived from a combination of the loan value rules of most large banks. Many banks limit loans to 50 per cent of the future discounted net revenues (total income minus royalties, production taxes and operating costs). Other banks will loan no more than 75 per cent of the undiscounted fair market when FMV is calculated by other methods. The assumption in this approach, however correct, is based on combining these two arbitrary rules-of-thumb. Thus, the fair market is two-thirds of the present worth of all future revenues.

Rate-of-Return Method. This approach calculates fair market value as the cumulative present worth of the future net revenue with a discount rate which is equal to an acceptable rate-of-return on investment. The rate-of-return used as the discount rate for this method might vary quite considerably, say from 15 to 40%, based on the opportunity rate, the risk associated with the reserve estimate and whether the effect of taxes have been fully considered.

For the purpose of bank loans the minimum rate-of-return utilized for this method is that rate equivalent to the average cost of capital that the borrower would pay the specific or similar investments. This "cost-of-capital" would be equal to the cost of debt plus cost of common equity plus the cost of preferred equity. The cost of debt for most borrowers would usually be in the range of 1 to 2 per cent above prime rate on a floating rate basis. The cost of common equity would be based on the cost of obtaining new capital using retained earnings (dividends plus anticipated rate of growth of firm's stock). The cost of preferred equity would be related by the cost associated with preferred stocks or bonds.

After establishing a fair market value most lending institutions will perform at least three loan soundness tests. The primary test is that 50 per cent of the future undiscounted revenue must be remaining at the anticipated payout of the loan. A second test requires that 50 per cent of reserves will still remain in-place at payout. The final test is the effect on the loan of a 50 per cent reduction in payments due to a reduced price, curtailed production or loss of production.

As would be expected, the fair market value devised by people or institutions will probably vary quite significantly. Consequently, the moral is - If you do not like the price or loan value, go submit the data elsewhere. Everyone will use the same general rules and approaches but my experience is that opinions of fair market value vary quite considerably due to the extent of uncertainty that always exists in studies of this nature.

Cash Flow-Compounding Frequency

Most of the equations in this chapter assume that the effective annual income is received in one lump sum at the mid-point of the year. No pattern has been shown for each annual income I. The purpose was to concentrate on equation principles, unencumbered by application details.

Income is really earned continually from an on-going project even though payment is periodic. Loans from a "bank" specify some compounding interval, but internal loans may be compounded at any convenient rate.

From a purely rational viewpoint, the choice of compounding frequency used usually will have less effect on the result of a time value computation than the inherent errors in the input data. Just pick one that satisfies your biases, prejudices, or company policy.

Appendix A contains three tables that may be used to solve the compounded value of each term of Equ. 5.9-5.13 using assumptions common to the petroleum industry. Examples showing the use of these tables are contained in the appendix.

Table A.1 - Annual compounding, lump sum mid-year cash flow

Table A.2 - Continuous compounding, continuous annual cash flow

Table A.3 - Continuous compounding, lump sum annual cash flow at any time during year

It is common to predict future income I_n from a decline curve analysis of the type summarized in Chapter 19. Continuous compounding is convenient for combining decline curve and interest computations. Continuous compounding is expressed in terms of nominal interest rate j (see Equ. 5.4).

When using continuous compounding, the equations of this chapter may be converted using the following equivalents. One merely substitutes the continuous compounding term for the periodic term shown.

<table>
<tr><td><u>Periodic Compounding</u></td><td><u>Continuous Compounding</u></td></tr>
<tr><td>$(1 + i)^n$</td><td>e^{jn}</td></tr>
<tr><td>$(1 + i)^{-n}$</td><td>e^{-jn}</td></tr>
<tr><td>$\dfrac{(1 + i)^n - 1}{i}$</td><td>$\dfrac{e^{jn} - 1}{e^j - 1}$</td></tr>
<tr><td>$\dfrac{i}{(1 + i)^n - 1}$</td><td>$\dfrac{e^j - 1}{e^{jn} - 1}$</td></tr>
</table>

Regardless of the equation or the compounding period, Appendix A is very useful for manual calculations. Along with a present value profit profile type plot, it may be used to minimize the trial-and-error aspect of Equ. 5.9 or 5.10. Simply use enough values of i or j to define the curve. Intersection with the zero axis provides i_r. NPV or PVP (depending on the plot used) may be read from the plot at the average opportunity rate.

Mis-Application of Equations

It is typical for a management to specify one or more methods by which projects will be screened. Cut-off points are provided to separate "favorable" from the "unfavorable" based on calculations made by the responsible division.

Not uncommonly, the division works the problem backward. If they preclude the project is favorable, cash flows and reserves may be juggled to produce a result meeting management criteria for a favorable response. The motivations for this practice are both rational and emotional.

If the process is arbitrary just to satisfy a pre-conceived notion or selfish motive, some element of professional prostitution is involved.

The mistakes caused by this practice are potentially greater than the inherent limitations of the equations.

RISK

The compound interest type equations are merely one input into the decision process. Assessment of risk is one other.

There are several ways that risk considerations may be imposed to some degree on the time value equations even though they are inherently no-risk measures of value.

Investment C may be an expected value - include costs of dry holes as well as producing wells to produce estimated cash flow, using some probability concepts.

Adjust the required rate of return, present value or other profit indicators in some manner to reflect relative risk involved.

Use present product prices, conservative production numbers, etc., to use a conservative posture to minimize risk.

Being merely conservative, arbitrarily, does not eliminate risk and is essentially self-defeating. You eliminate those marginal prospects necessary to produce a sound reserves situation. If too conservative, no project looks favorable.

The use of formal probability methods is beyond the scope of this book although basic risk and uncertainty is discussed in Chapter 8. References 4, 6, 17-19 consider these. Reference 6, and its accompanying training program, is particularly recommended. I only wish to point out that any approach that does not formally include risk does not represent complete input for decision purposes.

SUMMARY

This chapter summarizes the basic time value of money principles. Chapter 19 presents the applications using decline curves. There is some element of artificiality about these approaches. In effect one is injecting the fairly simple concept of money rental (interest) into a much more complex economic model. This requires judgment which, in turn, exposes the ideas to the biases and prejudices of the user. This is not necessarily bad; it merely imposes some arbitrary considerations.

The only logical answer is to understand the strengths, weaknesses and applicability of each approach. A blind recipe approach never achieves the desired goals. In addition, several value indicators should be employed. It should be apparent that rate of return or present value alone is not sufficient.

We do not choose to "argue" for one method versus another. We are only concerned that the final answer is consistent with the strategy and goals of the organization.

REFERENCES

1. Thuesen, H. G., Engineering Economy. Englewood Cliffs: Prentice Hall, Inc. 1957.
2. Campbell, J. M., Jour. of Petr. Techn., p. 708, July 1962.
3. Soc. of Petr. Engrs. of AIME, SPE Reprint No. 3. 1970.

4. Mcgill, R. E., Exploration Economics. Tulsa, Okla.: Petroleum Publishing Co. 1971.

5. Solomon, Ezra, The Management of Corporate Capital. Glencoe, New York: The Free Press, 1959.

6. Newendorp, P. D. and J. M. Campbell, Decision Methods for Petroleum Investments. Norman, Okla.: John M. Campbell and Co. 1970.

7. Wooddy, L. D. and T. D. Capshaw, J. Pet. Tech., p. 15, June 1960.

8. Lefkovits, H. C. et al., Proc. Fifth World Petr. Congress, IV, p. 67, 1959.

9. Phillips, C. E., J. Pet. Tech, p. 159, Feb. 1965.

10. Northern, I. G., Ibid., p. 727, July 1964.

11. Brons, F. and M. Silbergh, Ibid., p. 269, March 1964.

12. Garies, D. F., SPE Paper 1255. Oct. 1965.

13. Kaitz, M., J. Pet. Tech., p. 679 (plus discussion) May 1967.

14. Arps, J. J., Trans. AIME, Vol. 213, p. 337, 1958.

15. Hoskold, H. D., The Engineers Valuing Assistant. London: Longmans Green and Co. 1877.

16. Baldwin, R. H., Harv. Bus. Rev. 1959.

17. Grayson, C. J., Decisions Under Uncertainty. Boston: Harvard Business School. 1960.

18. Schlaifer, Robert, Probability and Statistics for Business Decisions. New York: McGraw Hill Book Co. 1959.

19. Weaver, Warren, Lady Luck. Garden City, N.Y.: Doubleday and Co. 1963.

20. Essley, P. L., J. Pet. Tech., p. 911, Aug. 1965.

21. Henry, A. J., J. Pet. Tech., p. 393, April 1972.

22. Capen, R. C., R. V. Clapp, and W. W. Phelps, J. Pet. Tech., p. 531-543, May 1976.

23. Campbell, Robert A. and J. M. Campbell, Jr., SPE Paper 6332. February Evaluation Symposium in Dallas, Texas. 1977.

24. Miller, R. J., SPE Paper 6080. Oct. 1976.

6

THE ECONOMICS DECISION PROCESS

Certain economic concepts are used in everyday life to explain, to support, and to justify current events. Terms like inflation, cost of capital, investment capital, competition, cartels, and federal debt invoke emotional reactions from many people, who oftentimes do not fully understnad the complete meaning nor all implications of these terms. We are constantly responding to changes in these areas, and our responses in turn change these forces.

Every mineral company in one way. or another reacts to or is impacted by these forces. Investments are selected based upon expectations about the state of the economy, the cost of production, demand for the mineral under consideration, and other factors. To adequately anticipate the effect of various economic forces, one has to understand the precise definitions of each and how they are measured. Common perceptions of many economic concepts all too often differ significantly from the working definition. Closing this gap is the objective of this chapter.

GENERAL ECONOMIC FORCES

Nine general economic forces are examined. Each concept is viewed in sufficient detail to identify the basic characteristics inherent in each, the empirical measurement, and areas where disagreement exists among professionals working in the area. The nine concepts are:

1. Inflation
2. Investment Capital
3. Cost of Capital
4. Trade-off between employment and inflation
5. Competition
6. Cartels
7. Gross National Product
8. Federal Debt
9. Supply and Demand

Every one of these concepts has been studied in great detail for years by experts in these fields. Debate still exists among professionals over the causes of these forces and the best way to rectify undesirable effects. We identify the points of disagreement where they are relevant.

INFLATION

Inflation is defined as the rise in the general price level in an economy. The key words defining inflation are general price level. Only when a great many prices increase together without offsetting price declines does inflation exist. Inflation is thus a general term defining the overall direction of the prices of all goods in an economy. The effect of inflation is obvious to most people. Annual rates of inflation ranging between 6% to over 100% have seriously eroded consumers ability to pay for commodities. Commodities like housing, health care, minerals, and others have risen significantly in price.

The undesirable consequences of inflation are easily demonstrated. A couple earning $20,000 a year spends $2,000 a year on food and $5,000 a year on housing. Inflation of 10% increases the amount spent on these items to an extra $700 to consume the same quality of food and housing. If income remains at $20,000, the consumer has a problem. An extra $700 expenditure on food and housing means consumption of other items must decrease in order to stay within their income. Inflation robs people of the ability to consume the same quality of goods as they did before, which is unacceptable in most countries.

The simple example brings out several important points. One is that all goods seldom increase in price at the same rate. Another is that calculations of inflation involve all price increases. For example, an increase in food prices of 5% and housing prices of 15% yields a simple average price increase of 10% = (5 + 15)/2. But consumption expenditures rise to (2,000) (1.05) = $2,100 for food and ($5,000)(1.15) = $5,750 for housing. Total expenditures are now $850 more than the previous year, instead of the $700 increase shown above. Greater expenses arise because the overall inflation rate (10%) understates the actual price rise in housing by $250 and overstates food costs by $100. General inflation rate measures do not always provide a good picture of how consumers are affected.

Averaging price increase of all goods to measure inflation produces periods when inflation is zero, even though the price of some goods rises significantly. Suppose the housing prices increase 10%, but food prices decrease 10%. General inflation is 0% for that period. But look what happens to the budget. Food consumption expenditures decrease $200 [$2,000 - (2,000)(0.90)] while housing costs rise $500 [($5,000)(1.10) - $5,000]. Expenses increase $300 = 500 - 200, even though inflation is zero, again forcing the consumers to change their consumption patterns. So, although inflation may be zero, this does not mean that consumers will always be able to buy the same quantity of goods.

The general price measures of inflation do not apply to all consumers. An existing homeowner, for example, is not affected as much by the rapid inflation in the cost of housing as a person entering the market for the first time. Maintenance costs and taxes may increase but the basic investment "floats" with inflation.

Consumption of food is a different story. Every consumer purchases food, and inflationary prices are borne almost equally. The lesson of inflation is that it strikes everyone differently. General inflation measures, although convenient and necessary, fail to tell the whole story about the consequences of inflation.

Let us examine how one measures inflation in practice and why policies based solely on measured inflation rates may produce bizarre results. Consider the following statements.

109

1) Measurements of inflation in a growing economy <u>overestimates</u> the decline in purchasing power
2) A constant inflation rate when the economy grows actually means that the purchasing power is <u>increasing</u>.

These statements cannot be observed in isolation from other events in the economy. In fact, the proper way to examine inflation is in relation to how much the economy is producing, often referred to as <u>productivity</u>. Suppose that a mineral company produces 5% more of its product and charges 5% more. Consumers pay 5% more for the mineral, but receive 5% more of the product return. Are they worse off? No, because productivity has risen at the same rate as have prices, leaving purchasing power unchanged. Even though an inflation rate of 5% is observed in the mineral industry, consumers are not harmed.

Most countries exhibit some sort of aggregate growth. In the U.S. the average growth rate over the last decade has been between 3 and 4 percent. If the inflation had been zero percent the real value of consumption would have increased. The basic premise of inflation analysis is that productivity does not change. Every percentage increase in productivity offsets a percentage increase in inflation.

With a historical growth rate of approximately 4%, inflation rates of less than 4% increase the purchasing power of consumers. A rate of inflation equalling 4% leaves purchasing power unchanged. Inflation rates in excess of 4% reduce purchasing power by the amount over 4%. Suppose productivity rose by its historical average of 4% in 1974 and inflation was 7%, the decrease in purchasing power is 7% - 4% = 3%. Expectations of a 3-4% increase in productivity is the reason national leaders are often heard to say that their objective is to reduce inflation to 4%. At 4% inflation and a 4% productivity increase, purchasing power is left unchanged. Consumers pay 4% more, but obtain 4% more in return.

Growing economies usually have some inflation. The objectives must be to keep inflation close to the growth in productivity. High inflation rates in the sixties and seventies greatly exceed productivity, reducing everyones standard of living.

Measures of Inflation

Indexes are employed to measure the severity of inflation. Some common measures of inflation include:

1) Cost of living index
2) Wholesale price index
3) Implicit GNP deflator

Table 6.1 summarizes historical changes in these measures of inflation over a 25 year period. Rising prices characterize each measure over these time frames. Inflation, as measured by the cost-of-living index, has risen from 72.1 to 161.2, with 1967 = 100. Inflation rose 67% between 1967 and 1975, with the biggest jump occuring after 1972. The wholesale price index and implicit price deflator follow similar patterns.

The severity of the inflation during the early seventies can be seen by observing the pre-1972 changes and the post 1972 changes. Inflation rose by 4% or less in almost every year before 1972. Productivity advances of around 4% in these years improved the purchasing power of consumers considerably. From 1955 to 1972, inflation rose about 45% or 45/17 = 2.64%. Yet in the three years after 1972 inflation rose 36% for an average over the three years of 12% per year, which is more than a 400% increase. The rapid rate of inflation was doubly troublesome because productivity rose only about 1-2% in this period.

Wholesale price indexes changed even more dramatically. From 1950 to 1972 wholesale price index rose 37% from 81.8% to 119.1%. After 1972 it jumped to 174.9% for an increase of about 54% in three years. This increase was caused primarily by the shift in mineral and some agricultural prices. Less dramatic inflation is, however, measured by the implicit price deflator, which rose 47% in the pre-72 period and 27% after 1972.

Table 6.1 Historical Changes in Inflation Measures

Year	Consumer Price Index	Wholesale Price Index	Implicit GNP Deflator
1950	72.1	81.8	53.6
1951	77.8	91.1	57.3
1952	79.5	88.6	58.0
1953	80.1	87.4	58.9
1954	80.5	87.6	59.7
1955	80.2	87.8	61.0
1956	81.4	90.7	62.9
1957	84.3	93.3	65.0
1958	86.6	94.6	66.1
1959	87.3	94.8	67.5
1960	88.7	94.9	68.7
1961	89.6	94.5	69.3
1962	90.6	94.8	70.6
1963	91.7	94.5	71.6
1964	92.9	94.7	72.7
1965	94.5	96.6	74.3
1966	97.2	99.8	76.8
1967	100.0	100.0	79.0
1968	104.2	102.5	82.6
1969	109.8	106.5	86.7
1970	116.3	110.4	91.4
1971	121.3	113.9	96.0
1972	125.3	119.1	100.0
1973	133.1	134.7	105.8
1974	147.7	160.1	116.4
1975	161.2	174.9	127.2

Presented with such different measures of inflation, a commonly asked question is which should be used. There is no easy answer since each represents different elements of an economy. Consumer price indexes, for example, are calculated by specifying goods purchased only by households - food, clothing, housing, medicine, transportation, and so on. Representatives of the government collect samples of the cost for each good on this list in about 50 cities each month. These groups are called a consumption bundle. If the bundle cost $700 per month in March of one year and $770 in March of next year, inflation was 770/700 = 1.10 or 10%. A rise in cost between March and April of $10 gives inflation for the month of April of 10/700 = 0.14%.

Wholesale price indexes are calculated using the same process. The only difference is that a different bundle of goods are considered. Instead of looking at costs from the consumers perspective, industry costs for minerals, labor, and capital are sampled. Changes in the cost of this bundle compose the wholesale price index.

Wholesale and consumer price indexes are related to each other. Since wholesale prices reflect the costs faced by stores, a wholesale price increase leads to a corresponding increase in consumer prices as stores pass the price increase on to customers. The relationship among prices is implicit, however. Stores may choose to absorb some of the price increase for competitive reasons. If wholesale prices eventually decline, stores often maintain the old price level as compensation for the period when they accepted less profit after the price rise. Examples of the inexact relationship in the seventies include sugar prices and coffee prices. But the two price indexes will in general move in tandem over time.

The implicit GNP deflator represents a different kind of index, as indicated by the use of the term implicit. Gross national product (GNP), which is explained in more detail later, measures total expenditure in an economy, including personal consumption expenditures, government expenditures, investment expenditures, and so on. Each of these categories possesses its own inflation index. Gross national product can be constructed in two ways. One sums actual expenditures, and the other sums these expenditures after they have been divided by their respective measures of inflation, referred to as constant dollars. The implicit GNP inflation measure is actual GNP expenditures divided by GNP expenditures in constant dollars.

Gross national product in 1975 was 1,516 billion. If personal expenditures were $900 billion, then all other expenditures would be $616 billion.

$$GNP = 900 + 616 = \$1,516 \text{ Billion}$$

These figures are in current dollars (uncorrected for inflation). If inflation was 15% and 5% respectively, the constant dollar value for GNP is

$$GNP = \$900/1.15 + \$616/1.05$$
$$= \$782 + \$586$$
$$= \$1,369 \text{ Billion}$$

The implicit price deflator (IPD) is

$$IPD = \$1,516/1,369$$
$$= 1.107, \text{ or } 10.7\%$$

The implicit price deflator, thus, is not actually used to deflate GNP, but summarizes the total effect of all inflation indexes. Note that it is actually a weighted average of each inflation index, with the weights being the proportion of GNP.

The question of which index should be used hinges on what you're trying to measure. A goal of capturing the purchase power of individual consumers dictates the consumer price index measure; for companies, the wholesale price index is best; and for overall approximation of inflation's impact on the economy, use the implicit price deflator.

112

Causes of Inflation

Perhaps the most interesting part of inflation analysis is the ongoing dispute of what causes it, with the obvious objective of curing it when it raises its ugly head. Remember, however, curing inflation means reducing it to a level equal to or less than productivity. Our understanding of the causes of inflation is clouded by conflicting explanations, making it difficult to derive and implement cures.

Sure-fired cures for inflation are not easy to come by. Inflation only means that the economy is sick, and, like personal illness, the sickness can be caused by many factors. Like doctors, the inflation analyst first has to diagnose the possible causes. Current theories identify three main sources of inflation. They are

1) Excess demand inflation
2) Cost-push inflation, and
3) Bottleneck inflation

Excess demand inflation occurs when total demand for products exceeds the capacity of industry to supply the products. Cost-push inflation arises when groups in society obtain the power to force wages and prices without increasing their productivity. Bottleneck inflation occurs as the demand for products increases, and industry has to take several years retooling their plants to meet these demands.

Excess demand inflation receives the most attention, because it represents the battle-ground where opposing philosophies attempt to establish superiority. The combatants are the Keynesians, who follow the principles of John Maynard Keynes, and the classicists, who are led by Nobel prize winner, Milton Friedman. Controversey centers around the cure, not the cause of excess demand inflation. Keynesians emphasize federal controls as the way to control inflation, while classical policy favors controlling the money supply.

Both groups recognize that demand-pull inflation occurs when demand increases faster than supply. Figure 6.1 illustrates the concept of demand-pull inflation. Aggregate supply represents the supply of all goods and services in the economy and aggregate demand is the same. When aggregate demand is $400 billion, the average price for all goods is $8. Now suppose aggregate demand rises to $1,800 billion; price goes to $15 per good. If we were just measuring price changes as all inflation measures do, the computed rate of inflation is 15/8 = 1.875 or 87.5%.

Keynesians look at this situation by subdividing demand into three basic components indicating where income may go--consumption by individuals (c), investment by corporations (I), and government (G). That is, a countries products are consumed by these three sectors, or

$$Income = C + I + G$$

The equilibrium in the economy occurs whenever the sum of three expenditures equals income because every dollar spent by one group yields a dollar's worth of income to the seller. The 45° line in Figure 6.2 represents this equilibrium. At the initial equilibrium $400 billion is spent and received as income. Now suppose that government, consumers and corporations increase expenditures to 1,800 billion, income also increases by the same amount.

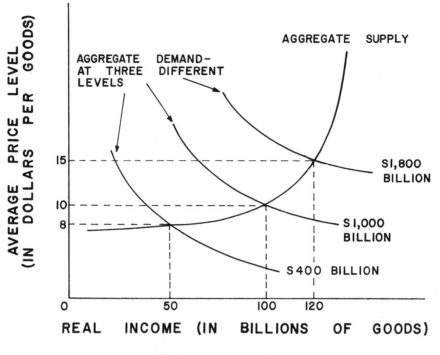

Figure 6.1 Demand-Pull Inflation

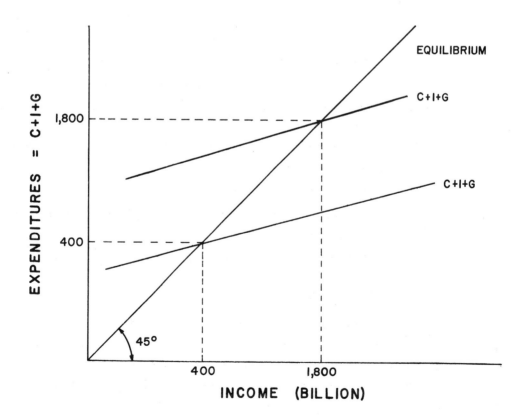

Figure 6.2 Keynesian Demand-Pull Inflation

114

Figure 6.2 converts to Figure 6.1 if we recognize that income and expenditures equal the price of goods times the quantity sold. At income of $400 = ($50)(8). To increase income to 1,800, quantity rises to 120 and price to 15, yielding the $1,800 estimate. Price has risen from 8 to 15, creating inflation as before. What is important in the Keynesian system is explicit consideration of the government sector as a consumer. The Keynesian approach argues that government can control inflation by changing their level of demand. When consumers and companies increase demand, government can reduce theirs to maintain equilibrium. Should demand be insufficient for products by consumers and corporations to help keep labor employed, government is recommended to step in an increase demand, thus creating additional jobs.

Events during the sixties and seventies illustrate both cases. Government demand rose rapidly during the Vietnam war. But without a corresponding decrease in consumer and corporate demand, high rates of inflation were created. In the seventies when unemployment was high, the government stepped in and increased their expenditures to create demand, which hopefully, would create more jobs. This was accomplished. The problem was that the additional jobs basically were unproductive, causing inflation with little change in output. Purchasing power therefore declined.

A classical analysis of inflation is based upon the quantity equation of money, which is

$$(M)(V) = (P)(Y)$$

where: M is the money supply, V is the velocity of money, or the number of times a dollar changes hands each year, P is the index of prices used to measure inflation, and Y is income measured in constant dollars. The quantity equation is true by definition.

Velocity of money is illustrated by comparing money supply to real income. If the money supply in a year is $300 billion and real income is $900 billion, every dollar has to change hands an average of 3 times each year in order to record an income of $900 billion. Money changing hands is consistent with our observations. A family earns a dollar which they record as income. The family purchases food from a grocery for a dollar, which the grocery also records as income. Next, the grocery purchases supplies from a wholesaler, and the dollar is again recorded as income. Every time a dollar changes hands, an extra dollar of income is tabulated.

Classical philosophy dictates that the money supply be used to control inflation in the following manner. Data for each variable are observed in a year

$$M = \$300 \text{ billion} \qquad P = \$1.00$$
$$V = 4 \qquad\qquad Y = \$1,200 \text{ billion}$$

Note that (300)(4) = (1.00)(1,200) by definition. Now suppose that real income is expected to increase to $1,500 billion next year. With a goal of zero inflation, or no change in P, the money supply has to be

$$M = \frac{1}{V}(P)(Y)$$
$$= (1/4)(1.0)(1,500)$$
$$= \$375 \text{ billion}$$

If an acceptable rate of inflation is 4% a year, money supply becomes

$$M = \frac{1}{4} (1.04)(1,500)$$
$$= \$397 \text{ billion}$$

What these calculations show is that changing the money supply and inflation are related to each other. Consider the common occurrence where the money supply is actually $420 billion, the inflation measure becomes

$$P = MV/Y$$
$$= (420)(4)/1,500$$
$$= 1.12, \text{ or } 12\% \text{ inflation}$$

Only by correctly matching the money supply with real income is inflation controlled.

Disputes over the proper level of the money supply are quite common. In 1977, President Carter replaced Dr. Burns, Chairman of Federal Reserve System, the comptroller of the money supply, because Burns was not increasing the money supply fast enough. According to the classical quantity equation, the net impact of a larger money supply is more inflation.

Several main issues divide the Keynesians and classicals. One critical factor is the velocity of money. The quantity equation assumes that velocity is constant over time, while Keynesian argue that it can be changed. This difference points up the primary area of dispute. Keynesians suggest altering federal expenditures and income to control inflation. Classicists, like Dr. Friedman, argue that the economy is so complex and government expenditures so hard to control that such a strategy is impractical; Friedman and his followers, instead, recommend increasing the money supply by a constant rate of 3 or 4% a year to control inflation.

Because of these different strategies we are constantly exposed to different recommendations for curing inflation. The question is how do we recognize which is the best given these conflicting philosophies. The answer is that none of them are likely to be perfect, all will contain mistakes. Economics of major nations are complex, integrated mechanisms whose size is so large that we seldom know the exact size of the money supply at any point in time, much less how much various sectors of the economy are consuming and producing. So, estimates of production and consumptions are the best that can be achieved. Developing specific cures from imprecise information about relationships between the various economic groups obviously yields inconclusive policies, and policies that are unlikely to be highly successful.

The low probability of success does not imply that efforts to reduce inflation are worthless, however. Every country in the world has adopted in different degrees the philosophy that government should intervene in the workings of the economy to stabilize economic fluctuations. Disputes certainly exist whether this should or should not be done, an issue which we leave for each reader to decide. As long as governments do intervene, the objective should be to do the most good (or the least harm). Recognition that the problem is complex only means that more effort has to be expended to understand the specifics of the problem in each country.

So we are back to the issue of how to analyze and develop policies for solving inflation. From our point of view both monetary and fiscal (government control of their income and expenditures), policies are necessary. Selecting the right combination of monetary and fiscal policy depends upon the exact nature of the causes of inflation, which, like every human being or mineral deposit, is never an exact replica of previous ones. Accomplishing this goal requires all elements of the economy, - the government, corporations, and consumers, - to agree on acceptable policies. Unlikely as this concensus is suggests that inflation will always be a highly political issue.

In the context of cost-push inflation, prices rise because suppliers of goods raise prices above the equilibrium levels shown in Figure 6.1. Cost-push inflation sometimes occurs because supplies of final goods increase their profit margins. Alternatively, labor unions or supplies of raw materials increase the cost of their products to manufacturers. For example, a labor union, which negotiates a substantial wage increase for its members contributes to cost-push inflation.

Examples of cost-push inflation in the seventies are plentiful. Coal miners in 1978 struck for over 100 days to obtain higher wages. Coal costs to other sectors rose because of the higher wages. Farmers, which historically contributed very little to inflation, also engaged in a strike designed to raise the price of agricultural products in the late seventies. Inflation occurred to the extent that they were successful.

Historically cost-push inflation contributed little to the overall inflation rate because consumers could simply shift their consumption to other items. By the mid-seventies modifications in the structure of the economy had changed all of this to the point where cost-push inflation became an integral part of most economies. The cause of this change was the practice of tieing labor union wages, non-union wages, social security benefits, government employee wages, and other cost factors to inflation rates. As things stand now a significant portion of the free world resources increase in price according to inflation. If inflation was 6% last year and labor costs, OPEC oil prices, coal prices, agricultural prices, and other factors rise by 6% in the coming year, it is virtually impossible to ever stop inflation. Tying compensation to inflation is perhaps the most discouraging force in today's world facing anyone concerned with fighting inflation.

Figure 6.3 illustrates cost-push inflation. Cost-push inflation occurs when the prices rise for any of the reasons above. The equilibrium price and quantity is $10 and $120 billion of real income. Should labor unions or an increase in profit margins raise price to $12, a disequilibrium appears where consumers want to spend $100 billion and producers desire to produce $130 billion. Excess supply of $30 billion results, and inflation is 12/10 = 20%. This graph is important because it captures an important element of inflation in the seventies. Higher prices were associated with excess production capacity. Everytime costs rose, excess capacity got larger, until many industries faced severe financial problems.

Cost-push inflation's duration was traditionally short-lived because the disequilibrium depicted in Figure 6.3 never lasted very long. Now with wages and salaries tied to inflation the time required to return to equilibrium has been lengthened and a valid question is whether cost-push inflation can ever be solved.

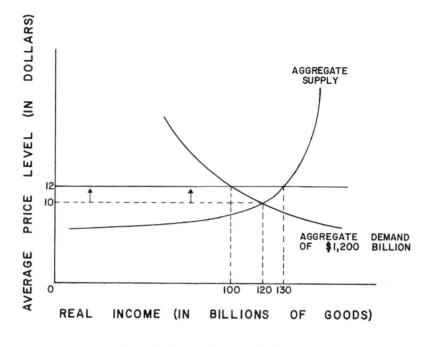

Figure 6.3 Cost-Push Inflation

Prospects for Effective Inflation Control

Controlling inflation is a difficult task as indicated in the discussion of excess demand inflation. But even more troublesome is how do we handle periods when excess demand inflation coincides with cost-push inflation. Since cost-push inflation is being mandated by law and excess demand inflation is complex and subject to competing views, the prospects for effective inflation control in an open economy are dim. In short, the trend in inflation control seems to be in the direction of reducing the impacts of inflation, rather than actually devising policies for curing it. Inflation is to economics what the common cold is to medicine. About all that can be done is to minimize the suffering of the patient, until the illness runs its course.

Complicating this is the matter of productivity, a resultant of man and machine working in tandem. This is not controllable by government although interference by government in the private sector tends to reduce it.

References to inflation are thus somewhat misleading. Inflation which indicates the rise in the general price level can be caused by a variety of economic forces. Formulation of cures, without agreement on the causes, is destined to fail.

INVESTMENT CAPITAL

One of the most interesting issues facing mineral industries, as well as all industries, relates to the attraction of investment capital. A frequently asked question is where does investment capital originate and how can a specific firm attract it? Investment capital falls into

118

two categories externally generated and internally generated. Externally generated investment capital comes from the sale of instruments like common stock and debt to investors, who receive title to a share of the companies assets in return for their money. Internal capital investment results from after tax profits that are not distributed to stockholders in the form of dividends.

External and internal generation of investment capital are functions of a variety of economic forces. A commonly heard statement is that the mineral industries cannot supply the necessary quantity of minerals because they are not able to generate sufficient internal capital. Such a statement usually supports recommendations for changing tax laws. Less restrictive tax laws are presumed to yield greater after tax profits, and thus a larger amount of internal capital. Repeal or changing the foriegn tax credit, intangible drilling tax credit, and the depletion allowance are examples of tax law changes that reduce after tax profits.

Analysis of investment capital sources is more complex than this because internal and external investment capital sources are directly tied to each other. As the profits determining internal generation of investment capital change, external investment capital is also altered. To see this relationship it is necessary to understand the motivation of the owners of mineral companies - the stockholders. The ultimate source of all investment capital is the individual. Private individuals purchase stocks or bonds in the hope that they will earn more money from this investment than they can from other investment alternatives. Institution investors, such as pension funds, mutual funds, and trusts, are nothing more than representatives of individuals that have been designated to make these decisions for them. For example when a bank purchases a corporate bond, that is actually owned by the stockholders of the bank.

For external capital sources the prime source of supply of investment capital is the discretionary income of families. Discretionary income refers to income which is not spent for basic items like housing, food, clothing, and so on. A family earning $2,000 a month before taxes might have $350 taken out for taxes, and spend $1400 on basics, leaving $2,000 - 350 - 1400 = $250 a month of discretionary income that could be used for investing. The interesting feature is how this money is used. Each consumer must decide whether to save (which is investing) or buying more goods with the discretionary income.

The absolute size of discretionary income changes as income shifts in relationship to families expenditures. If income, taxes, and basic expenditures all increase by 6% the family has 6% more discretionary income. Summing the discretionary income for all families yields the total amount available for investing. The total quantity of investment capital changes as income, taxes, and other expenditures vary. Increasing social security taxes, or federal income taxes lowers investment capital. Inflation, the prime source of higher basic expenditures, also reduces investment capital.

Actual investment capital seldom equals the sum of family discretionary income, however. To understand this it is necessary to ask why we, as consumers, would invest our discretionary income now instead of spending it. Like corporations which invest to increase future income, families do the same. The objective is to select investments that increase future purchasing power. But, since investing requires spending now and then waiting for uncertain profits to occur, investing is a risky undertaking. So the actual quantity of money available for investing

119

actually transformed into investment capital depends upon whether expected return is high enough to offset the risk of the future.

As a simple example, take a family with $500 of discretionary income. The option's are to spend the $500 for consumption goods or to place the funds in a savings account earning 5¼% a year. Investing in the savings account raises purchasing power next year to 500(1.0525) = $526.25. This amount is known with certainty since savings accounts are insured. What we don't know with certainty is how much we'll be able to buy with $526.25 in the next year. Suppose the consumer desires to purchase a stero costing $510. By saving the $500 for a year interest generated from the savings account is sufficient to purchase the stereo, thus making the investment desirable.

A price increase of 8.12% lifts the price to $540.60, leaving the consumer short of the required funds by $16.35. This example illustrates the principle that the promise of a positive return on an investment is insufficient. What is necessary is that the investment yield a return which allows the consumer to purchase a larger quantity of minerals tomorrow than he can today. Inflation exemplifies a good benchmark for comparing desired rates of return.

The general principal describing the quantity of investment capital is that the return from the investments must exceed the cost of investment capital. The greater return is over cost, the more the quantity of investment capital which will be made available. Figure 6.4 illustrates these principles. The historical relationship between return and the supply of investment capital shows a willingness on the part of investors to invest more at higher returns. Curve I_1, shows the $50 billion of investment capital is supplied at 6% return, assuming zero percent

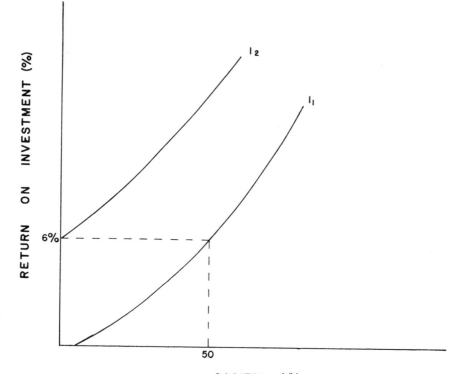

Figure 6.4 Supply of Investment Capital

120

inflation. If inflation rises to 6%, curve I_2, showing no investment capital would occur. Why?-because what advantages is there to supply investment capital returning 6% interest when the price of most goods also rises 6%. To stimulate investment capital supply borrowers must offer returns in excess of 6%.

The capital crunch in the seventies resulted from such a problem. Inflation reaching 8, 9, 10% exceeded returns from savings accounts, bonds, and stocks. Savings finding their way into investments, which historically averaged about 10% of personal income, declined to about 2% in 1974 and 1975. The result was the inability by corporations, governments and all borrowers to attract the necessary volume of investment capital until rates of return on the investment increased.

Allocation of investment capital among various industries also depends on the return offered relative to the riskiness of the investment. Investors given a choice of only two investments, a drilling company and a tax-free municipal bond, returning 8% each have an easy choice. They will invest in the bond because the return of 8% is less risky than the drilling company. Risks explains why some companies have a difficult time attracting capital even when the total supply of investment capital is adequate.

Figure 6.5 illustrates the effect of risk classes on the ability to attract investment capital. Curve R_1 represents investment like government sponsored assets that are relatively risk free. Curve R_2 is a similar curve for high risk enterprises like most mineral companies. If both investments offerred the same return of 9%, which is about the historical average in the petroleum industry, investment capital generated is $15 billion for the mineral industries and $35 billion for the less risky investments. Curve R_2 lies to the left of curve R_1 indicating that the same percentage return yields a smaller amount of investment capital.

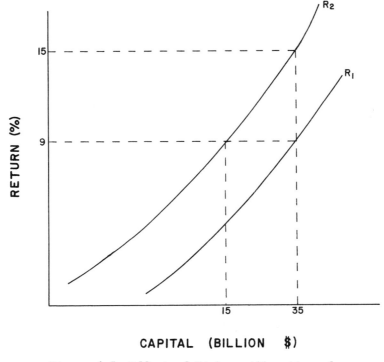

CAPITAL (BILLION $)

Figure 6.5 Effect of Risk on Allocation of

Investment Capital

121

Should the mineral companies decide that $15 billion was insufficient for their capital needs, return on investment should be raised. The question is by how much must returns be raised. A goal of collecting $35 billion requires a promise of a return of at least 15%, as Figure 6.5 indicates. In order to entice investment capital away from other inudstires, risky industries like minerals have to offer a higher return on investment. Periods when investment capital is in short supply are characterized by low returns from a mineral investment relative to returns in other areas.

Generating investment capital from internal sources is an alternative to seeking investment funds from external capital markets. Internal investment capital is equivalent to retained earnings on a corporation's balance sheet. The larger the retained earingins, the less the quantity of investment capital that must be borrowed directly. A typical mineral corporation income statement is presented below under column A. With sales of $6 billion the income statement denotes that $3 billion is used for producing the mineral, other operating expenses, like overhead and marketing are $500 million, depreciation is $300 million, leaving net income of $2,200.

Table 6.2   Income Statement
(In Million $)

Categories	A		B
Sales	6,000		
-Cost of Goods Sold	3,000		
Gross Profit	3,000		
-Operating Expenses	500		
Gross Operating Income	2,500		
-Depreciation	300		
Net operating income	2,200		
+Other Income and Royalties	100		
Gross Income	2,300		
-Other Expenses			
Interest	800		
Royalty Payments	300		
Net income before taxes	1,200		1,200
-Income Tax (50%)	600	(30%)	360
Net Income after Tax	600		840
-Dividend Payment	250		455
Retained Earnings	350		385

Subtracting other expenses from gross income, including federal taxes leaves $600 million of net income after taxes. The mineral company can retain the entire $600 million for investing or dispense part of it to the stockholders. Dividends are $250 million in this example, leaving $350 million worth of retained earnings, the internal source of investment capital.

Suppose the mineral company possesses investment opportunities costing $385 million. With retained earnings of $350 million, $35 million must be raised from the external capital markets by selling new stock or bonds. Mineral companies pressed federal agencies in the seventies and sixties for tax advantages that would enlarge their net profit after taxes. Column B illustrates a situation where tax measures reduce the effective tax rate to 30%. At the 30% tax rate, net income after tax income is $840 million, allowing the mineral company to pay dividends of $455 million and retain earnings of $385 million, which equals the desired quantity of investment capital.

122

The real world does not operate as simply as the income statement suggests. Retained earnings do not belong to the corporations; they are the property of stockholders. The mineral company is in essence borrowing funds from the stockholders by retaining the earnings. Retaining earnings is satisfying to stockholders only when the company earns a higher return on the use of these funds than the stockholders can earn if they had the money to invest. Mineral corporation management seeks a balance between their conflicting objectives by dispensing part of after-tax operating income in the form of dividends and keeping the rest as internal investment capital.

COST OF CAPITAL

Cost of capital is the name given to the price corporations have to pay to attract the desired quantity of investment capital. It is thus a composite of all forms of financing available to the mineral company - debt, preferred stock, and equity. Each financing form face different costs and risks that must be considered, and are discussed separately. Mineral companies cost of capital is the weighted average of the cost of the financing sources.

Cost of Debt

The cost of debt is the interest paid to bond holders. If a firm issues $10 million of debt at six percent interest, the before tax cost of debt is:

$$Kd = \frac{interest}{principal} = \frac{600,000}{10,000,000} = 6\%$$

Six percent is the return realized by the purchasers of the bond. The actual cost of debt to the mineral company is not 6%, however, because they get a tax credit for interest payments which protects part of their income. After tax cost of debt is the relevant cost in calculating the overall cost of capital, or:

$$Kd(1-t) = (before\ tax\ cost)(1.0-tax\ rate) = (6\%)(1-0.48)$$
$$= 3.12\% \tag{6.1}$$

when the effective tax rate is 48%. Always use $Kd(1-t)$ when calculating the overall cost of capital.

Cost of Preferred Stock

Preferred stock is a hybrid form of financing between debt and common stock. Like debt, preferred stock carries a fixed commitment to make periodic payments. Failure to make the dividend payments does not result in bankruptcy, as does nonpayment of interest on bonds. Preferred stock is thus more risky to the firm than common stock. To the investor, preferred stock is more risky than common stock and less risky than debt.

A mineral company offering to pay $5 interest on preferred stock selling for $50 per share has a cost of

$$Kp = \frac{dividend}{price\ of\ stock} = \frac{5}{50} = 10\%$$

for preferred stock. Since there is no tax write-off for preferred stock, 10% is the actual cost to the firm.

123

The actual cost of preferred stock is greater than 10% to the firm. Firms must pay a cost to brokers, called <u>flotation costs</u>, to get them to sell the stock. So even though investors pay $50 per share, the mineral company receives an amount less than $50 equal to the flotation cost. A brokerage fee of $2.50 per share raises the cost of preferred stock to

$$
\begin{aligned}
K_p &= \text{dividend}/(\text{price - flotation cost}) \\
&= 5/(50 - 2.50) \\
&= 5/47.50 \\
&= 10.52\%
\end{aligned}
$$

The higher cost of preferred is the result of flotation costs.

Cost of Retained Earnings

Internal capital sources like retained earnings are owed to stockholders and therefore cost the same as equity. The value of common stock depends ultimately on the dividends paid on the stock discounted over time.

$$
P_0 = D_1/(1 + K_r) + D_2/(1 + K_r)^2 + \ldots + D_t/(1 + K_r)^\infty \tag{6.2}
$$

Here P_0 is the current price of the stock, D_t is the dividend expected to be paid at the end of year t, and K_r is the required rate of return. This formulation differs slightly from the usual calculation, which is:

$$
P_0 = D_1/(1 + K_r) + D_2/(1 + K_r)^2 + \ldots + (D_t + P)/(1 + K_r)^n \tag{6.2a}
$$

The first equation deals with the case where the stock is held to infinity, and this equation recognizes that stock may be sold n periods later for P dollars. In either equation, we solve for K_r by the rate of return method explained in Chapter 5.

These equations yield the same conclusion if we recognize what is incorporated in the selling price, P, at time n. Capital gains represent expected cash flows which for the entire firm consist only of future dividends. Remember that we are viewing retained earnings from the perspective of the firm, not the investor. The cash flow of a firm is its dividends inasmuch as the firm itself never realizes capital gains, even though an investor might. In essence P represents the price paid an investor to forego future dividends from the company.

A simpler solution to finding K_r is to assume that dividends will grow at a constant rate g for an infinite time period.

$$
P_0 = \frac{D_0(1+g)^1}{(1+K_r)} + \frac{D_0(1+g)^2}{(1+K_r)^2} + \ldots + \frac{D_0(1+g)^\infty}{(1+K_r)^\infty} \tag{6.3}
$$

Since we cannot solve for values going to infinity, rewrite the equation as

$$
P_0 = D_0 \frac{1+g}{(1+K_r)} + \frac{(1+g)^2}{(1+K_r)^2} + \ldots + \frac{D_0(1+g)^n}{(1+K_r)^n} \cdot \tag{6.4}
$$

Multiply both sides by $(1 + K_r)/(1 + g)$

$$P_0 \frac{1 + K_r}{1 + g} = D_0 \left[1 + \frac{(1+g)^2}{(1+K_r)^2} + \ldots + \frac{(1+g)^{n-1}}{(1+K_r)^{n-1}} \right] \tag{6.5}$$

Subtracting equation 6.3 from 6.4 and assuming that the ratio of growth to retained earnings cost approaches 1 as n goes to ∞ gives:

$$P_0 = \frac{D_1}{K_r - g} \quad \text{or} \quad K_r = \frac{D_1}{P_0} + g \tag{6.6}$$

Equation 6.6 estimates the cost of retained earnings by dividing estimated dividends next year by current price and adding the growth rate. This is just an estimate since the model assumes that P_0 represents the actual value of the company. That is, the stock is neither under or over-priced.

Mineral company XYZ dividends are expected to grow at this historical rate of 4%. With stock currently selling at $30 per share and an expected dividend payment of $2.00 per share the cost of retained earnings is

$$K_r = \frac{2}{30} + 0.04$$
$$= .1066 \quad \text{or} \quad 10.66\%$$

Company XYZ's stock will rise by 10.66% only if the company earns 10.66% on the investments paid for with retained earnings. If the return on investment financed with retained earnings is less than 10.66%, the price of the stock should decline.

Cost of Equity

The cost of equity already issued is calculated the same way as retained earnings since they are equivalent forms of financing. To find the cost of issuing new common stock, K_e, flotation costs must also be considered. New stock issues cost more than existing equity as the result of flotation cost. Equation 6.7 solves for K_e,

$$K_e = \frac{\text{dividend}}{\text{price}} (1 - \text{flotation percentage}) + g \tag{6.7}$$
$$= \frac{D_1}{P_0(1-F)} + g$$
$$= \frac{2}{30(1-.10)} + 0.04$$
$$= .1104 \quad \text{or} \quad 11.4\%$$

when the values for D_1, P_0 and g are the same as before and 10% of the stock price is paid to brokers. New equity costs 11.4%, while current equity costs only 10.66%. The difference is due to the consideration of only costs forced by the mineral company. Investors still receive 10.66% return, but it cost the firm an extra 0.74% to issue the stock.

Composite Cost of Capital

Mineral companies employ various forms of financing to obtain investment capital, each with a different cost as we have shown. To find the overall cost of capital the relative cost of each form of financing is combined. The weighted cost of capital is given by

$$K_a = \sum_{i=1}^{n} (W_i)(K_i)$$

$$= (W_d)(K_d) + (W_p)(K_p) + (W_e)(K_r) + (W_e)(K_e) \qquad (8.8)$$

where K_d, K_p, K_r, K_e are the cost of each type of financing and W represents the percentage of each of total financing.

Suppose company XYZ utilized the following combination of financing

Type	Financing ($million)	Weight
Debt	$ 10	$10/24 = 0.416 = W_d$
Preferred Stock	2	$2/24 = 0.084 = W_p$
Retained earnings	6	$6/24 = 0.25 = W_r$
Equity	6	$6/24 = 0.25 = W_e$
	$ 24	1.00

Total financing of $24 million is converted to percentages by dividing every element by $24 million. The composite cost of capital is:

$$K_a = (0.416)(.0312) + (0.084)(.10 + 0.25)(.1066 + 0.25)(.1066)$$

$$= 0.0617 \quad \text{or} \quad 6.17\%$$

Combining relatively lower cost debt with stock and retained earnings the cost of capital for existing financing is 6.17%. The composite cost for new securities is

$$K_a = (0.416)(0.0312) + (0.084)(.1052) + (0.25)(.1066) + (0.25)(.114)$$

$$= 0.0769 \quad \text{or} \quad 7.69\%$$

which is the result of the extra flotation costs.

An often asked question is why don't mineral companies employ more of the lower cost debt as a way to reduce their cost of capital. Capital costs could be reduced to 3.12% by using 100% debt financing. Debt financing is not the best source of funds because of its inflexibility. If debt costs 6% and company issues $24 million worth, its annual interest expense is ($24 million)(0.06) = $1.44 million. For companies whose income before taxes and interest is close to $1.44 million, there is a high probability of bankruptcy since a reduction in income prevents the company from meeting their interest payments.

Debt, unlike stock financing, leaves companies little flexibility for surviving in periods when income before interest and taxes are depressed. Company financing objectives are to combine debt with equity in a way that keeps overall capital costs low, without increasing the danger of bankruptcy.

The risk associated with debt financing also affects the cost of debt and equity since investors require greater returns in order to compensate them for the risk of bankruptcy. Figures 6.6 and 6.7 illustrate how the cost of debt and equity vary, respectively, as more debt financing is utilized. The effect of _leverage_ (the percent of debt, W_d, in financing) shows an increase in cost in Figure 6.6. Lenders require higher returns to compensate them for the extra risk. Cost of debt is about 3.00% up to the point where debt comprises 40% of total financing. After this point, the cost of debt rises dramatically to 6% at 60% financing, which implies that investors are not sure that the mineral company can afford to continually meet the interest payments.

Figure 6.7 represents a similar situation for equity financing. Stockholders required return on equity has been simplifed in that the return has been broken down into three components-risk free rate of return, premium for business risk, and premium for financing risk. The risk free rate of return represents the expected return on assets like treasury bonds, and savings accounts which possess little risk. The premium for business is determined by the nature of the business. Financial risk, however, raises according to the amount of debt in the capital structure. The larger the interest payments, the less there are likely to be profits left over to pay stockholders. So as debt increases, stockholders demand greater returns to compensate them for risk.

Importance of the Cost of Capital

Capital cost represent the price mineral companies have to pay to obtain funds to operate. Like a bank, a corporation must earn more on its investments than it has to pay for the investment capital. Paying 10% capital while earning 10% from their investment leaves the company at a breakeven point. Company growth is measured by the excess of the return on the investment over the price paid for capital. This is the reason measures like capital cost are used to determine the time value of money in Chapter 5. Dividing by capital costs in present value calculations measures net cash flow after the cost of capital are taken into consideration.

MINERAL INDUSTRY COMPETITION

One of the issues of mineral economics that always provides a forum for heated discussion is the question of whether mineral companies really do compete against each other, or just go through the motions of competition to increase profits. Competition in the early years of the petroleum industry led to the bust-up of Standard Oil on the grounds that it was a monopoly. Anti-trust activity in the sixties and seventies turned to the question of whether the seven major petroleum companies held sufficient control over markets through vertical integration to constitute a vertical monopoly. Today the issue is whether acquisitions and mergers among mineral companies, such as coal-oil, oil-nuclear, gas-coal and so on retards the efficient flow of minerals to the consumer, while raising prices.

The cornerstone of these debates is the economic force called competition. If the mineral companies are competitive, it doesn't matter how large they are. But when increased size reduces competition, economic efficiency is retarded. Joint ventures in drilling and vertical inte-

127

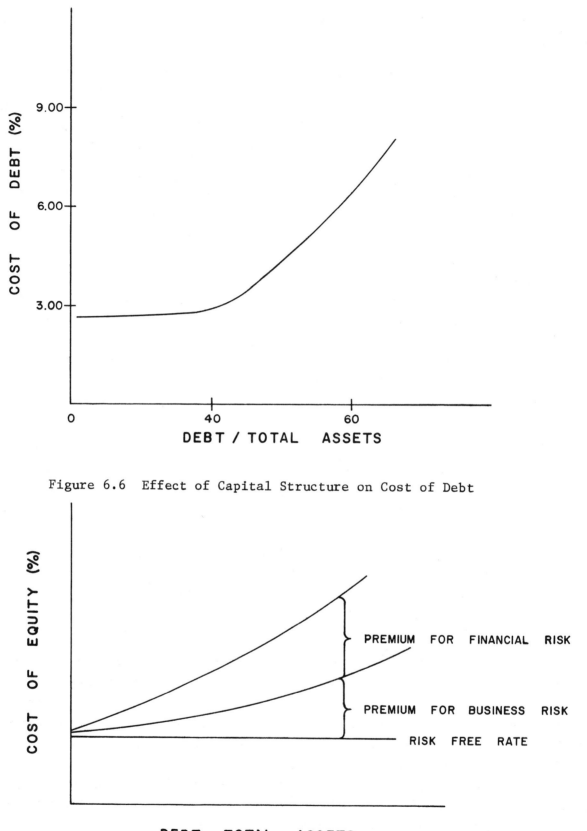

Figure 6.6   Effect of Capital Structure on Cost of Debt

Figure 6.7   Effect of Capital Structure on Cost of Equity

gration illustrate presumed anti-competitive behavior. Critics argue that mineral supplies are reduced by these activities, forcing consumers to pay higher prices. Supporters counter that these practices permit more efficient utilization of resources, allowing mineral companies to actually increase production.

Like we saw in the discussion of inflation, competition is a term more complicated than the general usage implies. As it stands now, competition implies a black and white world where mineral companies either compete or do not compete. There is little allowance for variations in the degree of competition, and whether and in what ways consumers may be affected. In this section a basic framework for looking at competition, and the opposite case, concentration, is presented.

Basis for Appraising Competition

Competition is presented as a goal because it is presumed to benefit consumers by increasing mineral supplies and keeping prices as low as possible. The ideal case, called perfect competition, represents minimum costs, maximum supply situations. Perfect competition, however, is an unobservable state because no industry ever has been, nor ever will be, perfectly competitive. It does define a benchmark for evaluating actual industry organizations, such as monopoly and oligopoly.

Evaluation of competition involves determining the number of firms in one industry and how they interact with each other. The effect of size on competition is illustrated by statements of several critics. The Special Committee on Integrated Oil Operations concluded that:

A substantial portion of the "observed" increased in concentration and market followed from a series of major mergers and acquisitions by large petroleum companies. Since 1960 alone, there have been more than fifty major corporate mergers and acquisitions in the industry. (Special Committee, 1974).

The trend toward larger petroleum companies is viewed as anti-competitive by the late John Blair because, once medium size firms become larger firms through mergers, the competitive or large firm is lost. According to Blair, attaining the status of a national major petroleum company makes the company realize that it has as much to lose from competition as older major firms. (Blair, 1972).

The validity of such assertions depends upon the definition of competition; competition means more than just "bigness is bad."

Perfect Competition

An industry is perfectly competitive only when the following conditions are met.

1) The number of buyers and sellers are extremely large.
2) No company or group of companies control the market.
3) Entrance and exit from the industry is extremely easy.
4) All buyers and sellers possess perfect information about the markets they deal with.
5) The minerals are all alike in quality.

The conditions for perfect competition are quite rigorous. It is easy to see why perfect competition is ideal when a large number of suppliers adjust immediately with perfect information and no dominant firm. Such a world has to be perfect (and does not exist).

Petroleum exploration easily satisfies part of the first condition with over 2,000 companies engaged in exploration and production and condition two with no dominant companies. Condition three is usually met since firms enter and leave production all the time. Condition four is never met since no one possesses perfect information about drilling prospects. All we have to do is to look at the record on bonus bidding and bid formations to observe the imperfections of the markets. Even petroleum explorations and production are imperfect by these conditions. Few products satisfy condition five - equality of mineral characteristics - because each producing site contains minerals with varying properties. So every producer cannot respond immediately to changes in demand for petroleum with specific properties.

Refineries on the other hand meet different conditions. There are few refineries (in relation to producers), building or shutting down an expensive refinery is easy, and the price wars in the past indicate that refineries do not have perfect information. Because of contract specifications refined products seldom possess exactly the same characteristics.

Marketing petroleum products is also imperfect. Only the condition of a large number of buyers is met. Even the condition of equal quality is not met if one believes petroleum advertising.

The net effect of these conditions is that no phase of the petroleum industry satisfies conditions for classification as perfectly competitive. But neither has any other industry now existing or ever known to have existed. The real issue is not whether an industry is perfectly competitive; since they are not, but how competitive are they, relatively.

Competition in Mineral Industries

Mineral industry critics try to apply the theories of perfect competition to evaluate any action, such as joint ventures, vertical integration, and acquisition of other mineral forms. Perfect competition presumably determines the pattern of mineral industry behavior that:

1) maximizes the total value of minerals
2) distributes minerals to consumers according to thier willingness to pay for the mineral, and
3) insures a fair return on mineral investments.

Joint ventures vertical integration, and acquisition of other minerals causes critics to argue that mineral companies are becoming too large. The claimed result is reduced valuation of minerals, unequal distribution, and excess returns.

Whatever the merits of the critics argument may be, the actions proposed to insure competition fail to serve the announced goals. The cost of regulation, entitlements programs and the like probably hurts the competitive posture of the smaller companies more than the majors. In many cases it is the classic case of the cure being worse than the disease. It is somewhat like the situation in the U.S. in 1976 when over $100 million dollars was spent to prevent a swine flu epidemic that never occurred.

Mergers. - Production is the most competitive area in the petroleum industry. Critics of the petroleum industry attack what they envision as declining competition in production as the result of several practices. Table 6.4 summarizes large crude oil company mergers between 1955 and 1970. Data on mergers is uninformative by itself. The Federal Trade Commission

Table 6.3

Large [1] Crude Oil Company Mergers, 1955-1970

1970 crude rank	Acquiring firm	Acquired firm	Year of acquisition	Production excluding royalty oil in barrels in year prior to acquisition	Percent of total U.S. production in year prior to acquisition
1	Exxon Corp.	1. Monterey Oil Co.	1960	5,491,708	0.2
2	Texaco, Inc.	1. Seaboard Oil Co.	1958	14,606,000 [2]	0.6
3	Gulf Oil Corp.	1. Warren Petroleum Corp.	1956	5,707,650	0.2
		2. Universal Consolidated Oil Co.	1962	2,562,216	0.1
6	Standard Oil Co. Ind.	1. Honolulu Oil Corp.	1961	14,990,000	0.6
		2. Midwest Oil Corp.	1964	8,951,941 [2]	0.3
		3. General Crude Oil Co.	1964	2,901,000	0.1
7	Atlantic Richfield Co.	1. Houston Oil Co. of Texas	1956	5,586,000 [3]	0.2
		2. Texas Pacific Coal & Oil	1958	7,397,088	0.3
		3. Argo Oil Corp.	1961	4,113,940	0.2
		4. Richfield Oil Corp.	1966	20,444,000 [2]	0.7
		5. Sinclair Oil Corp.	1969	59,368,710 [2]	1.8
8	Mobil Oil Corp.	1. Republic Natural Gas Co.	1961	2,783,646	0.1
9	Union Oil Co. of Calif.	1. Woodley Petroleum Co.	1960	3,459,000 [2]	0.1
		2. Pure Oil Co.	1965	27,260,738 [2]	1.0
11	Sun Oil Co.	1. Sunray DX Oil Co.	1968	33,136,000 [2]	1.0
		2. Mid-Continent Petroleum Corp.	1955	6,148,999	0.3
12	Continental Oil Co.	1. Pauley Petroleum, Inc.	1962	1,977,538	0.1
13	Marathon Oil Co.	1. Plymouth Oil Co.	1961	4,851,000	0.2
17	Tenneco Corp.	1. Middle States Petroleum Corp.	1958	2,605,081	0.1
		2. Wilcox Oil Co.	1964	1,270,392	0.1
20	Signal Companies, Inc.	1. Hancock Oil Co.	1958	8,294,291	0.3

[1] Large mergers are defined as mergers in which the acquired firm had a production of 1 million barrels or more of crude oil.
[2] Includes U.S. and Canada.
[3] Includes royalty.

Source: Moody's Industrial Manuals, various Years

converted these data to ratios showing the concentration of crude oil production by size of firm. Table 6.4 gives these figures. Oil production concentration rose from 21.2% in 1955 to 33.9% in 1975 for the four largest firms. The top twenty firms which accounted for about half (55.7%) of total production in 1955 increased their market share to 78.9% in 1975. Having approximately 80% of total production controlled by 20 firms is unacceptable to critics. What really has drawn the critics ire, however, is that over half of the increase to 78.9% is increased concentration by the four largest firms; 78.9 - 55.7 = 23.2% increase, of which 33.9 - 21.2 = 12.7 is due the four largest.

Table 6.4

U.S. Crude Oil Production Adjusted Concentration

Ratios 1955, 1960, 1965, 1970, 1975

Concentration Ratio	Year				
	1955	1960	1965	1970	1975
4 Firm	21.2	23.9	27.9	21.0	33.9
8 Firm	35.9	38.2	44.6	49.1	53.6
20 Firm	55.7	57.6	63.0	69.0	78.9

Such data do not prove that oil companies are anti-competitive. Greater concentration is defined by oil industry spokesman paritally on the grounds that it is necessary to allow them to explore in higher cost, higher risk regions where new oil supplies are likely to be found. Competition is just as keen among larger firms as smaller firms because more companies are able to play in these areas.

Trying to make bigness synonymous with lack of competition also ignores the very large capital needs which are beyond the capabilities of smaller companies. It also ignores the manpower and technological resources required to develop new mineral sources. Bigness is required to accomplish certain of the tasks involved. We must also recognize that joint ventures are not necessarily a device to control competition; they are usually a financial necessity. This is confirmed to a large degree by a detailed study of bidding patterns, patterns of day-to-day operations and the like. On a relative basis the minerals industry is far more competitive than most major manufacturing industries.

Joint Ventures. - Arguments over competition usually proceed to other areas at this point. Critics suggest that joint ventures especially among large firms, restrict this competition. Their reasoning is fairly straight forward. Company A and Company B bid together for a Gulf of Mexico sale, company B and company C join forces in the Gulf of Mexico, and A and C are together for a third sale. Three companies - A, B, and C - now share information and profits. And as the argument goes, the firms will be less competitive in production because what A does, affects B and C, who are partners. The critics fear is that interrelationship of profits creates an incentive to practice anti-competitive behavior.

The essence of the joint venture argument is that anti-competitive behavior does not always result; it just creates a forum by which anti-competitive behavior could occur. It is this possibility that is being attacked. Controversy over competition in the petroleum industry follows this dichotomy. Critics argue that growing concentration and joint ventures opens up avenues for reducing competition among petroleum companies that would like to increase their profits. Oil companies counter with the fact that petroleum company profits historically average less than 10%, indicating that they are not practicing anti-competitive behavior. What really divides the two groups is their time perspective. Reasonable critics agree with industry arguments. Their concern is over future competition. By allowing more mergers and joint ventures, is an environment being created where competition will be reduced significantly in the future? In essence the people in the middle recognize that it it easier to prevent anti-competitive behavior before it occurs rather than correcting it later. What they would like is an answer to the question of how much concentration should be allowed and how many joint ventures should occur.

Vertical Integrations. - In 1974 the Federal Trade Commission concluded that vertically integrated oil companies were a major source of anti-competitive behavior. The FTC reasoning behind the conclusion is exhibited in Figure 6.8. They reasoned that if one accepted that production was competitive, the small number of refineries (relative to producers) owned primarily by large producers limits the advantages of competition in production because everyone sells to the few refineries. If the refineries refuse to buy oil, or offer to buy the oil at uneconomic prices, then the ability to compete for future production is impaired.

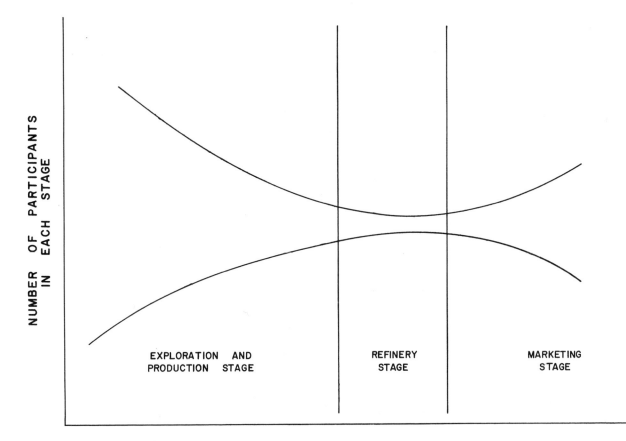

Figure 6.8 Illustration of Vertical Integration

Competition in marketing refined petroleum products also has been presumed to be impaired by refinery ownership. The refineries are hypothesized to control who they sell to in order to limit competition. Independent gas station operators provided ammunition for this position in the middle seventies when they charged that the major companies supplied gas only to company stations so that the independents would be forced out of business.

The issue of vertical integrations and its impact on competition lessened after the errors in the FTC report were identified, and federal controls on refineries began to allocate refined products to independents. The issue over the competitive effects of vertical integration is not dead, it is just lying dormant waiting for the right time to become a popular issue again. The continued increase in the marketing share captured by independents has been a telling factor in reducing demands for vertical divestiture.

Acquisition of Other Minerals. - Movement by the oil companies to acquire coal, geothermal, uranium, and other minerals elicits the same response from critics that, by incorporating several minerals into their companies inventory, mineral companies reduce the competition between minerals. Controlling outputs keeps prices high and permits excessive profits to be earned.

Anti-competitive behavior through acquisition of other mineral supplements the critics arguments about joint ventures. Companies A and B purchase coal companies, and C leases a geothermal deposit. If the three companies are tied together via joint ventures, critics argue that the incentive and the mechanism exists for controlling output in coal, oil, and geothermal. Again, as in earlier examples mineral companies seldom are actually accused by moderate critics of controlling output. The fear is that the potential for collusion exists. To date the fact is that participation in a joint venture has not really lessened general competition between companies in that venture. It has also spurred competition internally as different minerals divisions compete for limited available capital. It is difficult to explain to outsiders but large divisions of a major company are far more autonomous, characterized by major intra-company battles, than it may appear.

The upshot of these presumed forms of anti-competitive behavior reduces to the simple fact that competition, like inflation is a matter of degree, and is not measurable in precise quantities. Ignoring irrational mineral industry critics, moderate critics are left with the practice of joint ventures, mergers, vertical integration, and acquisition of other minerals which move the industry further away from the base reference, perfect competition. Every joint venture, every merger, every interaction among mineral compaines creates the potential for anti-competitive activity. The critics worry is that larger, more complex mineral companies prevent effective policing of anti-competitive behavior. Almost everyone bemoans the size and power of mineral regulatory agencies. But it is only fair to recognize that the growth in these bureaucracies is the direct result of the fear that the current size and operating practices of mineral companies retards competition.

It is the opinion of the authors that true competition does exist in the petroleum and other mineral industries at this time. But the fears of critics are real to them. It behooves the industry to perform in a manner that reduces, or eliminates, the credibility of such fears.

FEDERAL DEBT

The concept of a government borrowing money from its citizens began in the 1400's when the Dutch used such funding to finance their colonization of other countries. Proceeds from mining and trading were employed to pay off the debt plus interest. Government debt invested to return the principal plus interest is the same as bonds issued by private corporations. A bond issue is floated for a specific investment in both cases.

Over time, however, major changes in government borrowing occurred until today every governmental unit issues debt to meet expenditures. No longer do governments issue debt for investing in projects that return the principal plus interest, as corporations do. Instead many of their investments are in social projects, defense, social security, road construction, research and development, and education projects that never return a cent directly to the treasury. Debt principle and interest are paid from revenue sources like income taxes.

Debt at the governmental level differs from corporate debt in the types of investments selected, not in the way debt is issued. One can always question whether social investments undertaken by governments are valid, and whether debt should be issued to finance them. In free world countries these are political decisions and the overwhelming response appears to be that debt is an acceptable form of financing social investments. The consequences of such activity are less often understood by most people.

Between 1960 and 1974 the federal debt in the U.S. rose from $239 billion to $349 billion, an increase of 2.74% per year. Table 6.5 lists the federal debt figures for selected years. The deficit is the difference between revenues and expenditures. In 1976 federal government expenditures exceeded revenues by $71.2 billion. Total federal debt in 1974 was $349 billion and forecasted to reach $500 billion by the end of 1978 which is an increase of 9.4% per year. Total federal debt equals the deficit in a year plus the cost of refinancing old debts.

Table 6.5 Federal Deficits and Total Debt

(In $ Billions)

Year	Deficit	Total Debt
1960	-3.0	239.8
1965	-0.5	266.4
1970	-8.5	289.3
1974	-6.7	349.1
1975	-11.5	
1976	-71.2	

The controversy surrounding federal debt is an old one and will continue as long as there is a federal debt. At the center of this controversy is whether governments should balance their budgets every year (a zero deficit) or should be allowed to practice deficit financing.

Reasons for Deficit Financing

Governmental bodies sometimes incur debt rather than levy taxes to cover their expenditures for these reasons.

1) politicians are reluctant to raise taxes, especially in periods of rapid or extraordinary increases in expenditures, such as wartime. Approximately $200 billion of U.S. debt are expenses for World War II.

2) a position that capital improvements sponsored by the government ought to be paid over the life of the project by the taxpayers who benefit from the project. Bonds are used to pay for highways or schools rather than place the entire burden on the taxpayers in a specific year.

3) stimulation of the economy.

Critics of government policy argue that the debt cannot grow forever, and that the debt places an unfair burden on future generations.

Growth in Debt. - An often cited phrase is, "Our governments - like individuals and families - cannot continue to spend more than they take in without inviting disaster." Criticism of the national debt questions why governments should be allowed to "live beyond its means" when families and business do not. But families and businesses do borrow just like governments. Between 1946 and 1973, the national debt roughly doubled in size. During the same period corporate debt increased 9 times, home mortgage debt 11 times, and consumer debt 17 times. Compared to the other sectors of the economy, the federal government practiced restraint in its borrowings.

Borrowing is a perfectly responsible way of raising capital to finance major expenditures. Business does so to invest in projects that increase their earnings and families purchase homes and cars when they believe their earnings are sufficient to repay the interest and principal. Governemnts are in the same position; except that their earnings power - taxes, are much larger than family and business earnings sources.

If we are to apply the phrase "living beyond ones means" to governments, rationality dictates that we also refer to households and businesses. Every one of the largest 500 corporations in the world uses debt in its financing, often up to 60% of its total liabilities. Approximately 95% of all households incur debts which vary up to 30% of their earnings. In 1973 interest costs as a percent of GNP were 2.1%. It appears that households and corporations are "living beyond their means."

Government debt is analogous to business debt. When a mineral company needs funds or retires existing debt, they do so by issuing new debt. In most years governments pay off the holders of old bonds with funds raised by selling new bonds. Suggesting that governments should not incur debts, requires a general statement that no debt should be permitted by anyone to be consistant.

Burden on Future Generations. - President Eisenhower in 1960 referred to the national debt as "our children's inherited mortgage," which is certainly true. The question is whether taxes or debt is the appropriate source of financing. Households typically try to pay off their debt before they die so that their children will not have to face the debt obligations. Corporations, on the other hand, never pay off their entire debt, as do governments, which may be interpreted as "future customers inherited mortgage." Comparison of households to corporations or government is imprecise because households possess short, finite life, whereas corporations and government lifes extend beyond specific individuals.

There are two ways in which future generations might be burdened by national debt. First investment capital used to acquire capital is reduced. Borrowing money from the public bids capital away from private business seeking money to invest. Reduced capital flows to the private

136

sector means that future generations will have less goods producing capital available to them. Capital flowing to government was one cause of the "capital crunch" in the seventies.

Second, taxes in the future must be higher to cover interest payments on the debt. Higher taxes represents most peoples' view of the future burden of debt. The prevailing image is that people today are saddling the future with their unpaid bills. That the bills must be paid later is certainly true. But, at the heart of the matter, is the issue of whether the benefits obtained from the debt financed investments through higher taxes are unfair. Should the future generations receive better schools, better roads, and better health care as the result of the federal debt, the tax costs may not be unfair.

Future generations do not inherit the entire burden of the national debt. Because government expenditures take real resources away from the private citizens, the burden of the resource shift is on the current generation. During World War II, for example, labor and capital previously used to produce automobiles were employed to produce war supplies. It was that generation that sacrificed as much, if not more than future generations.

Total Impact of Federal Debt

Governmental decisions based upon the "buy now, pay later," principle are certainly consistant with practices by households and business, which few people question. In certain instances the national debt creates hardships on a nation's citizens, but not in all. The fundamental issue in appraising the merits of public debt in the end is just like the evaluation a business goes through in selecting investments. Do the benefits, cash inflows, exceed the cost of the investment adjusting for the time value of money? Costs in excess of benefits suggest dropping the investment. Favorable projects are justifiably financed with debt.

The simple analogy to corporate investment behavior is somewhat misleading, however. Identification of the benefits of public investments often are difficult to quantify. What do citizens gain precisely from expenditures for defense, health care, social security, low income housing, roads, regulatory agencies, and others? Whereas corporations value benefits in monetary units according to their profit motive, public debt financed projects involve weighing political trade-offs. Controversy over whether governments should use debt financing is generally more a dispute over the kind and quality of projects being financed by the debt rather than the debt itself.

New York City's financial problems add an interesting twist to the public debt issue. Until New York City became insolvent, many people presumed that public agencies could issue as much debt as they wished. The story of NYC is not that debt financing is bad, but that debt mismanagement is as bad for public agencies as it is for private groups. Issuing debt requiring interest and principal payments in excess of revenues produces insolvency, just as it does for households and corporations.

The real issue is deciding when public debt becomes too large. Suppose the federal government could only attract enough capital to pay its debts by offering tax free rates of 15%. As explained in the investment capital section, most people would favor investing in a government bond rather than mineral companies offering a historical return of 10%. The result - no investment capital for mineral companies.

It is probably true as some claim that the federal government could never become insolvent because of its ability to raise taxes and set interest rates. The real danger of uncontrolled public debt is that the size of the debt and the interest rates offered absorb all of the investment capital, leaving nothing or very little for the private sector. When that happens we will begin to observe the death of the private corporations, for without capital, it cannot survive.

GROSS NATIONAL PRODUCT

Gross national product (GNP) is the largest component of the national income accounts. Like an income statement for a firm, the national income accounts record the annual operations of the economy. Gross National Product, for example, corresponds to total sales of a firm. National income accounts, however, represent the annual flow of goods and services in the economy for all households, governments, and businesses.

Framework of National Income Accounting

Each of the accounts and their component parts (as defined by the Department of Commerce) are discussed below. We shall examine the relationships among them, working from the larger to the smaller-sized accounts.

Gross National Product. - GNP is the market value of the output of final goods and services produced by the nation's economy. It is "gross" because it does not deduct any charges for the depreciation of capital used up in the course of this production. However, all intermediate goods used up by business are excluded. GNP is made up of four components. They are:

1. Personal consumption expenditures, consisting of the market value of goods and services purchased by individuals and nonprofit institutions as well as the value of food, clothing, housing, and financial services received by them as income in kind.

2. Gross private domestic investment, consisting of newly produced capital goods - buildings and equipment - acquired by private business and nonprofit institutions, including the value of inventory changes and all new private houses and apartment buildings.

3. Net exports of goods and services, consisting of the amount of exports minus the amount of imports.

4. Government purchases of goods and services, consisting of general government expenditures for compensation of employees, and net purchases from business and from abroad.

National Income. - NI is the aggregate earnings of the factors of production which arise from the current production of goods and services by the nation's economy. National income is made up of five components.

1. Compensation of employees, consisting of wages salaries (including commissions, tips and bonuses, and payments in kind) and supplements to wages and salaries, such as employer contributions for social insurance and private pension, health and welfare funds.

2. Proprietor's income consisting of the earnings from current business operations of sole proprietorships, partnerships, and producers' cooperatives.

3. Rental income consisting of the monetary earnings of persons from the rental of real property, except for the earnings of persons primarily engaged in the real estate business.

4. Corporate profits, consisting of the earnings of corporations organized for profit which accrue to U.S. residents. These profits are measured before direct (income) taxes but after adjusting for the changing value of inventories.

5. Net interest, consisting of the excess of interest payments of the domestic business system over its interest receipts plus net interest received from abroad. Interest paid by consumers and by government is excluded.

Personal Income. - PI is the current income received by persons from all sources. It includes transfer payments from government (such as social securities payments) and business (such as corporate gifts to nonprofit institutions,) but excludes transfer among persons. The term "persons" includes not only individuals (including owners of unincorporated enterprises), but also nonprofit institutions, private trust funds, and private pension, health, and welfare funds.

Disposable Personal Income. - DPI is the income remaining to persons after deduction of personal taxes. Disposable Personal Income is made up of two components:

1. Personal outlays, consisting of the sum of personal consumption expenditures, interest paid by consumers, and personal transfer payments to foreigners.

2. Personal savings, consisting of the current savings of persons.

Figure 6.9 summarizes the national accounts for 1972. Personal savings and expenditures of $50 Billion and $747 Billion combine to give disposable personal income of $799 Billion. Continuing in this manner gives GNP to be $1,1555 Billion. Like income statements, careful analysis of GNP and its components is used to develop prescriptions for problems in the economy.

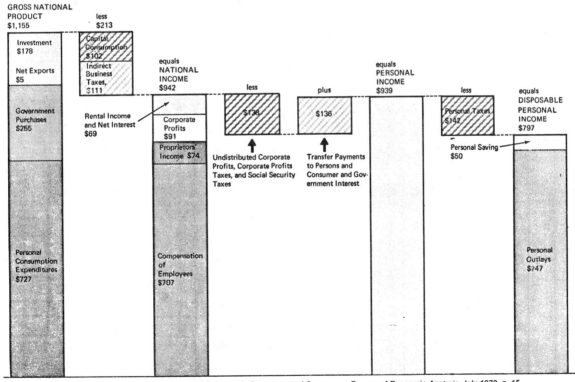

Source of data: *Survey of Current Business*, U.S. Department of Commerce, Bureau of Economic Analysis, July 1973, p. 15.

Figure 6.9   The National Income and Product Accounts: 1972 (all numbers are in billions of dollars)

During the 1970's almost every country was told by economists that the high rates of inflation that they were experiencing could not be reduced except by increasing the unemployment rate. Of course, most citizens responded by asking why do we have to increase unemployment to reduce inflation.

A precise relationship between employment and inflation was first observed by Prof. Phillips of Great Britain and later accepted as valid in the United States. The proposed relationship between price inflation and unemployment is called the Phillips curve.

Figure 6.10 illustrates the Phillips curve. At low levels of inflation, high levels of unemployment exist. Rising inflation is associated with lower unemployment. Inflation is 4% per year when the unemployment rate is 8%. Unemployment rates reduce to 4% only when inflation becomes 12% in this example.

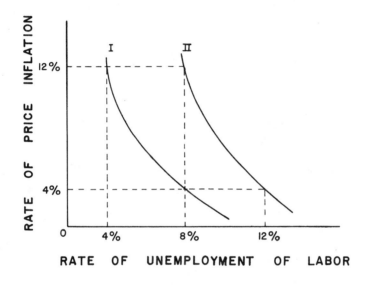

Figure 6.10   The General Form of the Phillips Curve

A 4% unemployment rate is called full employment, in spite of the seeming contradiction. Full employment occurs at 4% because at any time there are people recorded as unemployed who are in the process of switching jobs, on strike, or do not have any skills. It is the difference between 4% and the actual unemployment rate that economic policies are designed to cure. If the unemployment rate was 6%, the goal of administrations is to reduce it to 4%, not 0%.

Inflation measures the price of lower unemployment. In order to understand the trade-off between prices and employment, labor markets are evaluated-especially labor markets dominated by active unions.

If the rate of overall unemployment in the economy is low, labor markets are likely to be "tight"-that is, too few workers are likely to be available for the jobs that employers want to fill. Unions whose contracts are coming up for negotiation can be more aggressive in pressing their wage demands, because higher wages would be less likely to generate widespread unemployment for their members. At the same time employers will be especially interested in avoiding a strike, since it would prevent them from meeting the strong demand for the products of their firm. Employers have to raise wages to keep these employees. Higher wages, when reflected in the price of goods, means higher prices.

The coal industry strike in 1978 illustrates this trade-off. Because of growing coal demand due to expectations of hydrocarbon shortages and increased emphasis on coal called for in the U.S. energy plan, mine workers demanded and got significantly higher wages. Lower unemployment and higher coal prices resulted.

The actual location of the Phillips curve depends upon several economic forces. Curve I in Figure 6.10, for example, represents labor productivity--measured in output per man-hour. Variations in productivity affect employment and inflation. Curve II in Figure 6.10 illustrates how a decrease in productivity alters the location of the Phillips curve. A 4% inflation rate is associated with a 12% unemployment rate; and the minimum unemployment rate possible without excessive inflation is 12%.

To illustrate the trade-off in employment and inflation, consider a mineral company producing 500,000 tons (Q) and employing 100 workers (L), including management. Productivity (P) is calculated as

$$P = Q/(L)(2080) = 500,000/(100)(2080) = 2.40 \text{ tons per man-hour}$$

where 2080 is the number of hours worked per year. The total labor cost (LC) is the average hourly wage ($10) times the number of hours worked.

$$LC = (\$10)(100)(2080) = \$2.08 \text{ million}$$

A decline in demand by 10% to 450,000 tons reduces productivity to

$$P = 450,000/(100)(2080) = 2.16 \text{ tons per man-hour}$$

Management can reduce laborers by 10% to keep productivity at 2.40. But union contracts and unwillingness to fire executives and engineers often prevent such drastic measures. If only 3% of the laborers are laid off, productivity becomes

$$P = 450,000/(97)(2080) = 2.23$$

and if wages rose through union contracts to $11 per hour, wage costs are

$$LC = (\$11)(97)(2080) = \$2.21 \text{ million}$$

Company costs have increased by (2.21/2.08) = 1.06, or 6% and unemployment by 3%.

Had the same experience happened throughout the economy so that the unemployed could not find work elsewhere, inflation would be 6% and the unemployment rate 3%. This example points out that policies which accept a trade-off between unemployment and inflation are simplifications of reality. Other economic forces, like demand for minerals and productivity, also bear on inflation and unemployment rates.

During the seventies when inflation and unemployment increased together, some doubts were created about whether economic policies were adequate. Neglected in this controversy was the

141

shift in the productivity of labor and extreme upheavals in the demand for products like minerals. What occurred was misuse of a simplified relationship. Had productivity and mineral demands remained stable, trade-offs between unemployment and inflation rates would occur.

Any time someone states that we must go slowly in reducing unemployment in order to control inflation, our example points out that such a statement assumes no change in productivity, demand, and other economic forces. In dynamic economies the simplified trade-off between unemployment and inflation is unfounded.

ECONOMIC SUPPLY AND DEMAND

"It depends upon supply and demand" is a commonly heard phrase in discussions about a country's economy or specific minerals. Supply and demand are the end result, not the cause, of economic events, as this statement implies. In order to truly understand shortages of minerals, lack of investment capital, and unemployment (which imply the supply of minerals and money are inadequate and there are too many laborers seeking work) the causes of supply and demand have to be understood. Just the basic components of supply and demand are discussed here. Chapter 7 describes models for estimating the supply and demand.

Supply

The economic concept of supply measures producers output of minerals as economic forces. Observing that the natural gas industry produced 15 TCF in a year does not represent supply analysis. Economic supply explains why that quantity of gas was produced by describing the motive for production--profit. The objective is a better picture of how production might change as economic forces change. For example, would production of 15TCF of gas have taken place if gas prices were uncontrolled? Not likely, so the question is: Is it possible to develop a relationship which describes the impact of economic forces on supply?

Companies produce minerals and other products with the goal of making a profit. Profit (Π) is defined as

$$\Pi = PQ - rK - WL$$

where Q is the quantity of a mineral supplied, P is mineral price, K is the quantity of capital used to purchase plant and materials, r is the cost of capital as explained before, L is labor employed, and W is the wage paid to labor. Mineral companies combine capital and labor into their process to produce Q. The goal is to produce minerals at the point where profits are the greatest.

Mineral supply describes the change in production (Q) as price (P) and costs (rK, WL) change. In order to graphically depict a supply curve, rK and WL are held constant. Effects of changes in price on production are then observed. Figure 6.11 shows a supply curve for natural gas. At $0.50 per MCF, 10 TCF of gas is supplied. As gas prices rise to $1.00 and $1.50 per MCF, 15 and 20 TCF are supplied, respectively. After 20 TCF the supply curve S_1 which connects these points becomes vertical, indicating that about 20 TCF is the supply capacity in this example.

Curve S_1 has been constructed holding all other forces affecting supply, such as costs, environmental questions, lease bidding systems, and others, constant. Changes in these forces shift the entire supply curve to the right or left, depending upon whether costs increase of decrease.

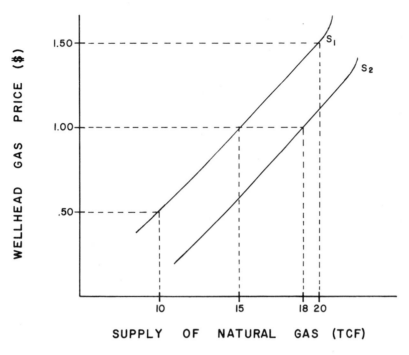

Figure 6.11 Natural Gas Supply Curve

Curve S_2 represents a case where costs decrease, by say a reduction in lease bid requirements. Producers can afford to supply more at the same price because reduced costs increase profits. With a price of \$1.00 producers are willing to supply 18 TCF with lower costs, instead of the 15 TCF suggested by curve S_1. Construction of actual supply curves are explained in Chapter 7.

Demand Curves

Mineral demand, like supply, responds to a number of economic forces, including price of the mineral, purchasing power, prices of substitute products, and others. Mineral demand is inversely related to price, because consumers prefer to purchase less at higher prices. Figure 6.12 depicts demand curves for gasoline and natural gas. Gasoline demand is almost vertical, indicating that price changes have little effect on demand. Consumer response to gas price increase is greater, as the flatter curve shows, because more substitutes exist for gas than for gasoline.

Demand curves are established holding other economic forces constant. A change in income might, for instance, shift the demand curve for gasoline to the right. In a dynamic world where economic forces constantly change, identification of the relative impacts of each economic force on demand permits accurate analysis. Such a procedure is the basis for governmental and business positions on mineral policies. In the United States the energy policy presented in 1977 was designated to raise prices of gasoline and natural gas to reduce demand. Price increases proposed were designed to more than offset forces that would increase demand, like income increases.

Techniques for establishing actual demand curves are explained in Chapter 7.

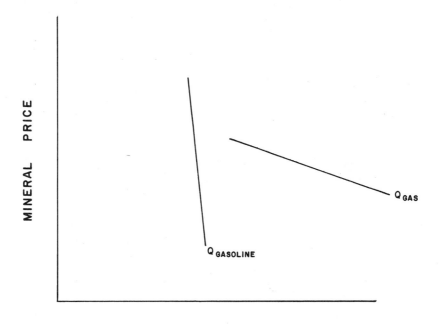

Figure 6.12   Demand Curves for Gasoline
and Natural Gas

Equilibrium

The intersection of demand and supply curves defines the point of equilibrium. The price consumers are willing to pay for a mineral equals the price producers require to supply that quantity of minerals. Any other price creates a _disequilibrium_ in the mineral market. Point A in Figure 6.13 identifies the equilibrium position for natural gas. At an average wellhead price of natural gas of $1.00, consumers demand and producers supply 15 TCF of gas. This is the optimal position shown in the example (which is intended only to illustrate the concept).

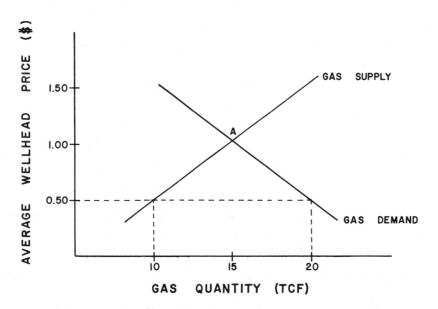

Figure 6.13 Equilibrium Price of Natural Gas

In actuality, markets for minerals are never in equilibrium. But when supply and demand forces are free to operate, price tries to move toward equilibrium. Natural gas is a case in point. At an average wellhead price of $0.50 per MCF, producers desire to supply 10 TCF and customers want to consume 20 TCF. If gas prices were free to change, the price would rise to $1.00 per MCF, the point where demand equals supply. Controlled gas prices, like those in the U.S., are not able to rise to clear the market. Under normal circumstances consumers only would be able to buy 10 TCF of gas. Additional controls, however, force suppliers to produce the 20 TCF of gas demanded by customers. The result is faster depletion of gas reserves at a price that is too low to generate sufficient profits to replace the gas consumed.

Harmful effects like the excess consumption and insufficient profits encountered in the gas industry illustrate the importance of careful analysis of the forces determining demand and supply. A supply and demand framework is utilized to evaluate other aspects of mineral industry behavior.

Remember, Figure 6.13 only shows the equilibrium concept. The actual supply and demand curves need not be straight lines. As noted previously, these lines must properly reflect all factors affecting their position.

Cartels

Creation of the OPEC cartel at an effective level in 1973 illustrates control of oil supply using political, as opposed to economic, pressures. By agreeing to limit production the cartel shifted the supply curve to the left, which effectively raised prices. Figure 6.14 shows the effect of cartel imposed supply restrictions. Oil price rose from $4.00 to $12.00 per barrel, the effect being a reduction in demand from Q_1 to Q_2 barrels. The goal of cartels is to restrict supply in order to achieve the goal of higher prices.

Success of a cartel depends on the willingness of its members to restrict supply. Should one or more cartel member cheat on the supply restrictions, the supply curve shifts to the right, forcing producers to sell at a lower price in order to sell their excess oil. The success of OPEC has stimulated producing nations of other products to investigate the formation of a cartel. Examples include coffee and copper.

Impacts of cartels carry over to other mineral industries as well. Figure 6.15 shows demand and supply curves for coal prior to the cartel. Coal, a close substitute for oil, shows an equilibrium position of $6.00 per ton at Q_1. Higher oil prices and uncertain supply made coal a more attractive fuel source. The effect of the cartel was to shift the demand for coal to the right, increasing the price to $12.00 per ton and consumption to Q_2. Switching to coal placed additional pressure on the cartel because it effectively reduced the demand for oil as consumers substituted coal for oil.

Extending Figures 6.14 and 6.15 to include other minerals, that are substitutes or used in conjunction with oil, points out the complex nature of the economic system. The total impact of the cartel in any country can be determined only by careful analysis of the supply and demand relationships in every segment of the economy.

Pseudo-cartels. - Mineral industries, especially the oil industry, have been accused of acting in a cartel-like manner themselves. Programs like import quotas and prorationing effec-

145

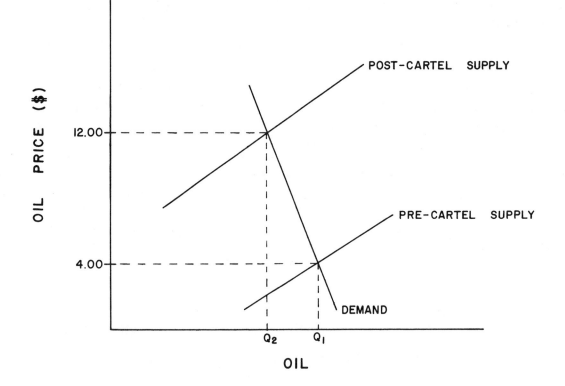

Figure 6.14   Effect of OPEC Cartel on Oil Demand

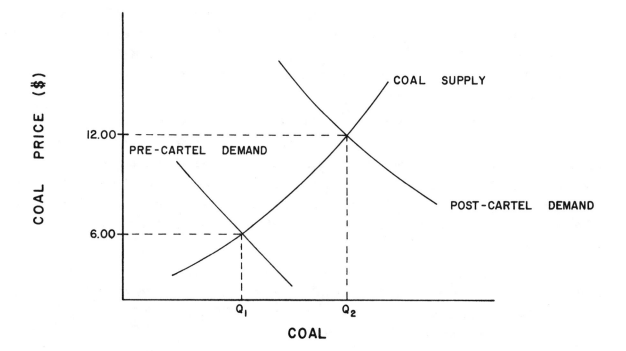

Figure 6.15   Effect of OPEC Cartel on Coal Demand

tively shift supply curves as shown in Figure 6.15. One of the arguments stated by OPEC members in justifying the cartel is avoidance of excessive depletion of their natural resources, an argument which closely resembles the arguments for prorationing. Policies or programs that shift supply curves, regardless of their origin, are referred to as <u>pseudo cartels</u>.

We are neither defending nor critizing any of these programs. We are only pointing out that programs which shift supply curves, regardless of whether a group of countries or one country is involved, raise prices and reduce mineral consumption. Valid reasons exist for such programs. But from a pure economic viewpoint the effects are equivalent.

SUMMARY

Economic terms used in the course of everyday conversation often are thrown around as if the economic system can be reduced to simple statements. The complexity of simple sounding terms, such as inflation, investment capital, cost of capital, supply and demand, and others have hopefully been demonstrated. Proper accounting for complexities of the economic world requires at least a basic understanding of what these terms really mean, or, more importantly, do not mean.

REFERENCES

1. Blair, John, <u>Concentration in the U.S. Petroleum Industry</u>, New York: McGraw Hill Co., (1957).
2. Moody's Industrial Manuals, various years.
3. Mulholland, J. and Webbink, D., <u>Concentration Level and Trends in the Energy Sector of the U.S. Economy</u>, Bureau of Economics, Federal Trade Commission, Washington: Government Printing Office, (1974).
4. Survey of Current Business, Department of Commerce.
5. Van Horne, J., <u>Financial Management and Policy</u>. Englewood Cliffs, New Jersey: Prentice Hall Inc., (1977).
6. Dernbourg, S. and T. McDougall, <u>Macroeconomics</u>. New York: McGraw Hill, (1971).
7. Wannacott, P., <u>Macroeconomics</u>, Homewood, Ill.: Irwin Publishing Co., (1978).

7

ANALYSIS OF COMMON ECONOMIC MODELS

Evaluation of mineral investments compares net cash flows over time among alternative investments in order to maximize returns to the public or private treasury, as explained in Chapter 5. Careful and thoughtful appraisal of the economic elements comprising these cash flows are often nonexistent in many mineral evaluations. The failure to recognize economic forces determining profitability in detail is perplexing when one considers the quantitative manner in which the engineering and geological forces are considered. To spend the time and effort applying the material presented in Volume 2 to reserve estimation and production, and then simply "making-up" estimates of inflation rates, mineral prices, environmental expenditures, and other items affecting the time value of money often distorts the potential value of implementing these procedures. Extremely accurate engineering calculations combined with poor estimates for economic forces yields mediocre mineral evaluations. Accurate appraisal of mineral reserves economic value follows only when both the engineering and economic parameters are estimated accurately. Further complicating mineral evaluation is the fact that recoverable reserves and production schedules are partially dependent upon economic values, generally increasing as the economic value increases.

In order to improve the accuracy of mineral evaluations, it is necessary to explicitly understand the relationship between engineering and economic parameters, and how these relationships convert into cash flows for calculating the time value of money. Transforming engineering based estimates to economic value is based upon two general concepts familiar to everyone - supply and demand. Economic valuation, however, differs from the conception most people have of supply and demand. Over the life of most mineral reserves, which may last more than 20 years, producers and consumers alike adjust their patterns of usage as economic change. Supply and demand analysis measures changes in the pattern of mineral production and consumption as economic forces vary over time, and thus, is more than just a casual review of historical movements in supply and demand.

The importance of careful study of supply - demand patterns cannot be understated in the mineral industries. Review of mineral industry economics clarifies the dynamic nature of the profitability of past investments in mineral reserves. Coal, for example, dominated the world energy scene until the onset of liquid hydrocarbons. During the period when other energy forms were preferred, coal investments declined in desirability, reducing the level of supply and demand. Faced with declining discoveries, higher exploration and production costs, and the fourfold price increase for these minerals in the sixties and seventies, coal suddenly became a sound investment again. A reverse situation existed for liquid hydrocarbons. Other minerals, such as geothermal and synthetic hydrocarbons, which were once poor economic investments, have also risen in desirability as the world economic situation evolves.

Integrating engineering and economic evaluations allows the mineral valuation process to supply needed minerals at a point in time where demand is sufficient to make mineral investment profitable. Ignoring either engineering or economic reality makes for bad investment decisions, and bad investment decisions creates hardships for producers and consumers alike. Likewise, both producers and consumers benefit from sound investment decisions. Consumers obtain the quantity of minerals they desire at a price that earns the producers a "fair" return on their investment.

Supply and demand models provide one mechanism for improving the accuracy of mineral investment evaluation. Many of the elements essential to complete supply and demand analysis are considered subjectively. But, like the interaction between recoverable reserves and production schedules with economic forces, the elements of supply and demand analysis also interact with each other. Subjective consideration of supply and demand is limited unless such interactions are accounted for, in much the same way that reserve estimates which ignore economics are limited. Basic inputs for constructing supply and demand models and for considering these interactions are developed and explained in this chapter.

RELATIONSHIP BETWEEN ECONOMIC AND ENGINEERING MODELS

Constructing and implementing economic models proceeds along the same lines as engineering models - reservoir models, gas processing plant design models, and refinery models. In all cases the model building process follows similar steps.

1. Specification of limiting assumptions.

2. Accounting for relationships between physical and economic forces.

3. Testing individual and combined relationships for results consistent with real events.

4. Revision of model inconsistencies when discovered.

Representative economic and engineering models generally follow this outline.

Perhaps the greatest difference between the two modelling efforts results from the subject matter of each discipline. Engineering, the practical application of physical principles, endeavors to discover and apply new physical ideas to meet the needs of society. Being based upon physical principles, engineering models can be developed and refined to some extent in the laboratory, where the elements reacting with a physical problem can be carefully controlled. Applying principles learned in the lab to mineral production or processing usually points out areas where the controlled experimentation fails to measure real world relationships. Returning to the "drawing board" with actual results allows for further refinements in the model.

Scale effects form one of the major obstacles to successful application of experimentally derived or pilot project principles. In general, experiments and pilot projects by necessity are smaller in scale than their implementation. Simply expanding the size introduces uncertainty into the analysis, especially when physical relationships are extremely nonlinear or limited to certain ranges in value. Newendorp (7.21), for example, discusses ways of measuring joint probabilities in exploration simulations.

Economic models suffer from slightly contrasting problems. For one thing, economics does not possess a subject matter that can be observed in a laboratory. Its subject is the human decision-making process that results in mineral production, mineral demand, and the other economic forces acting on mineral supply and demand. Controlled experiments are difficult, and usually impossible. Economic models therefore must rely upon explaining the causes of mineral supply and demand based upon actual transactions. This is a looking backward (induction) process. Only by explaining the causes of historical mineral activity can future mineral activity be anticipated. Economic models, in essence, employ historical patterns to reveal cause-effect relationships underlying mineral supply and demand. The accuracy of economic models, like engineering models, requires a long trial-and-error process as theories are developed and discarded when they fail to explain behavior in the real world. Sophisticated empirical testing in economics is, unfortunately, less than a half century old, hardly enough time to produce totally acceptable, highly accurate models in a dynamic world.

Another problem plaguing economic models relates to their general applicability. Models are designed to capture the interrelationship among a specific mineral supply or demand and its causes. Transferring a model built for a specific purpose to another use distorts the entire analysis. As an extreme example, suppose a model constructed to forecast coal supplied from deep shaft mines in West Virginia was used for a similar purpose at a strip mine in Wyoming. The model would certainly be judged inadequate. But is this any more farfetched than assuming a single reservoir, gas processing, or aquifer model exists which explains all situations in their respective fields adequately? Of course not.

Misuse of economic models in the extreme example cited above or, as is more commonly the case, the model's outputs are misinterpreted, leads to an interesting paradox. Economic modelling, although continually improving in sophistication, still is inadequate. To recommend usage of models with some deficiencies represents an improvement over subjective evaluations of economic forces or no such evaluation at all. Misuse of economic models, on the other hand, by well meaning, but improperly trained professionals offers the potential for poor evaluation as well. So do we go with an approach which improves on subjective evaluation, but may be misinterpreted, or do we stick with subjective evaluation?

Answering this question relies on several elements. One, subjective and formal modelling complement each other; they are not substitutes, as Chapter 9 explains. Second, economic models offer the potential for learning through experience. As models face the acid test of explaining mineral economic events, efforts to discover the sources of any deficiencies lead to improved models that are better equipped to explain the causes and effects of mineral economics.

BASIC ECONOMIC MODELS

Mineral economic evaluation comprises nothing more than a thorough, detailed analysis of two concepts familiar to every professional - supply and demand. Terms like supply and demand are difficult to explain because everyone has heard them and uses them in conversation (but seldom correctly).

Common application of the terms supply and demand is not wrong; it is merely incomplete. Statements like 500 tons of coal were supplied or 1,000 Mcf of gas were demanded in a given time period exemplify the usage of the supply-demand terms. But more important than observing exact values of supply and demand is the explanation of why a particular supply and demand emerges. Explaining the causes of supply and demand patterns represents the extension from existing terminology.

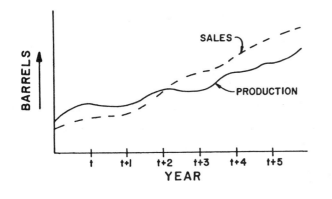

The figure at the left is a typical supply demand curve. But what do such graphs really tell us? Basically, only that supply and demand have shifted historically, nothing more and nothing less. What is not contained in these graphs is an understanding of why these graphs occurred. Something caused these patterns. Without a detailed understanding of the economic and engineering forces working on supply and demand, the graphs are virtually meaningless. Engineering statements often are made about the efficiency of equipment. Actual efficiency is a function of operating procedures, system design and other features. Graphing efficiency levels of a piece of equipment over time is equivalent to graphing supply and demand; it tells us nothing about the causes of efficiency changes. About the only approach one can follow is to draw upon knowledge of physical principles and operating procedures to infer causes. For improving future system design or current efficiencies loose approaches like this are difficult to implement in practice. Similiar problems hold in supply-demand analysis.

Supply-demand analysis emphasizes the price received by producers and price paid by consumers, respectively, as the primary vehicle for describing supply-demand behavior.

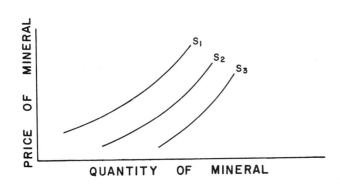
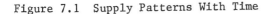

Figure 7.1 Supply Patterns With Time

Figure 7.1 represents a typical graphical analysis of supply patterns over time. Supply in economic models defines output of minerals as a function of economic resources used in the production process. These resources include capital equipment, labor, raw material feedstocks, distribution, and other factors. To produce greater quantities of a mineral, one or more of these economic resources must increase. Whether mineral production occurs depends upon the relationship between the total cost of producing minerals and the price producer's receive. Costs in excess of price obviously retard production. Prices in excess of costs make production profitable.

A typical mineral supply curve with a positive slope to the right, indicates an increasing supply of minerals as prices rise. Such a simplistic statement is obvious to every mineral professional. Natural gas, coal, synthetic crude, and other mineral industries have argued for some time that higher prices are necessary to stimulate production for meeting demand. What is important is the need to describe the factors influencing the supply curve based on graphical patterns of historical supply. This is the mineral supply modelling framework that fully accounts for the cause-effect relationship between mineral price and supply.

Presentation of mineral supply curves at this point is somewhat like placing the cart before the horse, since the derivation of such a curve has not been explained. We choose to get ahead of ourselves in order to emphasize the importance of understanding the value of supply curve analysis. Several techniques for actually deriving supply curves are explained in the next section. At this stage it is sufficient to note that each curve in Figure 7.1 describes only the relationship between mineral price and supply. Other resources, like the price of inputs used to produce minerals and their availability, determine whether curve S_1, S_2, or S_3 is relevant.

Analysis of mineral demand follows a pattern similar to that of mineral supply. Graphical representation of demand describes the interaction between mineral demand and price. The only difference is the slope of the demand curve. Demand curves slope downward and to the right instead of up and to the right. Figure 7.2 shows several demand curves. Downward sloping curves imply that a decrease in price increases demand, or, conversely, rising prices decreases demand. Lowering mineral prices from P_1 to P_2 increases demand from Q_1 to Q_2.

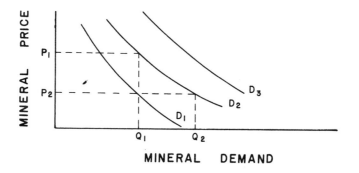

Figure 7.2 Example of Demand Curves.

A single demand curve describes the relationship between mineral price and demand, holding all other economic forces affecting demand constant. Other economic forces impacting demand are consumers' ability to purchase the mineral (income), the price of minerals that can be substituted for a specific mineral, shortages of minerals, and so on. These economic forces determine whether curve D_1, D_2, or D_3 is relevant at any point in time.

152

Looking at mineral demand curves in this way illustrates the error in comparing graphs of a mineral demand, say natural gas, and gas prices, and reaching the conclusion that using gas prices raise gas demand. Figure 7.3 uses hypothetical demand curves to show why such analysis is spurious. Curve D_t represents demand for gas in year t. At a price of $1.10 per MCF, gas demand is 500,000 MCF (Point A). A price increase to $1.20 in year t+1 reduces gas demand to 480,000 MCF (Point B), if none of the other economic forces influencing gas demand change. One of the major determinants of gas demand is temperature. Year t might be characterized by moderate temperatures. Year t+1, however, might have an extremely cold winter. As a result the relevant demand curve is D_{t+1}. Gas demand in year t+1 actually is 505 MCF at a price of $1.20 per MCF (Point C).

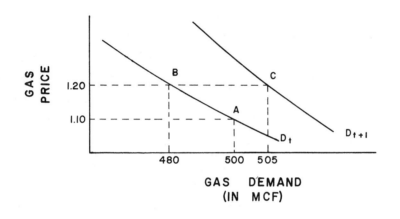

Figure 7.3 Example of Shifting Gas Demand

Comparing gas demand and price between points A and C suggests that increasing price by 20¢ increases demand by 5 MCF. Can it be concluded that the gas price increase caused demand to rise? If it does, this implies that consumers like to spend their money to purchase natural gas. The relationship between price and demand is divided and cannot be observed directly because other forces affecting gas demand have also changed. Rational people seldom consume more of a mineral just because prices increase.

Judging the relationship between price and demand, while ignoring other forces, is inconsistent with techniques employed in engineering and geological analysis. In evaluating enthalpy, for example, of a light hydrocarbon at a vapor pressure of 2000 lbs/sq inch abs and a constant molecular weight of 20, we might observe at a temperature of 400° a total heat, of 490 Btu/lb. Now suppose that molecular weight changes to 30, giving a value of 445 Btu/lb at the same temperature. If the change in molecular weights is not accounted for, a simple graphical analysis like that used to explain demand would show that as temperature rises, the total heat generated by light hydrocarbons at 2000 lbs/sq inch absolute decreases, an obvious distortion of what is really happening.

153

Mineral demand, and mineral supply have to be treated with the same kind of careful analysis that is applied to enthalpy problems and other engineering and geological situations to avoid spurious conclusions. As Chapter 3 indicates one reason for conducting detailed engineering appraisals of mineral reserves is that these are the calculations the professional knows how to make. Economic analysis, on the other hand, may do the same thing - ignoring the complexities of the economic calculations in order to perform the arithmetic one knows how to do. Sound mineral evaluation obviously requires detailed understanding of both areas.

Micro and Macro Economic Models

Confusion often reigns supreme when explaining supply and demand models over precise interpretation of their results. The interpretation varies according to the number of producers and consumers being analyzed. Looking at the supply curve for one mineral producer is micro-economic analysis, while measuring the supply curve for all producers of the mineral combined represents a macroeconomic analysis. Analysis of some proportion of customers between one and all usually is considered a macro analysis. Macro models are basically a combination of several micro models aggregated into one family of curves. Macro and micro demand models are interpreted in the same manner.

Even though both types of supply-demand analysis address the same types of questions, each has advantages. Micro models are ideally suited for mineral evaluations within individual firms since it identifies resources, production processes, engineering constraints, demand patterns, and other forces specific to that firm. Macro models on the other hand facilitate analysis of general policy changes, such as taxes, prices, international agreements, and so on, on a particular mineral industry.

Evaluating the impact of such policies on a mineral industry could proceed based on individual supply and demand curves for all producers and consumers. But in most mineral industries where several thousand producers and hundreds of thousands of customers operate, detailed analysis of individual supply and demand patterns is infeasible. Table 7.1 illustrates how macro supply curves are constructed from micro supply relationships using the following data for two firms.

For prices ranging in value from $10/ton to $15/ton, Firm 1 produces 10,000 to 15,000 tons and Firm 2, 30,000 to 60,000 tons. Summing the production of the two firms yields the macro supply curve. Figure 7.4 illustrates these curves graphically. Note that each firm's supply curve reflects a differential response to price changes, indicated by different slopes. Macro supply curves for a thousand firms are constructed in the same way, but are more complex.

Almost all evaluations of governmental policies are predicated upon similar macro analysis of supply and demand. One danger inherent in any macro analysis results from making inferences about each company's response to price changes. In essence, knowing only the macro relationship, analysts work backwards to make inferences about the impact of price changes on each company's supply patterns.

If we did not possess the information on Firms 1 and 2 supply curves given in Table 7.1, but did know the macro supply curve, would it be possible to infer what the supply curves for the individual firms would look like? Yes, if anything is known about the two firms. Suppose

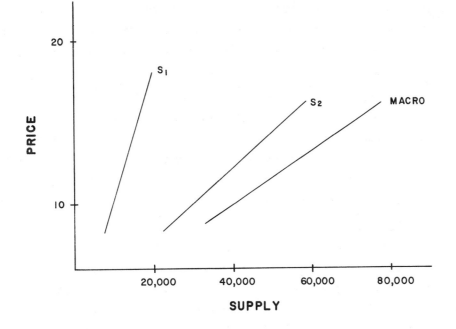

Figure 7.4 Micro and Macro Supply Curves

	Production		
Price	Firm 1	Firm 2	Macro
$10.	10,000	30,000	40,000
$11.	11,000	35,000	46,000
$12.	12,000	39,000	51,000
$13.	13,000	44,000	57,000
$14.	14,000	51,000	65,000
$15.	15,000	60,000	75,000

Table 7.1 Construction of Macro Supply Relationship
from Micro Supply Curves

that in year t mineral price was $12 and we know that Firm 1 produced 12,000 tons and Firm 2, 39,000 tons. Price is expected to rise to $13 under a new policy in year t+1. From Table 7.1, 57,000 tons are expected to be produced, but information about the relative contribution of each firm is desired.

One approach often employed is to assume each firm's production increases by the same percentage as total production, which is (57,000-51,000/51,000) = 11.7 percent. Firm 1 is estimated by this procedure to produce (12,900)(1.117) = 13,414 tons and Firm 2, (39,000)(1.117) = 43,583 tons. Had we had the advantage of the figures in Table 7.1 the obvious distortion would.

155

be readily apparent. First, observe that total production from the two firms is approximately equal to total production, or 13,414 + 43,583 = 56,997 tons. Production of Firm 2 is understated, being 43,583 tons, compared to an actual figure of 44,000 tons. The production of Firm 1 is overstated by 414 tons (= 13,414 - 13,000).

Errors of this kind are unimportant when we are concerned with anticipating the effect of a price change on total production. Economic evaluation emphasizes measurement of impacts on respective nations, states, companies, and citizens, as well as just the total supply increments, however. Distributional impacts of this type are discussed in greater detail in Chapter 10. But, the thrust of our discussion is to emphasize that distributional impacts cannot be evaluated accurately using macro models of either supply or demand.

Suppose that a panel of experts is meeting to decide on whether to increase mineral prices from $12 to $13 per ton as part of a governmental policy to stimulate production. Also, a part of the objective is encouragement for small producers to increase their share of total production to offset growing criticism of increased concentration in our mineral industries. Projections based upon the macro model support this objective. The smaller firm maintains its share of total production at about 23.5% = 12,000/51,000 = 13,414/57,000, suggesting that raising prices benefits large and small firms equally. Such a conclusion is erroneous because Table 7.1 shows that actual production by the small firm is 13,000 tons, giving a market share percentage of 22.8% = (13,000/57,000) x 100. A policy resulting in a price increase raises the large firm's market share at the expense of the small firm.

Declining market shares for small firms, or for large firms, are not bad per se. They are bad only if there is a valid reason. But, in many areas of mineral activity bigness and "badness" are considered synonymous terms. In the example employed here, which does not represent all situations, the large firm benefits relatively more from the price increase because its supply curve is more responsive to price changes. Several explanations may exist for this situation--better management, better planning, anti-competitive controls of marketing outlets, and so on. Without a detailed understanding of the contribution of these and other forces in the formation of a supply curve, definitive statements about the desirability of a price increase or mineral industry composition are impossible.

Macro supply and demand models' inability to capture microimpacts accurately can be illustrated by putting this example into the framework most often used in economic analysis. Data in Table 7.1 is analyzed using the concept of the representative firm. A hypothetical supply curve representing the characteristics of all companies is calculated by merely dividing total production at each price by the total number of companies. This average supply curve is representative of all firms, hence the name representative firm. Like any average, as explained in Chapter 8, some distortion is introduced into the analysis since an average measures neither firm's supply curve exactly. The representative supply curve is:

Price	Supply
$10	20,000
11	23,000
12	25,500
13	28,500
14	32,500
15	37,500

Every dollar increase in price yields the specified increase in production. At $12 per ton the representative firm's supply curve implies each company produces 25,500 tons. This is 13,500 tons more than Firm 1's actual production, and less than Firm 2's production by the same amount. Reaching conclusions about the distributional impact of any economic force on a specific firm or all firms in a mineral industry is impossible based on the representative firm concept. Policies or mineral programs designed around the representative firm approach are thus open to the criticism of favoritism.

Macro models force binding constraints in their use as these examples illustrate. But it cannot be inferred that macro models therefore cannot be used for mineral economic evaluations. Macro models are valuable tools for mineral evaluations, and the extent of this value is determined by the purpose of the evaluation. In summary, macro models have the following characteristics.

1. Reduce time and monetary costs of conducting mineral evaluations;

2. Accurately account for total supply and demand changes in the entire mineral industry; and,

3. Fail to capture the impact of decisions on specific mineral companies, except when the company supply-demand curve corresponds exactly to the macro supply curve or the representative firm supply curve.

Point 3 reemphasizes the problem of using mean or expected values. Means are a convenient way to summarize complex data series like mineral supply and demand patterns. Anytime data is summarized, however, a certain amount of information is lost. Mineral firms should only utilize representative supply or demand models when their firm may be characterized as average.

It is interesting to note that the same limitations affect the evaluation process with regard to the engineering elements in the mineral industry. Reservoir models, depicting the relationship between various engineering forces and fluid dynamics, may be considered applicable for typical reservoirs (a representative reservoir). For reservoirs possessing unique characteristics, reservoir models often fail to adequately measure recoverable reserves. Just as care is utilized in applying these models, so it should be used in supply-demand model usage.

MINERAL SUPPLY MODEL

Constructing a logical framework for analyzing the determinants of mineral supply begins by listing the major forces, economic and engineering, which affect supply. Every element on this list is composed of two items, the physical quantity used and the cost per item, if it is possible to obtain such detailed data. Analysis of coal supply, for example, would specify a certain quantity of laborers, say 10 men per shaft or strip mine. Each laborer may cost on average $15,000 per year working full-time. Total cost for labor is $150,000 per year working full-time. Since a mineral company only produces their product when they can recover their costs, finding the total cost for all inputs tells them how much they must receive in revenue to be profitable. Unprofitable productions are closed. The objective is to find the level of production that maximizes corporate profits.

Economic and engineering parameters that act to determine these costs include:

Parameters	Price Measure
Lease	Cost of Acquiring
Exploration	Expense
Transportation Equipment	Expense
Labor	Wage rate
Investment capital	Interest payments
Mining Equipment	Expense
Managerial Services	Wages
Marketing	Wages

For most mineral production processes lists of necessary inputs are quite long. To simplify the exposition of constructing supply curves, the inputs are often separated into two or three categories. The most common categories are given in Equation 7.1 Cost (C) is the sum of

$$C = (W \cdot L) + (K \cdot r) \qquad (7.1)$$

the product of labor (L) utilized times the wage paid labor (W) and capital equipment purchased (K) times the return on the investment that has to be earned (r). Capital equipment is priced at the necessary return on investment since firms use either portions of stockholders' equity (retained earnings) or other borrowed funds to purchase the capital. The amount borrowed plus the return investors must have to lend this capital represents r, the cost of acquiring capital.

Most mineral companies cannot control wages paid labor or investor's desired return to any great extent. Quantities of labor and capital employed in the production process, however, are subject to some manipulation by the company. Total cost of production is controlled by combining labor and capital equipment in the proportion that keeps costs within acceptable limits.

Calculation of total cost for different levels of inputs are shown in Table 7.2, assuming annual wages are $15,000 and the required return on investment to investors is 15%. Total costs increase as the quantity of inputs for producing a mineral increase, as everyone knows. Changes in wages or required investment returns will cause the total costs to vary.

Quantity of Labor (1)	Labor Cost (1) x 15,000 (2)	Capital Investment (3)	Investment Cost (3) x 1.15 (4)	Total Cost (5) = (2)+(4)
10	150,000	50,000	57,500	207,000
15	225,000	60,000	69,000	294,000
20	300,000	70,000	80,500	380,500
25	375,000	80,000	92,000	467,000

Table 7.2 Calculation of Total Cost

Knowing total costs for different input combinations is valuable information. But this information does not explain directly the choices open to a mineral company in seeking the objective of profit maximization, or to public agencies in apprasing the impact of mineral policies. This is especially important since supply analysis concerns explaining why a total cost figure occurs as illustrated by the historical graphical approach. Just adding numbers as we have done so far fails to explain the causes; and, without understanding the causes, planning for the future is difficult.

Numbers like those in Table 7.2 might represent growth in the size of a mineral plant over a four-year period. But, again, we need to know why each cost occurs. Focus on the cost figure of $294,000, which is the cost in year 2. After hiring 15 employees and spending $60,000 on capital equipment, the company selects the quantity of production that yields the largest return. Production can vary from zero output to the engineering capacity of the mineral deposit plus the capacity of installed equipment. The easiest principle to follow is to produce that quantity of mineral that yields the lowest average cost, defined simply as total cost divided by production. If production in year 2 was 30,000 tons, average total cost is

$$ATC = 294,000/30,000$$
$$= \$9.8 \text{ per ton}$$

Careful inspection of actual mineral company operations shows that a company's actual cost seldom equals the minimum average cost per unit of output.

Average and minimum cost seldom coincide. Minimum average cost occurs at the point where output is the largest for a given cost. Critics of mineral companies often cite this discrepancy as evidence that mineral companies are controlling output as a way to increase profits. This is an erroneous interpretation of the data.

To illustrate the fallacy in simple conclusions, an even more detailed analysis of costs is necessary. Figure 7.5 shows typical cost curves for a mineral company as a function of the quantity of minerals produced. Average total cost (ATC) is represented by an inverse shaped

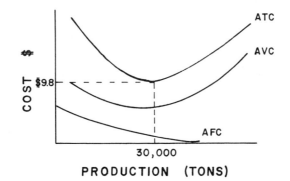

Figure 7.5 Types of Cost Curves

function. An ATC curve like this implies that a single minimum cost point exists for each mineral deposit. Minimum cost is $9.8 for a production level of 30,000 tons. The inverted U-shaped curve occurs because at a lower production level, say 15,000 tons, cost per unit output is higher.

Total cost curves are developed by reference to the two main components of cost-fixed (FC) and variable cost (VC). Fixed costs represent any cost that does not vary with output. Bonus costs, property taxes, lease costs, interest payments, etc. are examples of fixed costs. Since fixed costs represent a significant portion of total cost for petroleum refineries, synthetic hydrocarbon and geothermal plants, they play an important role in choosing the optimal quantity of a mineral to supply. Why?--because the mineral company has to pay these costs even if no minerals are produced. Total fixed cost divided by the quantity produced is average fixed cost, AFC = TFC/Q.

Variable costs reflect costs incurred only when minerals are produced. Royalty payments, wages, and other items characterize variable costs. Average variable cost is found by dividing total variable cost by production, AVC = TVC/Q.

The shapes of the average cost curves in Figure 7.5 illustrate the fallacy behind assuming output is controlled to increase profits. Total fixed costs are constant, so the AFC curve slopes toward the X-axis as the constant TFC value is divided by successively larger outputs. Average variable cost is U-shaped for most production functions, indicating that an optimal production rate exists. Such is the case in most extractive industries. Suppose a well has the capability of producing 30,000 Mcf per day with present capital equipment. For a given set of variable costs, say $10,000, average variable cost is $10,000/15,000 MCF = $0.66/MCF for a production rate of 15,000 Mcf. Increasing production to 30,000 MCF at a cost of $15,000 gives an average cost of $0.50 = 15,000/30,000. To go beyond 30,000 MCF/day requires additional manpower, capital equipment, and work on the reservoir itself, all of which tend to increase costs per unit of output. So average variable costs tend to rise beyond some point in most circumstances.

Adding AVC to AFC yields ATC. Since AFC declines at higher production levels the AVC and ATC approach each other as AFC → 0. A question often asked by mineral personnel is how these curves are derived in practice. To do this two additional terms require explanation--<u>returns to scale</u> and <u>input substitution</u>.

Input Substitution

Any relationship describing the different possible ways in which the inputs in the production process can be employed is termed the <u>elasticity</u> of substitution. Figure 7.6 shows how different combinations of capital and labor can be employed to produce mineral output. At point (A) 15 units of labor and $69,000 worth of capital are employed to produce 30,000 tons of mineral.

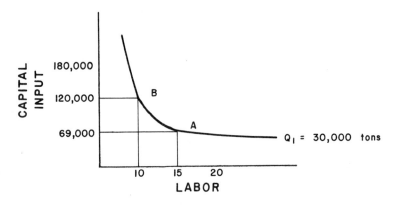

Figure 7.6   Example of Input Substitution

Could another combination of capital and labor produce the same amount of mineral. Point B signifies such a point. Connecting all possible combinations of inputs for a specific output level is the <u>input substitution curve</u>.

160

The average cost curves dealt only with variable and fixed costs for different output levels. Implicit in that framework is the assumption that capital and labor are combined in the least cost manner. The total cost of capital and labor was $294,000 in Table 7.2. If 10 labor units and $120,000 worth of capital are employed, total cost is $270,000. Management saves $24,000 in costs and still produces 30,000 tons by substituting capital for labor. Selecting the right combination of inputs for each output level insures minimum cost production process.

The point of this exercise is to demonstrate the need to recognize that supply models which add costs, without considering the effect of alternative combinations of inputs on cost (as we have done so far) may be misleading. The extent to which they are misleading is a function of the degree to which inputs may be substituted. Large refineries or processing plants are highly capital intensive. Adding or reducing the labor force does not alter output drastically. Coal production, on the other hand, could employ only labor to mine and transport coal as it did in its early days and still does in some developing countries. The capability for substituting inputs, and the resulting impact on cost curves, is obviously greater in this situation.

A measure of the degree to which inputs are substitutable is termed the elasticity of substitution. Values of zero indicate that the inputs cannot be substituted for each other at all. The larger the value of elasticity of substitution, the greater the degree of substitution. For example, a coal mine with 10 trucks for hauling coal needs 10 drivers. Reducing the number of drivers cannot be compensated for by adding trucks since there would be no one to drive them.

Several types of supply models which possess different values for the elasticity of substitution are discussed later. Explanation of returns to scale is necessary before these examples are presented.

Returns to Scale

Input substitutability explains changes in cost of production at a given output level as inputs are varied. But mineral companies also must consider alternative production levels. Returns to scale measures the change in output as capital input levels change. Curve Q_2 shows an increased production from 30,000 to 50,000 tons. At this latter production level the optimal combination of capital and labor is point B, 30 units of labor and $138,000 worth of capital. By doubling the quantity of inputs--capital and labor--production increases 20,000 tons.

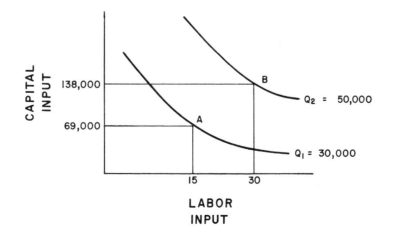

Figure 7.7 Returns to Scale

161

Returns to scale are classified into three categories--increasing, decreasing and constant. Doubling inputs results in a 66% [=(50,000 - 30,000)/30,000] increase in output in this example. This is a decreasing returns to scale case. Had output risen to 60,000 tons instead of 50,000 tons, a constant returns to scale situation would exist because output increases at the same rate as the labor inputs (100%). Increasing returns to scale situations exist whenever the percentage increase in output exceeds the percentage increase in inputs.

Evaluation of return to scale are commonplace in mineral industry engineering studies. Pilot projects for coal gasification and secondary recovery projects face the problem of verifying that the physical relationships uncovered on small projects will also hold for plants of a commercial size. Returns to scale is the counterpart of the same problem in economics. It is an effort to measure changes in economic relationships as the size of the operation increases. Highly non-linear relationships tend to distort both engineering and economic relationships presumed to be constant as the level of activity increases.

Two factors, substitutability of inputs and returns to scale, classify supply functions. Some techniques for determining supply functions are more flexible than others. That is, a greater range in the elasticity of substitution and return to scale are considered. Several techniques for estimating supply relationships, judged in relation to these considerations, are discussed below.

ESTIMATION OF SUPPLY CURVES

Supply curves come in a variety of forms, from the very simple to the very complex. Four supply relationships, representative of this range, are discussed with examples. These are:

1. Break-even analysis;

2. Zero substitution among inputs;

3. Unitary substitutability between inputs;

4. Constant substituability of inputs with values other than zero or one;

Break-Even Analysis

The precise definition of break-even analysis is the volume of production necessary to equate revenues with the fixed and variable costs of production. In some cases, particularly in the petroleum industry, break-even is sometimes defined as that point in time when income equals cost of the operation, meaning net income is zero. Observe that this definition is inconsistent with our objective in analysis of supply relationships. It tells us how many years before a project recovers its fixed and operating costs. But it does not explain what economic forces caused production to occur at a specific level in each year. The precise definition of break-even analysis is used for this reason.

The break-even point is defined as

$$Q = F/(P - V) \qquad (7.2)$$

where Q is the quantity of mineral produced, F is the fixed cost, V is variable cost per unit of output ($/ton, $/Mcf, or $/bbl), and P is the price received for each unit of sales. Given fixed cost equals $1 million, variable costs equal $250/unit, and the price is $500/unit, the break-even production rate is

$$Q = 1,000,000/(500 - 250) = 1,000,000/250 = 4000 \text{ units.}$$

During any period in which fixed cost, variable costs, and price possess these values, 4000 is the minimum level of production that should occur; producing at a lower rate results in an operating loss for that period. Of course, one can produce at a higher rate and make a profit. The most profitable production level is infinity, which obviously cannot be achieved.

By defining variable costs and prices as unchanging over all production levels, maximum profit occurs at maximum production. OPEC's lid on production in the seventies indicates an obvious example that not everyone can produce at the maximum rate without altering prices and costs. Inability to determine anything but the minimum production level is the primary weakness of break-even analysis.

Figure 7.8 graphs the break-even relationship. A break-even chart shows the relationship between operating profits and volume in a given time period. The greater the ratio of price to variable costs per unit, the greater the sensitivity of profits to production level changes. To show how variations in each force affects the break-even point, they are changed one-at-a-time.

In year t+1, variable and fixed costs are expected to remain constant, with price rising to $750/unit. The minimum production level for next year is then

$$Q = 1,000,000/(750 - 250) = 2,000 \text{ units}$$

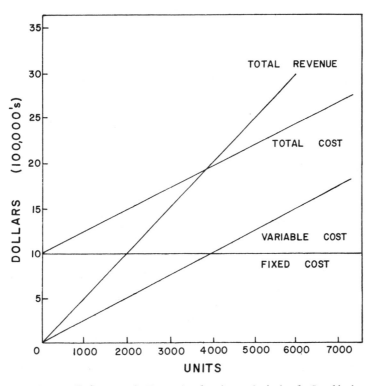

Figure 7.8 Break-Even Analysis: Original Conditions

163

Raising prices lowers the minimum break-even production level. Common sense, however, suggests that producers would desire to produce more.

This counterintuitive finding exemplifies much of the supply analysis applied to mineral policies. The reasoning is as follows: with a higher price of $750/unit, producers can still produce 4,000 units and increase their profits by the following amount:

$$\text{Profit} = PQ - F - VQ$$
$$= (750)(4,000) - 1,000,000 - (250)(4,000)$$
$$= \$3,000,000 - 1,000,000 - 1,000,000$$
$$= \$1,000,000$$

Profit increases to $1,000,000 in year t+1 since zero profits were made in year t. Inferences are often drawn that this is a windfall profit to producers.

Because mineral production depletes resources, it is not necessarily true that a specific mineral reserve could produce 4,000 units in that year. Mineral reserves typically decline from one year to the next. If the maximum producible quantity of mineral in year t+1 is 3,000 units, profit is = (750)(3,000) - 1,000,000 - (250)(3,000) = $500,000. More typical, however, is the case where producers, expecting higher profits, plow back these funds into increasing recoverable reserves. Since increasing recoverable reserves increases both fixed and variable costs as less economical reserves are produced, the profit picture changes. Raising fixed costs to $2,000,000 and variable costs to $500/unit might yield an extra prodcution of 500, or 3,500 units in year t+1. Profit is expected to be

$$\text{Profit} = (750)(3,500) - 2,000,000 - (300)(3,500)$$
$$= -\$425,000$$

or, $425,000 deficit occurs even with a price of $750/unit, simply because costs have offset the price rise.

Break-even analysis, as these examples illustrate, suffers from several unrealistic assumptions. These are:

1. Price and variable cost are constant at any output level; and,

2. Classifying costs as variable or fixed is difficult, since many costs could be either (intangible drilling costs, for example).

In spite of these limitations, break-even analysis has several offsetting advantages, including:

1. Ease of preparation when time or data limitations do not allow for more detailed analysis, since analysis is based on historical relationships;

2. An easy to comprehend standard of performance is obtained--if the break-even point cannot be achieved, production should cease;

3. Risk can be incorporated into the analysis by considering the probability distributions of costs and price.

164

When a quick, computationally cheap, fairly accurate supply relationship for short-run analysis is required, break-even analysis is acceptable.

What is required for a more accurate analysis is a supply relationship like that shown in Figure 7.9, where total revenue and total cost vary at different output levels. Curvilinear

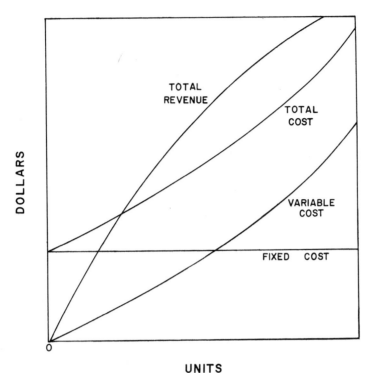

Figure 7.9 Break-Even Analysis: Curvilinear
Relationship

relationships are also helpful in that specific production of minerals are estimated for profit maximizing producers. Maximum profits occur at the point where total revenue and total cost are farthest apart. In order to find the profit maximizing production rate, more sophisticated supply relationships have to be developed.

Zero Substitution of Inputs

Supply relationships which allow for more detailed analysis involve several types. One example are the fixed coefficient class of models, like input-output models. Two primary assumptions are involved in input-output analysis. One is the zero substitutability of inputs as prices change and the other is that output is directly proportional to changes in inputs, doubling inputs doubles output. These assumptions are graphically shown in Figure 7.10. At an output level of 100,000 units of the mineral, 10 units of labor and 15 units of capital are optimal. Why?--because according to our assumption capital and labor cannot be substituted for each other. Adding labor, say 5 units, to the same quantity of capital does not increase output. Labor costs go up at the same time. So, in input-output analysis only one possible combination of inputs exist. Doubling inputs to 20 and 30 doubles output to 200,000.

Originally developed by W. Leontief (18) to help the Soviet Union plan their economy, input-output analysis has been modified to let free world countries obtain a better picture of

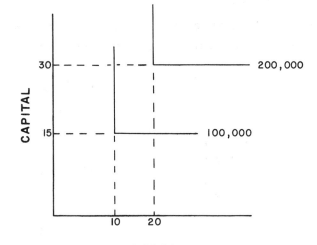

Figure 7.10 Substitutability and Constant Return
to Scale in Input-Output Supply

the operation of their economies. Essentially, input-output analysis attempts to measure the input needed by various industries to produce a given level of output. The primary usage of input-output analysis has been at the national level or the so-called macroeconomic forecast. In this instance, the input-output analysis breaks down the Gross National Product (GNP) of the country into the various components of the National Income Accounting scheme.

A series of input-output tables are published regularly by the Survey of Current Business for 87 intermediate industries. One industry covered is petroleum and natural gas. In a recent table for this industry the total inputs were $5,338 million. Of these expenses the largest were

Real estate and rental - $2,246 million (bonus and royalty payments)
Transferred imports - $1,046 million (value of imported products)

The same table shows the receipts from the sale of crude petroleum and natural gas totalled $12,265 million. The difference between input and output, called value added, is $12,265 - 5338 = $6,926 million.

An output from one industry will be an input for another industry. Of the $12,265 million above, $12,237 million was shown as an input by other industries, $16 million was an increase in inventory and $12 million was exported. So, the input-output table also describes total revenue received and how it was dispersed among other industries. In the case being cited $9,813 million was sold to petroleum refineries and related industries, $1,948 million to electric and gas utilities and $297 million to itself, with the remainder allocated to other industries.

For analyzing supply changes it is convenient to convert money amounts into relative percentages. Many of these percentages stay relatively constant with time and can be used to estimate future supply for a particular industry. One also needs the effect of final and intermediate demands on supply since most production ends up as a consumer product. Input-output tables are transformed through a process called matrix inversion.

166

To illustrate technique consider a simplified analysis of two interacting industries--petroleum and steel. The purpose is to show how matrix inversion and the multiplier table are derived. Input-output analysis is based upon the double entry form of bookkeeping. Table 7.3 shows one possible interaction between the two industries and final consumption. Inputs into each firm are shown in the left column and outputs in the right column.

Table 7.3
Two Industry Production Accounts

Petroleum Industry

Inputs		Outputs	
1. Purchases from Petroleum Industry	15	Sales to Petroleum Industry	15
2. Purchases from Steel Industry	50	Sales to Steel Industry	40
3. Wages Paid	30	Sales to Consumers	55
4. Depreciation	5		
5. Profits	10		
Total	110	Total	110

Steel Industry

Inputs		Outputs	
1. Purchases from Petroleum Industry	40	Sales to Petroleum Industry	50
2. Purchases from Steel Industry	15	Sales to Steel Industry	15
3. Wages	50	Sales to Consumers	55
4. Depreciation	10		
5. Profits	5		
Total	120	Total	120

Petroleum companies purchase $15 million of petroleum from their own industry and $50 million from the steel industry. Adding labor cost (wages) of $30 million, depreciation of $5 million, and profits of $10 million, the total cost of production is $110 million. Output valued at $110 million is sold to other petroleum companies ($15 M), to the steel industry ($40 M), and to consumers ($55). The steel industry is interpreted in the same way.

Comparable figures to those for the national input-output model are given in Table 7.4. Here the figures have just been rearranged to show intermediate sales between the two industries. Wages plus depreciation and profit equals value added. Consumers consume $55 M of both products as shown under the consumer expenditures column. Table 7.4 can be in either physical units or dollar terms depending on the preferences of the user. All of the numbers come directly from the T-account sheets. Rows 1, 2 and columns 1, 2 represent what is called intermediate demand. Intermediate demand represents the purchase of raw materials for transformation into final goods. This process is exemplified by a refinery purchasing crude oil and turning it into gasoline, where crude oil is the intermediate product and gasoline is the final product. In this example the value added, or cost of product transformation plus profit, is $45 M by the petroleum industry and $65 M by the steel industry.

Table 7.4

Two Industry Flow Table

($000)

	PI	SI	Consumption Expenditure	Total
1. Petroleum Indus. (PI)	15	40	55	110
2. Steel Indus. (SI)	50	15	55	120
3. Wages + Dep. + Profit	45	65		110
Total				

Another way of looking at the input table is to figure the direct requirements needed to produce a certain dollar amount of output. What is needed is information on how much input from each industry is needed to produce each unit of output. To calculate these numbers, merely divide the values in each column by the total at the end of each column. From Table 7.4, if the steel industry wants to increase their output by a million dollars, then 40/120 = 0.334, or $334 thousand of petroleum products is purchased, 15/120 = 0.125, or $125 thousand of steel input is needed, and 65/120 = 0.541, $541 thousand of wages, depriciation, and profits are required. The complete table is given in Table 7.5.

Table 7.5

Two Industry Direct Requirements

	PI	SI	Cons.
1. Petroleum Indus.	0.136	0.334	0.5
2. Steel Indus.	0.454	0.125	0.5
3. Wages + Dep. + Profit	0.410	0.541	
Total	1.00	1.000	1.0

Now we know that if the steel companies want to increase production, they will have to increase purchases by $334 thousand from petroleum companies. In order for the petroleum companies to supply this output, they have to increase their own production.

Because the steel industry needs more inputs from the petroleum industry, increased oil production has to be forthcoming. Since the petroleum industry has to supply $334 thousand of petroleum products and value added is 41% of the petroleum industry's output, $334 x 0.41 = $136 thousand increase in value added in the petroleum industry. Also, because the steel industry has to produce $125 thousand of output, total value added attributable to the increased production is $125 x 0.542 or $67.7 thousand. The total impact on value added in the two-sector model is the summation of the value added of the petroleum industry, steel industry, and the value added

168

required for the original production of the million dollars of output. For the sake of clari-
fication, remember that to produce a million dollars worth of steel, 12.5% of the steel produced
is consumed in the production process by the steel plant itself. In actuality, the steel companies have to produce $1,125 million of steel in order to produce a million dollars worth for sale
to consumers. Now the total impact on value added is $136,000 + $6,770 + 542,000 = $745,700.
Alternatively, one can say $136,000 + ((0.542)([1,000 + 125])) = $745,700.

Unfortunately, the direct requirement table doesn't deal with all of the impacts. Remember that the petroleum industry has to consume 13.6% of all oil production to supply the inputs
desired by the steel industry. Because the steel industry needs to produce $334,000 for steel
and $136,000 to consume itself in producing the petroleum needed by the steel industry. Further-
more, some additional inputs from the steel industry may be required. The reader can see that
the simultaneous relationship between the steel and petroleum industry can become complicated
rather fast in determining the total impact of a million dollar increase in sales in the steel
industry. Instead of grinding through these calculations by hand, manipulation of the matrix of
intermediate inputs will yield the same results.

Before discussing the matrix operation, the inverse of a matrix needs to be defined. In
matrix algebra terminology, an inverse matrix is defined as any matrix which when multiplied by
the original matrix results in an identify matrix, or matrix with ones in the principal diagonal
terms. Conceptually, the inverse of a matrix is equivalent to the scalar operation $(2)(1/2) = 1$
where $1/2$ is the inverse of 2. Let A represent a 2 x 2 matrix, A^{-1} represent the inverse of the
A matrix, and I be the identify matrix. By definition, the equation is

$$(A)(A^{-1}) = I \qquad\qquad (7.3)$$

Now assume, $A = \begin{vmatrix} 4 & 3 \\ 7 & 1 \end{vmatrix}$. For a 2 x 2 matrix, A^{-1} is found by switching the elements on the principal diagonal, and switching the elements on the off diagonal and multiplying each off diagonal
element by -1. To complete the inversion process, divide these elements by the determinant of A,
which gives

$$A^{-1} = \begin{vmatrix} -1/17 & 3/17 \\ 7/17 & -4/17 \end{vmatrix}$$

Check for errors by multiplying A by A^{-1}

$$(A)(A^{-1}) = \begin{vmatrix} 4 & 3 \\ 7 & 1 \end{vmatrix} \begin{vmatrix} -1/17 & 3/17 \\ 7/17 & 4/17 \end{vmatrix} = \begin{vmatrix} 1 & 0 \\ 0 & 1 \end{vmatrix}$$

where

$$4(1/17) + 3*(7/17) = 21/17 - 4/17 = 17/17 = 1$$

$$4(3/17) + 3(-4/17) = 12/17 - 12/17 = 0$$

$$7(1-1/17) + 1(7.17) = 7/17 - 7/17 - 7/17 = 0$$

$$7(3/17) + 1(-4/17) = 21/17 - 41/7 = 17/17 = 1$$

Although the use of the inverse technique is fairly simple here, larger matrices require computers for easy calculation of the inverse matrix. In addition, inversion of large matrices by hand usually leads to errors in calculation.

Computation of the total impact of our input output table is found by inverting the elements of the direct requirements matrix, or Table 7.5. First, however, the elements of the intermediate demand matrix on the main diagonal are subtracted from one and the off diagonal elements are made negative by subtracting them from zero. In matrix notation, this series of events is given by (I-A), or

$$
\begin{vmatrix} 1 & 0 \\ 0 & 1 \end{vmatrix} - \begin{vmatrix} 0.136 & 0.334 \\ 0.454 & 0.125 \end{vmatrix} = \begin{vmatrix} .864 & -0.334 \\ -0.454 & 0.875 \end{vmatrix}
$$

The reason for subtracting each element is an example of mathematical manipulation to achieve desired results. Because numbers ≥ 1 in the main diagonal are desired in the total requirements matrix, the elements of the direct requirements matrix are made less than one. If the element on the main diagonal equals one, this means all of their production goes to final demand. When the main diagonal term is greater than one, the excess of the value over one is used internally. To determine the total requirements matrix, invert (I-A), or $(I-A)^{-1}$. From the previous example, we have

$$
\begin{vmatrix} 0.864 & -0.334 \\ -0.454 & 0.875 \end{vmatrix}^{-1} = \begin{vmatrix} 0.875/0.609 & 0.334/0.609 \\ 0.454/0.609 & 0.864/0.609 \end{vmatrix}
$$

$$
= \begin{vmatrix} 1.4477 & 0.5526 \\ 0.7511 & 1.4295 \end{vmatrix}
$$

Table 7.6 puts these results in tabular form.

Table 7.6

Two Industry Total Requirements

	PI	SI
Petroleum Industry	1.4477	0.5526
Steel Industry	0.7511	1.4295

The interpretation of Table 7.6 is equivalent to Table 7.5, although the values of the numbers are numerically greater in Table 7.6. According to Table 7.5, the steel industry has to produce $1,429,500 of output to meet the demand of $1,000,000 for steel, compared to a value of $1,125,000, calculated earlier. As is always the case, the total requirement table is equal to or greater than the direct requirement table, because the indirect effects have to be either zero or positive.

The total requirements table for the national economy, can be computed in the same manner as this example. As a company representative, suppose that the only expected change in customer demand is expected to occur from the electrical and gas utilities, which had total final demand sales of $12,513 million. They are forecasting a doubling of sales to $25,026 million. Crude petroleum and natural gas sales must rise by 0.432% for each dollar increase, or ($12,513) (0.09432)= $1,180 million. Not only can we tell the dollar increase in sales, we can also tell how the extra funds will be spent. The table may show that 56.475% of expenses are value added. So value added after the increased demand will be $6,926 million + ($1,180)(0.56475) = $7,592 million. Expected increases in other inputs are analyzed in the same manner.

Input-output modelling represents a detailed appraisal of specific industries' interaction with other industries. Supply curves can be generated from these tables given forecasts of final demand of every industry using your mineral, as these examples illustrate.

Input-output supply models thus simultaneously makes forecasting more realistic and more expensive. In return for more information about the final use of a product and necessary inputs, the time and expense of construction expand accordingly. Of all supply models, I-O models are the most complete in determining the relationship between various economic variables.

Assumptions play a critical role in determining its value for analyzing mineral supply changes. Situations where the substitutability of inputs is non-zero and returns to scale are not constant may not be adaptable to the IO supply model approach described here. More detailed analysis is required to handle these problems.

Aside from these supply relationship problems, several practical problems arise in I-O models. The first step in constructing an input-output model is to classify your firm and the companies your product is sold to by the types of products manufactured. The first question thus becomes one of how to classify all the firms your company does business with. For instance, in today's economy, most companies produce more than one line of product. In national calculations, secondary products are lumped into the dollar volume of output with the primary product, although the secondary output is listed as being purchased by proper industry. At the firm or industry level, the various products can be explicitly broken out. The Survey of Current Business lists petroleum industry activity in two categories - crude oil and natural gas production, and refinery and allied products, respectively. Major refineries can enlarge on this category by explicitly considering the production of gasoline, jet fuels, fuel oil, naptha, etc. and who uses them. The same can also be done for crude oil and natural gas production. Emphasizing only the major purchases of their product allows the firm to reduce the size of the input-output study.

Procedures for calculating technological coefficients are probably the biggest source of worry to users of input-output models. The simplest and most common step is to merely divide the input to one industry from another by total industry output. In Table 7.4, the interindustry flows are given in dollar terms. Since total demand for steel and petroleum are also known, division of each intermediate and final component by the total values results in the direct requirements data in Table 7.5. This division process may be misleading in some cases. For instance, industrial firms often use multiple fuels in their heating systems. During slack periods of residential demand for natural gas, the price of natural gas/BTU becomes less than the price of coal/BTU, motivating the firm to switch to natural gas as a source of energy. During peak periods of natural gas by residential customers, the reverse occurs, and the firm switches back

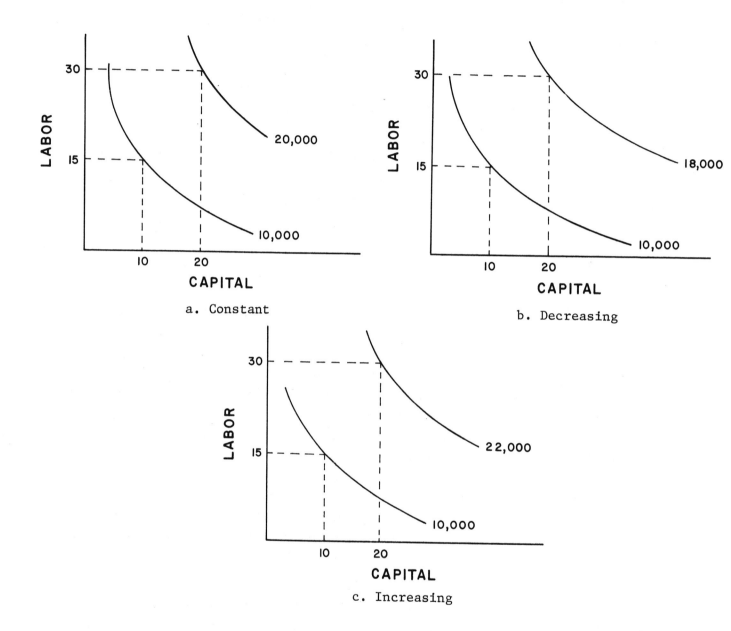

Figure 7.11   Returns To Scale In Mineral Supply Models

172

to coal. However, suppose the price of natural gas and coal remains in the same proportion. In this instance, input-output analysis says the two sources of energy will be used in the same ratio, in essence, input-output analysis ignores economic and technological forces which might make one fuel better than the other. If dramatic shifts in the economics of mineral prices or shortages are expected to occur, modifications in the technical coefficients have to occur.

Non-Zero Substitutability

Shifting to a mineral supply model where inputs are substitutable involves statistical estimation techniques because the proportion of inputs used to produce a given quantity of supply obviously change over time in response to economic and technological forces. The simplest models capturing substitutability are called <u>Cobb-Douglas models</u>. This model is expressed as

$$Q_S = AK^\alpha L^\beta \tag{7.4}$$

where Q_S is mineral supply, K, L are the quantity of capital and labor used respectively, A represents technological change over time, α measures the proportion of capital used in the production process, and β is the proportion of labor inputs.

Various alternatives exist with Equation 7.4. Constant returns to scale characterize mineral supply when $\alpha + \beta = 1$. Doubling inputs in turn doubles output. When $\alpha + \beta < 1$, doubling inputs results in less than a doubling of output, called decreasing returns to scale. Increasing returns to scale occur when $\alpha + \beta > 1$, so that doubling inputs more than doubles output. Figure 7.11 illustrates these alternatives under a, b, and c, respectively.

Selecting among the three alternatives is an empirical problem. By collecting actual mineral production and inputs for several years for a mineral producer, the specification of $\alpha + \beta$ can be established using the statistical analysis described in Chapter 9. Suppose the following historical data is acquired for this year and the previous four years.

Year	Coal Supply	Capital	Labor
t-4	10,000	2,083	40
t-3	12,000	2,497	48
t-2	14,000	2,940	54
t-1	13,000	2,719	51
t	18,000	3,954	58

Estimate the equation $Q_S = AL^\beta K^\alpha$ in double log form as

$$\log Q_S = (\log A)(\beta \log L)(\alpha \log K) \tag{7.5}$$

Finding the value for β and α tells us what kind of returns to scale exist. To simplify the statistics, presume that $\alpha + \beta = 1$. If this is true, then $\beta = 1 - \alpha$ by definition so Equation 7.3 becomes $Q_S = A(L^{1-\alpha})(K^\alpha)$. Dividing through by L, gives $Q_S/L = A(K/L)^\alpha$, which is a simple univariate regression with output per unit of labor input a function of the capital-labor ratio. The double-log linear regression results are

$$\log Q_S/L = \log 2.38 + 0.8 \log(K/L)$$
$$= (10.59)((K/L)^{0.8})$$

The rate of substitution of capital for labor in models where $\alpha + \beta = 1 = 0.8 + 0.2$ has been proven to be 1, meaning every unit increase in capital is exactly offset by a reduction in a unit of labor used.

Models of this kind allow for substitutability of various inputs, which the input-output did not allow for. Suppose that a mineral company states that it will increase labor used to 64 units and capital utilization to 4,100 units in year t+1. Production is forecasted by plugging these values into

$$Q_S^{t+1} = (\log 2.38) + 0.8 (\log 41,100) + 0.2 (\log 64)$$
$$= 19,278 \text{ tons}$$

Production of 19,278 tons is expected in year t+1.

An alternative way to calculate future production is to explain the growth in capital and labor directly with the model. Functional forms like $Q = AL^{\beta}K^{\alpha}$ assumes that technology remains constant because A, the technological parameter is constant. Technology can have three effects: a) labor saving, b) capital-saving, or c) neutral. Labor saving technology reduces the amount of labor needed to produce the same quantity of output. For example, a new technology that produces 10,000 tons of mineral output with 40 laborers, instead of 50 labor units, is referred to labor saving technology. Capital saving technology reduces capital inputs required to produce the same output. Neutral technology reduces both capital and labor necessary to reach the same mineral output. The equation for each are

Capital Saving	Labor Saving	Neutral
$Q = K^{\alpha}(e^{mt}L)^{1-\alpha}$	$Q = (e^{mt}K)^{\alpha}(L^{1-\alpha})$	$Q = (e^{mt}K)^{\alpha}(e^{mt}L)^{1-\alpha}$
$= (e^{m(1-\alpha)t})(K\ L^{1-\alpha})$	$= (e^{m\alpha t}K\ )(L^{1-\alpha})$	$= (e^{mt})(K^{\alpha})(L^{1-\alpha})$
$= (A^t)(K^{\alpha})(L^{1-\alpha})$	$= (A^t)(K^{\alpha})(L^{1-\alpha})$	$= (A^t)(K^{\alpha})(L^{1-\alpha})$
where $A = e^{m(1-\alpha)}$	where $A = e^{m\alpha}$	where $A = e^{m}$

Determining the kind of technological progress in a specific industry is a statistical question. Note that the statistical formulation of this problem can be solved by applying the same data used earlier to the double-log equation: $\log Q_S = (\gamma)(t + \alpha)(\log t) + (1 - \alpha)(\log L)$, where t is time, represented by the values t-4, t-3, ..., t, and $\gamma = m\alpha$. Note that the value for m is solved for by $\gamma/\alpha = m$, where γ and α are taken from the statistical model.

Given statistical findings like,

$$\log Q_S = (0.001)(t) + 0.8 (\log K) + 0.2 (\log L)$$
$$Q_S = (e^{(0.001)(t)}(K^{0.8})(L^{0.2})$$

and the assumptions that technology has been increasing labor inputs, n, the increase in labor utilization, per year is $0.001/(1-0.8) = 0.05$ or 5% per year.

Non-zero substitutability models certainly account for changes in inputs into the production process in mineral industries better than break-even or input-output analysis. With this greater detail, however, comes the need to convert the information back to a simpler supply framework that can be compared to these other supply models. This is easy. Equation 7.1 defined profits as

$$\pi = (Q_s)(P) - (K)(r) - (L)(W)$$
$$= TR - TC$$

or total revenue minus total cost equals profits. Production (Q), labor inputs (L), and capital inputs (r) are known from the supply relationships. Multiplying these values by the sales price (P), price of capital (r), and labor price (wages = W), respectively, yields an estimate of profits. Given that mineral companies desire to maximize profits, we can use detailed supply relationships to estimate supply curves described as a function of price, holding capital and labor costs constant.

Suppose a mineral company has 58 employees costing an average of $15,000 per year, 3,954 units of capital costing $10 per unit, and expects to receive $55 per ton for its mineral production of 18,000 tons, what is the expected profit? Profit is calculated as

$$\pi = (18,000)(\$55) - (58)(15,000) - (3,954)(\$10)$$
$$= \underline{\$80,456.}$$

The break-even production level is given by

$$\text{Break-even point,} \quad 0 = (Q)(55) - (58)(15,000) - (3,954)(10)$$
$$Q = \frac{(58)(15,000) + (3,954)(10)}{55}$$
$$= \underline{16,537 \text{ tons}}$$

Such analysis is comparable to the break-even procedure.

The advantage of these supply models relates to the following question. What happens if prices rise to $60/ton and you want to increase output to 21,000 tons. Substituting these values into the total revenue π equation does not account for extra labor and capital inputs. We can calculate extra inputs needed by recognizing that if α measures the percentage increase in output due to capital and the desired production is 21,000 tons, this is a production increase of 3,000 tons, or $(3,000/18,000)(100) = 16.6\%$. Producers have a choice of using only new capital, only new labor, or both. Presume two labor units are added, then needed capital requirements are

$$\log K = \frac{\log 21,000 - \log 10.59 - 0.2 \log 6.0}{0.8}$$
$$= 8.466, \text{ or, anti-log of } 8.466 = 4,750 \text{ units}$$

Profits are

$$\pi = (21,000)(60) - (4,750)(10) - (60)(15,000)$$

$$= \$312,500$$

By upping production to 21,000 units at the higher prices, mineral companies make a greater profit.

Using these models, like the break-even analysis, lead to the conclusion that the greatest profits occur when production is infinite. Why?--because the inputs into the production process do not cost more at higher production levels. This is unrealistic. As mineral companies push into the intensive margins or use secondary recovery techniques, costs will exceed total revenue.

To illustrate the profit maximizing producing region, the preceding analysis is synthesized by employing a production relationship showing decreasing returns to scale. Adding further inputs thus increases output at a slower rate than the increase in inputs. The following data and equations are given

$$\text{Labor cost } (W) = \$15,000$$

$$\text{Capital } (r) = \$10$$

Supply relationship $\log Q_s = \log 6.09 + 0.2 \log K + 0.5 \log L$

$$\text{where } \alpha + \beta = 0.5 + 0.2 \leq 1$$

Using the following data, the profit maximizing position can be found:

Price	Quantity (tons)	Capital Inputs	Labor Inputs	Profit
$40	8,000	2,300	25	-78,000
45	10,000	2,500	28	5,000
50	15,000	3,000	45	25,000
55	18,000	4,100	60	49,000
60	21,000	4,200	80	18,000
63	23,000	4,400	95	-20,000

The largest profit (49,000) is earned when 18,000 tons are produced and sold at $50 per ton. Production of more than 21,000 tons or less than 10,000 tons reduces profits because labor and capital costs offset total revenues.

Production below 10,000 tons is uneconomical because the cost of basic inputs cannot be covered. Above 10,000 tons revenues cover cost for a time. As companies begin to reach the intensive margin, adding inputs fails to increase production at a rate sufficient to cover the extra costs. Total costs, therefore, exceed revenues at some point. The relationship between total costs and total revenues is shown in Figure 7.12. In order to be profitable the mineral company has to produce more than 9,500 tons (Point B) and less than 21,500 tons (Point A).

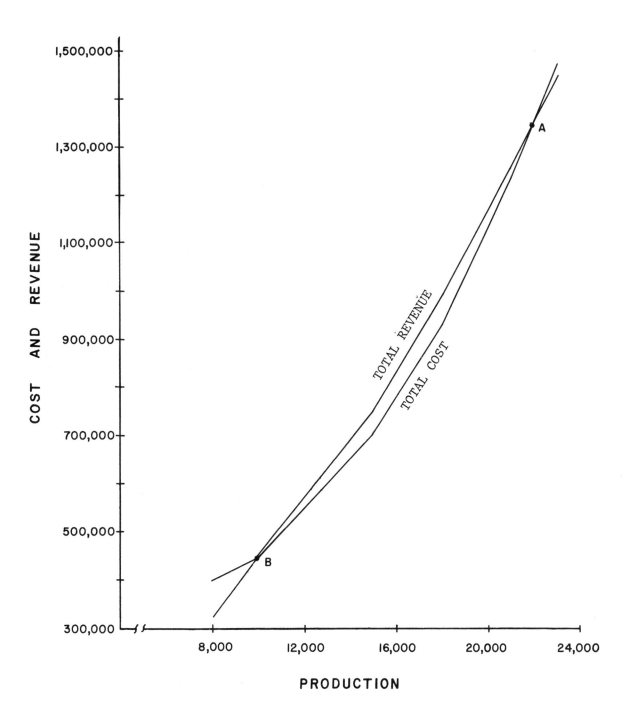

Figure 7.12 Total Cost and Total Revenue.

177

The area where total revenue exceeds total cost represents the mineral companies' supply curve. Instead of dealing with Figure 7.12, it is easier to graph just the relationship between price received and supply, as done in Figure 7.13. The supply curve, S_1, represents the quantity and cost of inputs. Curve S_1 shifts in the 2-dimensional graph as costs change. For example, if labor negotiated an average salary of $20,000 for the year, curve S_1 might become S_2, indicating less output is produced at each price. At $60 per ton 15,000 tons can be profitably produced at $20,000 per labor unit instead of $21,000 tons as before.

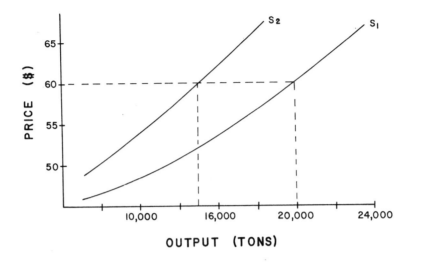

Figure 7.13 Mineral Supply Curve

Supply analysis, like engineering relationships, holds certain forces constant so that we can focus on forces felt worthy of emphasis. Since mineral prices are the mechanism by which mineral supply and demand are allocated, price is placed on the Y-axis for later comparision to demand changes which are also a function of price. Interactions between supply and demand can be compared in this way.

DEMAND MODELS

Mineral demand is divisible into two categories--derived demand and final demand. Derived demand corresponds to minerals used in the production process, such as fuel, to produce other goods for consumption. The input-output model referred to derived demand in its direct requirements table (Table 7.5). Final demand is that portion of mineral supply going directly to customers for their use directly. Minerals, thus, follow two paths to its end use, one in processed form and one directly.

Figure 7.13 illustrates the alternative path for minerals to final markets. The supply analysis just discussed developed the relationship for determining the profitable supply of minerals. For processors these minerals are inputs into their production process, and similar supply curves could be developed using the same techniques by adding an additional parameter in the functional form $Q_s = f(K,L,M)$, where M is the mineral used. We will not deal with this process again since the procedure is the same as the supply analysis, only more complicated.

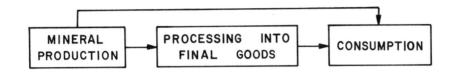

Figure 7.14   Flow of Mineral Supply to End Uses

Direct sales of minerals to consumers usually provide heating. Natural gas, coal, and oil fall into this category. Demand analysis concerns explaining how customers decide to choose among alternative minerals, again holding certain forces constant. In explaining demand for a specific mineral, say natural gas, economics concentrates on the price of natural gas. Economic forces held constant include income, the prices of other fuels, weather, mineral availability, and tastes. The end result of demand models is a graphical depiction of mineral demand as a function of price. Supply and demand curves can be combined to determine the market clearing price in this way.

Underlying the monetary aspects of demand is the concept of utility. Utility determines how consumers respond to different prices for minerals.

Utility Basis of Mineral Demand

Demand for a commodity in an economic sense, encompasses more than just the observed quantity of minerals purchased at some price. This limited definition describes demand after the fact and, more importantly, fails to provide any clue to what demand would be at other prices. The concept of demand is a powerful analytical tool because of the ability to explain or predict the quantities of mineral products purchased at alternative prices. More importantly, managers would not know how to set prices if they could not anticipate how customers would respond. Setting prices at a level where profits are maximized is, after all, the main objective of companies. The ability of a good to satisfy a desire is called utility. Those goods with the highest utility, i.e., those which provide the most satisfaction, are the products the consumer will spend most of his income on. For any individual, utility will depend on the inherent qualities of the goods, and his tastes and preferences. The existence of a multitude of goods and a limited income requires a comparison of the relative utility of different goods. When more than one good is compared, the problem arises of how to compare the level of satisfaction obtained from good X to the satisfaction of good Y, since utility is unobservable.

The near impossibility of determining actual utility led to the development of the indifference curve. Construction of indifference curves begins by asking consumers to compare different combinations of goods and services. Corporate managers, for example, must choose among different mineral investments by asking the manager to state which mineral investments are equally attractive. Connecting investments of equal satisfaction yields the indifference curve.

Assume a manager of a major integrated company has a choice of allocating his budget between petroleum or coal exploration. The following table gives the results of rankings by one

179

manager. Combination A represents 12 petroleum and 10 coal leases, and B represents 11 petroleum and 12 coal. The manager gave each a ranking of 1, indicating combination A and B are equally satisfactory. Investment combinations A, B are preferred to the other investment possibilities because they have been assigned a rank of 1. Investment combinations C, D, E are preferred over G, and G is preferred over H, I.

Ranking of Investment Combinations

Combination	Petroleum	Coal	Rank
A	12	10	1
B	11	12	1
C	6	9	2
D	7	7	2
E	10	4	2
F	6	8	3
G	2	8	4
H	4	4	4
I	6	2	4

Connecting the investment combinations with the same rank, as shown in Figure 7.15, yields three indifference curves, ignoring investment F. Point A on indifference curve I_1 (the subscript 1 represents the rank assigned A by the manager) is put at (12, 10), B at (11, 12), and so on. Indifference curve I_2 combines the points C, D, and E. The higher the indifference curve, the more preferred each combination is. These indifference curves are used to construct the demand curve.

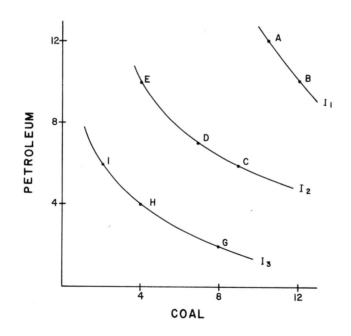

Figure 7.15   Indifference Curves

Derivation of Demand Curves

Demand is determined by the interaction of several forces, including the amount of money for spending, the price (or cost) of each investment, the price of equipment and other items necessary to develop the investment, and preferences for investments by the manager. Demand for petroleum and coal investments expressed in equation form are

$$Q_p = F_p(P_p, Y, P_c, T, P_{d,p}), \text{ and}$$

$$Q_c = F_c(P_c, Y, P_o, T, P_{d,c})$$

where Q_p, Q_c are the demand for petroleum and coal investments, respectively; P_p, P_c are the prices for petroleum and coal investment, Y is income or budget available for investing, T is preference for one type of investment, and $P_{d,p}$, $P_{d,c}$ are development costs for each.

In order to graph a relationship of several variables, we assume that all of the variables on the right hand side of the equation remain constant at first. Demand curves are constructed by relaxing this assumption for one variable at a time. Assume for the moment that $10 million dollars is available for investment, petroleum leases cost, on average, $500,000, and coal leases average $666,000.

The first step is calculation of the maximum number of coal or petroleum leases that can be consumed given this budget. Maximum possible consumption is given by the equation

$$Y = (Q_p)(P_p) + (Q_c)(P_c) \tag{7.7}$$

which is called the budget constraint. The budget constraint says that the budget, $10 million, can be spent for either petroleum or coal. Consumption of only petroleum leases specifies the quantity of leases purchased by

$$Y/P_p = Q_p + (0)(P_c)$$

$$\$10 \text{ million}/500,000 = 20 \text{ leases}$$

or, if only coal is consumed, the maximum possible purchase is

$$Y/P_c = Q_c$$

$$\$10 \text{ million}/666,000 \cong 15 \text{ leases}$$

With a budget of $10 million the manager can buy 20 petroleum leases or 15 coal leases. But as indicated by his rankings of combinations (Page 180) he prefers to put some money in both, instead of putting "all of his eggs in one basket."

How do we establish exactly what combination of each he will choose. Since indifference curves measure his relative preferences for each, we can combine the indifference curves with the budget constraint. The objective is to find the combination that maximizes the manager's satisfaction. Figure 7.16 uses indifference curves similar to those in Figure 7.15 with the budget

constraint added. Drawing a straight line between 20 and 15 gives a budget line (Y₁, Y), representing all possible combinations of coal and petroleum leases he can afford. The point of
tangency between the budget line and the indifference yields the combination of petroleum and
coal leases giving maximum satisfaction. Point R on indifference curve I_5 is that point.

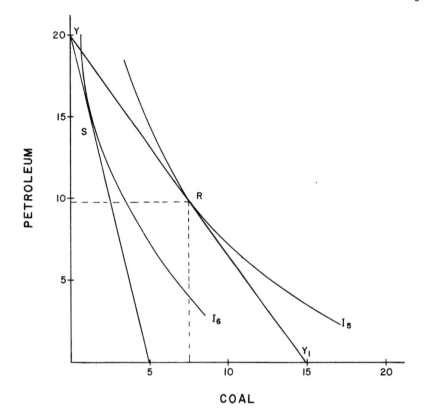

Figure 7.16   Indifference Curves with the Budget Line

Dropping straight lines to each axis gives the optimal investment combination. Approximately 7-1/2 coal leases and 9-1/2 petroleum leases are selected, at a total expense of
(7.5)(5000,000) + (9.5)(666,000) = $10 million. Fractional shares result from joint ventures and
farmounts. Point R is optimal because no other investment combination that fulfills the budget
constraint allows him to reach the highest level of satisfaction, the highest indifference curve.

Suppose now that the price of oil leases and income remain constant, but the price of
coal leases rise to $2 million each. The budget line revolves downward to signify that fewer
coal leases can be bought with the same income to S = $10/$2 million. The new tangency is at
Point S, indicating a preferred consumption of 14.5 petroleum and 2 coal leases. Moving from
point R to point S determines the demand curve. Note that the cause of the move to point S was
the rise in the coal lease price. Figure 7.17 plots these points as a function of coal lease
prices.

At a price of $2 million, two coal leases maximize satisfaction, while 15 coal leases
could be purchased at $666,000. Coal lease demand increases as prices decrease. The inverse
relationship between coal lease demand and price is logical if only for the fact that the manager
has more money to spend as prices decrease. The same is true for petroleum leases, at a lower
price, he can purchase more leases. Demand curves are expected to be inversely related to price
for this reason.

182

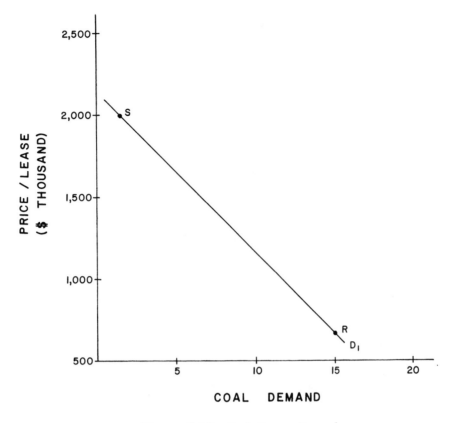

Figure 7.17 Coal Lease Demand

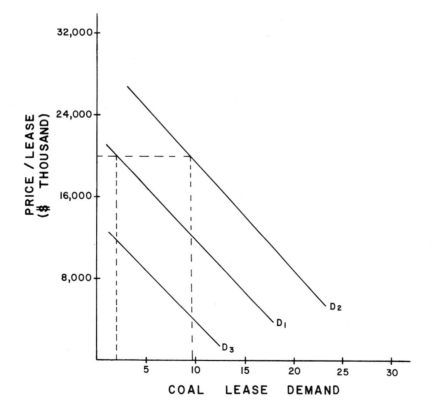

Figure 7.18 Effect of Budget Changes on Coal Demand

A question which arises quite often is how do these curves accomodate shifts in budgets (Y) or price of petroleum leases, which occur often in the real world. Changing any other economic force shifts the entire demand curve to the right or the left. Figure 7.18 reproduces a previous coal demand curve, D_1. Doubling the investment budget to \$20 million simply doubles the number of petroleum and coal leases that can be bought at the original prices.

$$Q_c = \$20 \text{ Million}/666{,}000 \qquad Q_p = \$20 \text{ Million}/50{,}000$$
$$\cong 30 \qquad\qquad\qquad \cong 40$$

Finding the point of tangency with this new budget line at these prices that gives the optimal combination, repeating the rise of coal lease price to \$2 million for a second point, produces a coal lease demand curve of D_2. Coal lease demand with a \$20 million budget lies above the initial curve D_1 because more leases of both kinds can be afforded at any price. Conversely, if budgets decrease, the demand curve looks like D_3 for the same reason. At a price of \$2 million per coal lease, 9.5 leases would be consumed at the higher budget level (see D_2). Budget decreases indicated by curve D_3 suggest that no coal leases would be invested.

The similar process takes place when petroleum leases change in price. For example, increasing the price of petroleum leases reduces the quantity of affordable leases. By repeating the analysis presented for income changes, the effect petroleum lease price on coal lease demand can be calculated. If such an analysis is conducted, shifts in the demand curve occur.

What we have completed is a framework for evaluating the _individual_ effects of economic forces on mineral demand. Instead of simply tracing historical demand patterns over time, demand curves permit us to control these individual forces so that these historical patterns can be broken down into finer components. The advantage: expected changes in each component can be utilized to realize better forecasts of future demand.

Impacts of mineral price, income, and other mineral prices on demand are seldom available graphically. To avoid the limitations of graphical analysis which can handle only two dimensions, the impacts of these forces are measured quantitatively in terms of percentage changes, called _elasticities_.

Price elasticity. - Measures the percentage change in demand resulting from a percentage change in price, as expressed by

$$E_p = \frac{(Q_1 - Q_2)/[(Q_1 + Q_2)/2]}{(P_1 - P_2)/[(P_1 + P_2)/2]}$$

where E_p is the price elasticity of demand, Q_1, Q_2 are the two values of demand, and P_1, P_2 are the corresponding prices. In the coal demand example, the two quantities are $Q_1 = 15$, $Q_2 = 2$ and the two prices are $P_1 = \$666{,}000$, $P_2 = \$2$ million. Thus, E_p is

$$E_p = \frac{(15-2)/[(15+2)/2]}{(\$.66M-\$2.M)/[(.666M+\$2M/2]}$$

$$= \frac{(13/8.5)}{[(-1.333M)/1.333]}$$

$$= 1.5/(-1) = \underline{-1.52}$$

184

Every percentage increase in the price of coal leases reduces the demand for coal leases by 1.52 percent. Negative signs indicate a demand curve sloping downward and the right (Figure 7.17).

The price elasticity of demand can take on five possible forms. The three most common forms are <u>inelastic</u>, <u>unitary</u> and <u>elastic-price elasticity</u> of demand. The remaining two are <u>perfectly elastic</u> and <u>perfectly inelastic</u> demand curves and normally are not encountered.

The importance of elasticity is its ability to show how the total revenue of a firm will change as price (P) and quantity demand (Q) vary. Since total revenue (TR) is the product of the quantity demanded of a good and the good's price $(TR = PQ)$, any change in price and quantity sold will affect the total revenue of the firm. Figure 7.19 depicts the five possible shapes of the demand curve.

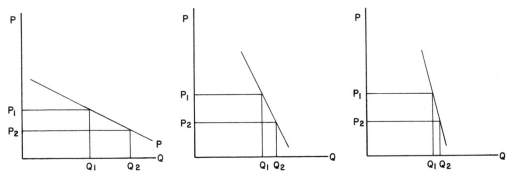

a. Elastic   b. Unitary   c. Inelastic

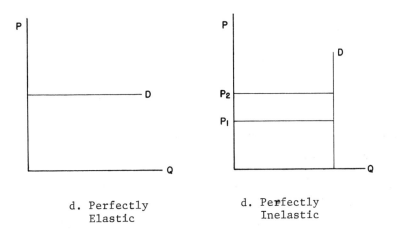

d. Perfectly
Elastic

d. Perfectly
Inelastic

Figure 7.19 Various Demand Curves

In the case of the <u>elastic price</u> elasticity demand curve, the drop in the price from P_1 to P_2 causes the quantity demanded to increase from Q_1 to Q_2. Since the change in quantity demanded is greater than the decline in price $(\Delta Q > \Delta P)$, $Q_q P_2 > Q_1 P_1$. Substituting from $TR = PQ$ gives $TR_2 > TR_1$ and shows that a firm with an elastic demand curve can increase total revenue by lowering prices.

For the <u>unitary</u> elasticity of demand, the change in price is exactly offset by the change in quantity demanded. The net effect is no change in total revenue, or $P_1Q_1 = P_2Q_2 = TR_1 = TR_2$. In this situation, the pricing decisions of the firm will have no impact on total revenue.

The third case is represented by the increase in quantity demanded outweighing the decrease in price. In this instance $P_1Q_1 > P_2Q_2$ and, thusly, $TR_1 > TR_2$. When an <u>inelastic price</u> elasticity of demand faces the firm, the lowering of prices will reduce total revenue. The impact of a change in price on total revenue for each of the three cases is summarized in Table 7.7.

Table 7.7

Impact of Price Decline on Total Revenue

Price Elasticity	Price	Demand	TR = P x Q
1. Elastic	↑	↓	↑
2. Unitary	↑	↓	—
3. Inelastic	↑	↓	↓

<u>Income Elasticity of Demand</u>. - Measures how changes in income alter demand. The equation for calculating the income elasticity of demand is the same as that for price elasticity, after income is substituted for price. In earlier discussions, it was shown how demand changes when income is increased. Coal lease demand jumped from 2 to 9.5 leases when income doubled to $20 million. The income elasticity equation and calculated value for this example are

$$E_Y = \frac{(Q_1 - Q_2)/[(Q_1 + Q_2)/2]}{(Y_1 - Y_2)/[(Y_1 + Y_2)/2]}$$

$$= \frac{(2 - 9.5)/[(2 + 9.5)/2]}{(10 - 20)/[(10 + 20)/2]}$$

$$= \frac{(-7.5/5.75)}{(-10/15)}$$

$$= \frac{-1.30}{-0.66} = 1.97$$

Coal lease demand increases 1.97% for every 1% increase in the budget. A 10% budget increase leads to a 20% rise in coal lease purchases.

Income elasticities fall into three categories. Values between zero and one, exclusive of one, are termed <u>inelastic</u>; values equal to one are <u>unitary elastic</u>; and values exceeding one are <u>elastic</u>. Note that for coal lease demand to be elastic, the elasticity for the alternative investment, petroleum leases, must be inelastic, indicating that the manager substitutes coal for petroleum leases. Substitution of one lease for another arises because if income increases 10% and coal lease demand jumps 19.7%, the manager must increase his demand for petroleum leases by less than 10% in order to stay within his budget constraint.

<u>Cross-price Elasticity</u>. - The cross-price elasticity captures the relationship between coal lease demand and the price of petroleum leases. The same type of equation is employed again, only petroleum lease price is substituted for income. Suppose that an increase in the average price of a petroleum lease from $500,000. to $1,000,000. per lease increases coal lease demand from 7.5 to 14 leases. The cross-price elasticity is

186

$$E_{po} = \frac{(Q_1 - Q_2)/(Q_1 + Q_2)/2}{(P_1 - P_2)/(P_1 + P_2)/2}$$

$$= \frac{7.5 - 14/(21.5/2)}{500 - 1000/(1500/2)}$$

$$= 0.604/0.66 = 0.916$$

indicating that a 10% increase in oil lease price raises the demand for coal leases by 9.16%.

The sign of the cross-price elasticity tells how two products relate to each other, unlike the price and income elasticities that are expected to be negative and positive, respectively. Positive cross-price elasticities mean that two products are substitutes. As the price of petroleum leases goes up, the manager will substitute the relatively lower cost coal leases for them. A price rise of one good followed by a decline of other good demand (a negative sign) denotes complementarity between the two goods. Complementary goods are typically used together in some way. Examples of complementary demands include drilling activity and the demand for tubing. If the price of drilling increases enough to reduce drilling activity, less tubing will be needed, so demand will decline.

The major shortcoming in constructing demand curves is the inability to observe indifference curves for one manager, much less all managers. So, how do we identify demand practically? By making the assumption that, even though most consumers are not always aware of their performances, consumers choose among alternative products in a way which is consistent with these preferences. Demand analysis is consistent with our discussion as long as consumers' selection of goods for purchase is consistent with their preferences. It is hard to imagine a person continually going against these preferences when faced with harsh realities of mineral economics.

Examples of Demand Models - The Case of Natural Gas

One of the hottest issues in the seventies centered around the price of natural gas. Basic to the arguments was the fact that natural gas reserves were badly depleted, and needed to be preserved or improved. Consumer groups, of course, resisted higher gas prices, believing in part that some miracle would occur. As always, debates on the issues became lost in emotional charges and counter-charges.

What is central to bringing back some sense of order to such situations is a detailed understanding of the determinants of natural gas demand. To illustrate such an analysis, we start with the basic model outlined earlier:

$$Q_{N \cdot G} = f(P_{NG}, Y, P_S, X)$$

where Q_{NG} is the quantity of natural gas demanded each year, P_{NG} is the price of natural gas, Y is the income of customers, P_S is the price of fuels competing with natural gas, and X is any other set of forces believed to affect gas demand. Chapter 9 explains that the statistical approach would be to collect annual data for these variables and estimate the relationships. Before that is done, let us recognize that there are different uses of natural gas.

Natural gas is used in different ways. Residential customers use it primarily to heat their houses, cook meals, and heat water. Industrial customers employ gas as both a boiler fuel and as a feedstock. Commercial customers utilize it in a manner similar to residential customers. What is apparent is that different customer classes use gas in different ways, and therefore likely will respond differently to the same economic forces. More importantly, each customer class faces different prices. The net result is that demand curves may not be the same for each customer class. The following model extends the general gas demand equation to take account of these differences.

$$Q_{NG} = Q_{NG}^R + Q_{NG}^C + Q_{NG}^I$$

$$Q_{NG}^C = F_R(P_G^R, P_E, P_{FO}, Y)$$

$$Q_{NG}^C + F_C(P_G^C, P_E, P_{FO}, IP)$$

$$Q_{NG}^I + F_I(P_G^I, P_E, P_{FO}, IP)$$

where the variables are defined as

Q_{NG}^R = residential gas demand (trillion BTU/customer);

Q_{NG}^C = commercial gas demand (trillion BTU/customer);

Q_{NG}^I = industrial gas demand (trillion BTU/customer);

P_G^R = real price of residential gas (¢/BTU);

P_G^C = real price of commercial gas (¢/BTU);

P_G^I = real price of industrial gas (¢/BTU);

P_E = real price of electricity (¢/BTU);

P_{FO} = real price of fuel oil (¢/BTU);

Y = real income ($/person);

IP = index of industrial production (1967 = 100).

Each dollar value is defined in real terms, meaning that actual prices are deflated by the cost-of-living index. Real gas prices are measured relative to the price of other items. For example, if the price of gas was $50 in year t and t+1, and the cost-of-living rose 10% in year t+1, the effective price of gas is $0.50/1.10 = $0.45. With a downward sloping demand curve, gas prices would increase, even if there is no change in the actual price of gas, the exact increase depending on the elasticity of demand.

Total gas demand equals the sum of the three gas demand components. Determination of the demand curve involves collecting actual data on the specified variables, and then estimating the relationships statistically. Suppose we did just that and obtained the following results.

$$Q_G^R = 24.20 - (10.72)(P_G^R) + (10.28)(P_E) + (4.02)(P_{FO}) + (0.031)(Y)$$
$$(12.4) \quad (1.81) \qquad (20.6) \qquad (2.25) \qquad (3.87)$$

$$R^2 = 0.994$$

$$Q_G^C = 73.64 - (242.1)(P_G^C) + (25.8) - (0.911)(P_{FO}) + (2.711)(IP)$$
$$(3.0) \quad (1.7) \qquad (3.0) \quad (0.1) \qquad (2.3)$$

$$R^2 = 0.978$$

$$Q_G^I = 36{,}741 - (1{,}258)(P_G^I) + (784)(P_E) + (3{,}630)(P_{FO}) + (138.7)(IP)$$
$$(5.73) \quad (0.2) \qquad (2.2) \qquad (3.1) \qquad (3.8)$$

$$R^2 = 0.971$$

Variation in the dependent variable, Q_G^R, Q_G^C, and Q_G^I, explained by the model is measured with the <u>coefficient of multiple determination</u> - R^2. The value for R^2 in the Q_G^R equation, 0.994, means that 99.4% of the variation in the residential demand for natural gas is explained by the equation. The closer the value of R^2 is to unity the better the correlation fits what are considered the pertinent variables. It should be pointed out, however, that a satisfactory value does not <u>automatically</u> insure that the correlation is meaningful; it could be mere coincidence. The point - the model must be developed by one who is truly knowledgable about how the "real world" system operates.

In the equations Q_G^C, Q_G^I, the index of industrial production is added to capture the impact of production levels by firms on the demand for gas. Firms closed for business do not need gas at any price, but as the level of production increases they will need more gas. The values in parenthesis are standard errors (See Chapter 9).

Gas price has the expected negative relationship with demand in each equation and is statistically different from zero. Coefficients for substitute fuels also possess the correct, statistically significant signs, with the exception of fuel oil prices in the commerical gas demand equation. The proxy for productive activity, IP, is positive and statistically significant, supporting our feeling that gas demand is directly related to business activity. Only the real income coefficient explaining residential gas demand is disappointing because it implies that income's effect on gas demand is statistically equal to zero.

With coefficients of determination (R^2's) exceeding 0.971 for the three equations, the models are fairly accurate, and acceptable for the following demand analysis. The information contained in these equations can be converted to graphs of demand by utilizing the coefficients. For residential demand, holding the price of other fuels and income constant, the intercept value 24.2 (trillion Btu/customer) provides the starting point on the Y-axis for price equal to 0. This says 24.2 trillion Btu's/person would be used each year if the price of natural gas was zero. If the price of gas was $0.50 per MBtu, expected demand is 24.20 - 10.72 x 0.50 = 18.74 TBtu/person. Raising the gas price to $0.50 reduces gas demand by 5.64 TBtu/person. Should gas prices go to \$1.00 per MBtu, gas demand becomes 24.20 - (10.72)(1) = 13.48 TBtu/person. Figure 7.20 connects these points to construct the demand curve, D_1. Gas demand declines as prices increase.

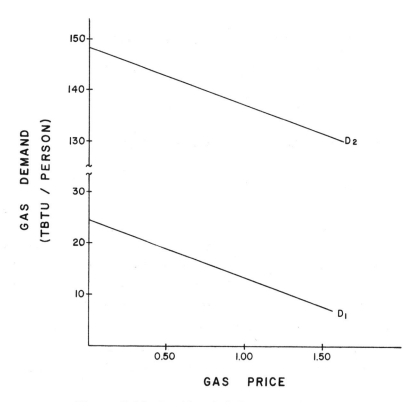

Figure 7.20 Residential Gas Demand Estimated
from the Statistical Model

Gas demand models oftentimes are improperly interpreted, leading some to erroneously question their validity. How would we verify this demand curve? One approach is to look at the mean demand per customer over a period of years. For example, mean gas demand over a 25-year period was 104.58 trillion Btu/customer. Does is seem likely that the demand curve is valid given the significant variation in these values? It does if you remember that other forces affecting gas demand were presumed to be zero in constructing D_1. Since income, electricity price, and fuel oil price have positive coefficients, the demand curve shifts upward as values are plugged into the equation. For example, if income has a value of \$4,000/person, the new demand curve is found by connecting the following values.

Intercept: 24.20 - (10.72)(0) + (0.031)(4,000) = 148.2

Price = \$0.50: 24.20 - (10.72)(0.5) + (0.031)(4,000) = 142.8

Price = \$1.00: 24.20 - (10.72)(1.0) + (0.031)(4,000) = 137.4

Demand curve D_1 has been shifted upward by the value of (0.031)(4000) = 124 trillion Btu/customer, a value more consistent with historical demand per customer. Curve D_2 represents these values. Note that the curves D_1, D_2 would be parallel if the scales were the same.

Adding specific values for other fuels along with their coefficients creates different demand curves. Demand curves for commercial and industrial sectors are generated in the same way. But what is important is the observation that demand represents a level of consumption that depends upon a variety of economic forces. As these forces change, so will demand. To properly understand demand, which suppliers constantly work at, we have to know what determines demand, and the relative importance of these forces.

Elasticities are an excellent way to determine relative importance. Calculation of elasticities using data over several time periods cannot use Equation 7.8 because this equation does not control for the effects of other economic forces on gas demand. The solution is to multiply the coefficients for each variable by the ratio of the means for Q and the independent variable. This is equivalent to the other elasticity equations.

$$E_P = \frac{\dfrac{Q_1-Q_2}{(Q_1+Q_2)/2}}{\dfrac{P_1-P_2}{(P_1+P_2)/R}} = \frac{Q_1-Q_2}{P_1-P_2}\frac{(P_1+P_2)/2}{(Q_1+Q_2)/2} \tag{7.9}$$

For a linear system like $Q_{NG}^{R} = 24.20 - (10.72)(P_G^{R})$, we can take partial derivatives.

$$Q/P = 10.72 = \frac{Q_1-Q_2}{P_1-P_2}$$

Since partials represent small first differences, Q/ P is equivalent to the first part of the elasticity equation. For the second part observe that this is nothing but the mean of P divided by the mean of Q, $\overline{P}/\overline{Q}$. Combining these gives

$$E_P = (\partial Q/\partial P)(\overline{P}/\overline{Q})$$
$$= (-10.72)(\overline{P}/\overline{Q})$$

If $\overline{P} = 1.35$, $\overline{Q} = 104.58$

$$= (10.72)(1.35/104.58)$$
$$= -0.138, \text{ gas demand is price inelastic}$$

With a mean value of income of 3,121 over the sample period

$$E_Y = (0.031)(3,121/104.58)$$
$$= (0.031)(29.94)$$
$$= 0.925, \text{ income inelastic}$$

A mean value for fuel oil price of 1.65 gives

$$E_{FO} = (4.02)(1.65/104.58)$$
$$= 0.063$$

The means and elasticities for all variables are summarized in Table 7.8.

191

Table 7.8
Variable Means and Elasticities

Variable	Mean	Equation Elasticities		
		Q_{NG}^R	Q_{NG}^C	Q_{NG}^I
P_G^C	1.35	-0.138		
P_G^C	1.05		-.576	
P_G^I	0.44			-0.016
P_E	10.13	0.998	0.592	0.235
P_{FO}	1.65	0.063	0.0	0.173
Y	3,121	0.925		
IIP	81.56		0.501	0.327
Q_{NG}^R	104.58			
Q_{NG}^C	441			
Q_{NG}^I				

Residential gas demands increase of 9.25% for every 10% increase in income, and 9.98% increase when electricity price rises 10% have the biggest impacts. Natural gas prices have the greatest impact on commercial demand, reducing it 5.76% for every 10% rise. Also observe that industrial demand responds very little to gas prices.

Changing gas prices in the three sectors affects each differently as was hypothesized. Gas pricing policies which raise prices across the board reduce commercial demand the most, have little impact on residential demand, and almost no effect on industrial demand.

SUPPLY-DEMAND INTERACTIONS

Establishment of supply and demand curves provides a means of analyzing the impact of a multitude of economic forces. What remains to be determined is the price that can be expected to exist. This is easy. Supply is a function of price, as is demand. Combining supply and demand curves yields the equilibrium price where these two curves interact. Figure 7.21 illustrates equilibrium price for natural gas supply and demand in the residential sector. Supply curve S_1 and demand curve D_1 intersect at price P_1. At this price Q_1 units of gas are supplied and consumed.

Equilibrium in a mineral market is simply the point where supply and demand interact. The equations for computing equilibrium are

$$Q_{NG}^D = f(P_{BG}, X)$$

$$Q_{NG}^S = f(P_{NG}, Z)$$

$$Q_{NG}^D = Q_{NG}^S = Q_1$$

The first two equations explain demand and supply, respectively. The equality $Q_{NG}^D = Q_{NG}^S$ states that price adjusts to bring quantity traded into equilibrium with supply. Value Q_1 is the equilibrium point in Figure 7.21

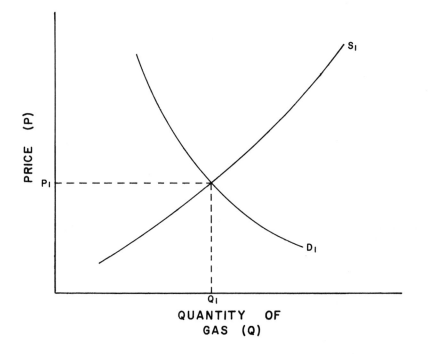

Figure 7.21 Equilibrium Price of Natural Gas

Finding the equilibrium quantity and price requires solving the three equations for three unknowns--Q_{NG}^D, Q_{NG}^S, and Q_1. Because of the equilibrium conditions equating supply and demand, this reduces to finding two unknowns P_1 and Q_1. Take the models

$$Q_{NG}^D = (\alpha_0 - \alpha_1)(P_{NG})(\alpha_2)(Y)$$

$$Q_{NG}^S = (\beta_0 + \beta_1)(P_{NG} - \beta_2)(Z - \beta_3)(C)$$

where C is the cost of producing and Z is another force reducing supply, the solution is

$$P_1 = \left[\frac{\alpha_0/\beta_0}{\alpha_1 + \beta_1}\right] + \left[\frac{\alpha_2}{\alpha_1 + \beta_1}\right](Y) + \left[\frac{\beta_2}{\alpha_1 + \beta_1}\right](Z) + \left[\frac{\beta_3}{\alpha_1 + \beta_1}\right](C)$$

193

$$Q_1 = \frac{\alpha_1\beta_0 + \alpha_0\beta_1}{\alpha_1 + \beta_1} + \left[\frac{\alpha_2\beta_1}{\alpha_1 + \beta_1}\right](Y) - \left[\frac{\alpha_1\beta_1}{\alpha_1 + \beta_1}\right](Z_t) - \left[\frac{\alpha_1\beta_3}{\alpha_1 + \beta_1}\right](C)$$

which is found by substituting one equation for another.

The equilibrium quantity and price of gas depends upon the values taken on by ·the exogeneous economic forces--Y, Z, and C, the coefficients. As the economic forces change value, so will the equilibrium point. To find the expected equilibrium values, we have to possess information about these values. If the model coefficients were

$$Q_{NG}^D = 24.4 - (10.3)(P_{NG}) + (0.031)(Y)$$

$$Q_{NG}^S = (100) + (40)(P_{NG}) - (0.10)(Z) - (32)(C)$$

and the values for the economic forces were Y = $4,000, Z - 100, and C = 500, the equilibrium quantities are

$$P_1 = \frac{24.4 - 100}{-10.3 + 40} + \frac{0.031}{-10.3 + 40}(4,000) + \frac{-0.10}{-10.3 + 40}(100) + \frac{-.32}{-10.3 + 40}(50)$$

$$= -2.54 + 4.17 - 0.33 - 0.538 = \$0.762$$

$$Q_1 = \frac{[(-10.3)(100)]+[(24.4)(40)]}{-10.3 + 40} + \frac{(0.031)(40)}{-10.3 + 40}(4,000) - \frac{(-10.3)(40)}{-10.3 + 40}(100)$$

$$- \frac{(-10.3)(0.32)}{-10.3 + 40}(50)$$

$$= -1.81 + 1.670 + 1,387 + 5.54$$

$$= 3,060 \text{ trillion Btu of gas per customer.}$$

These are the P_1, Q_1 values in Figure 7.21.

A common question is how does the effect of a change in any of these forces affect the equilibrium values. Calculate these impacts by incorporating the new values into the equations and simply recompute. For example, suppose income is expected to be $4,100 per capita next year, with no change expected in C and Z. Since only. income changes value, we find

$$P_2 = \frac{0.31}{-10.3 + 40}(4,100 - 4,000) + 0.762$$

$$= \$0.104 + 0.764 = \$0.868$$

$$Q_2 = \frac{(0.031)(40)}{-10.3 + 40}(4,100 - 4,000) + 3,060$$

$$= 5.177 + 3,060 = 3,065.2$$

that equilibrium price and income have risen by 10.4 cents and 5.177 trillion Btu/customer.

At the outset of this chapter we stated that simply tracing historical patterns of actual supply and demand of minerals inadequately described the causes. Observing the changes in price and quantity caused by income indicates why. Casual inspection of the change in this change would suggest that gas demand and gas price as positively related. But, in fact, even in a model where gas demand and price are inversely related, we observe a positive relationship. Graphical analysis of the type commonly practiced in many areas certainly distorts the real mineral economic relationships, and should be avoided in all mineral economic analysis when accurate results are desired.

Equilibrium has been assumed to always exist up to this point. The interaction of supply and demand, equilibrium, only occurs when producers and consumers bargain with each other fairly and openly. However, because any bargaining process takes time a market for minerals is unlikely to be exactly at the equilibrium position at all times. Every time an economic force deviates from a previous value, such as income, the equilibrium position changes. A certain amount of time must pass before complete adjustment is completed. Should we observe mineral price and quantity on the path to equilibrium, the false conclusion that the market is in equilibrium would be reached. Markets are seldom actually in equilibrium; they are always just moving toward equilibrium; often they are close enough to be treated as if they were in equilibrium.

Quite a different problem plagues economic analysis. That is the intervention of governmental agencies, nations (cartels), and firms banding together to control prices and quantities of minerals. The advantage of controlling prices and quantities is illustrated in Figure 7.22. Points P_1, Q_1, as usual, represent the equilibrium prices of \$0.764 and demand of 3,060, respectively. Now a governmental agency steps in and sets prices at \$0.40. Producers would supply 2,800 trillion Btu of gas and consumers would seek about \$3,500 trillion Btu's. Consumers would bid against each other until price was driven back to the equilibrium point.

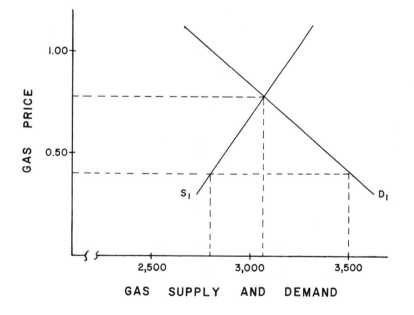

Figure 7.22 Effect of Gas Price Controls on
Gas Demand and Supply

Movement toward equilibrium prices does not occur in controlled markets. In controlled mineral markets we expect producers to determine the quantity of minerals actually sold since they will supply only so much, creating excess demand for gas. Gas price controls, unfortunately, also contained provisions guaranteeing that customers receive all of the gas they desire. Gas companies supply $3,500 trillion Btu instead of 2,800 trillion Btu. Consequences of such a policy are twofold. First, gas was consumed at a level that exceeded what producers could profitably supply. Poor profits made it necessary to supply the gas out of current gas inventories. Second, high demand combined with low profitability prevented economic replacement of depleted inventory, thus, reducing gas reserves for future customers. Government regulation, in essence, stated by such a policy that gas reserves for future customers was not as important as current gas needs. "The future is now" philosophy was being followed. Remember that these supply demand curves fully account for major economic and technical forces affecting supply and demand.

Equilibrium at the intersection between supply and demand defines the <u>market efficient</u> allocation of minerals and the efficient price. Any other combination of price and quantity is <u>inefficient</u>, regardless of supporters' justification. Economic efficiency, as such, provides a sound basis for evaluating policies, as shown in Chapter 10.

Efficiency plainly supports gas price decontrol. Adherence to efficiency is a two way street, however. Mineral companies historically supported policies like prorationing and import quotas for a variety of reasons. Critics of such programs pointed out that inefficiencies resulted from these programs. Curves S_1, D_1 in Figure 7.23 illustrate the effect of prorationing on market efficiency. Curve S_1 is vertical, representing that every well has a maximum production rate. Equilibrium occurs at a price of $4.50 and a quantity of production of 1,000 bbl/day. Now suppose the state steps in and interferes with the efficient price and quantity by restricting output to 80% of capacity.

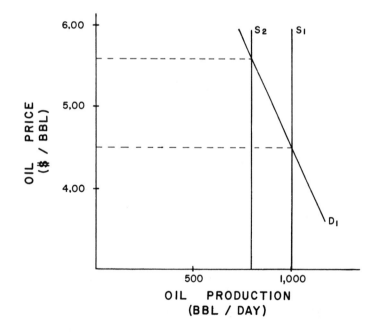

Figure 7.23 Effect of Proration on the Market Price
and Demand for Oil

196

Reducing output has the effect of raising prices to $5.62 per barrel, as the result of the downward sloping demand curve. Interestingly enough total revenue from the well is the same (4.50)(1,000) = 4,500 = ($5.62)(800), allowing the producer to argue that he is not making any more money than before. Consumers, furthermore, still spend $4,500 for the oil. Consumption is reduced by 200 barrels per day because of prorationing, and the reduced consumption, because of prorationing's interference in the market place, is what critics object to.

Valid reasons existed for prorationing (and import quotas), but, given our definition of economic efficiency, distortions in the economic market place occur. Like natural gas price controls, such distortions are viewed as being unfavorable by some members of society. Whether deviations from economic efficiency merit attention is difficult to determine easily because subjective evaluation of the benefits of national security, reduced or controlled inflation, preservation of adequate mineral inventories for future generations, is required. Supply-demand analysis tells us what the costs and benefits of moving away from market efficiency are likely to be. Professionals, other than technical personnel, have to decide when such deviations are in the national interest.

Example of Supply-Demand Analysis

Natural gas supply and demand patterns provide an excellent example of how to use this knowledge to analyze market efficiency because of past price controls that have insured that market efficient allocation of gas quantities and prices did not exist. Equations for gas demand in residential, commercial, and industrial sectors can be used to explain total gas demand. If we take as our objective the maintenance of some fixed reserve of gas for future use with depletion of gas inventory from demand being matched by new gas reserve additions, the determinants of gas reserve additions must be known.

One set of equations explaining gas reserve additions are presented in Chapter 10. For simplicity, we employ the following statistical equation to represent gas reserve additions.

$$Q_{NG}^S = (3,000) + 450(PW) + (882)(DV) - (1,270)(SR) - (5.41)(\$/FD)$$
$$\quad\quad (3.18) \quad (2.41) \quad\quad (1.91) \quad\quad (1.26) \quad\quad (2.74)$$

$$R^2 = 0.787$$

where Q_{NG}^S equals revisions plus extensions plus new pool wildcats measured in trillion cubic feet, PW is the wellhead price of gas, DV is a dummy variable indicating a significant new reservoir discovery, SR is the success rate, and $/FD is the cost per foot drilled.

Wellhead price captures expected revenue per unit volume from a commercial well, while cost per foot drilled accounts for the outflows incurred in drilling. The difference between receipts less disbursements represents gross revenue from a well. Rising receipts should lead to greater reserve additions since more drilling capital is available. Success rate is a proxy for the kind of drilling prospects being undertaken - higher success rates indicate that smaller, less risky deposits are being drilled, while lowering the success rate suggests riskier areas are being drilled in. Finally, DV is a variable taking on values of 0 or 1, with 1 entering when a significant find, like the North Slope of Alaska, is expected.

The sign of each coefficient is correct. Wellhead prices induce greater reserve additions as prices rise, thus the positive sign. Greater costs make drilling more expensive, so $/FD is negative; likewise, successive rate is negative. The elasticities for these variables are

$$E_{PW} = 0.497, \quad E_{\$/FD} = -1.38, \quad E_{SR} = -0.814$$

These elasticities explain the depressed state of natural gas drilling during the sixties and seventies. A ten percent price increase yields only a 4.97% increase in reserve additions. Should cost per foot drilled rise by 18% also, the net effect on reserve additions is a 9.83% = -13.8% - 4.97% reduction in reserve additions. Wellhead gas prices must rise more than costs in order to preserve gas inventories.

Gas inventories may also be preserved by considering the effect of higher wellhead gas prices on the demand for gas in the residential, commercial, and industrial sectors. The following identity represents the change in gas inventory, accounting for the variation in both new reserve additions and demand.

$$GR(t) = GR(t-1) + Q_G^S(t) - Q_G^D(t)$$

$$Q_G^D(t) = Q_G^R(t) + Q_G^C(t) + Q_G^I(t)$$

Gas inventory in year t (GR(t)) is the sum of gas reserves at the end of the previous year plus new additions and minus production. Reserve additions are explained by the equation for Q_G^S, and Q_G^D is the sum of the demand in the three sectors. Since gas reserves are known for the previous year, only changes in Q_G^S and Q_G^D have to be estimated from the respective statistical equations to find gas reserves in year t.

One other identity is necessary to complete the analysis. Demand in each sector was measured in million Btu's/customer, while GR is measured in trillion cubic feet. The units of measurement are incompatible. Gas demand is converted to trillion cubic feet by assuming 1,000 Btu's per cubic feet and calculating

$$Q_G^R(t) = (Q_G^R(t))(C^R(t))(0.001)$$

$$Q_G^C(t) = (Q_G^C(t))(C^C(t))(0.001)$$

$$Q_G^I(t) = (Q_G^I(t))(C^I(t))(0.001)$$

where C^R, C^C, C^I are customers in the residential, commercial, and industrial sectors in year t, respectively. The problem of transforming Btu's to cubic feet is solved by the assumption that one cubic foot equals 1000 Btu. Multiplying the number of customers by gas demand per customer gives total demand in the sector.

Finally, the last identity relates to how prices at the wellhead turn into prices facing consumers. There is no easy solution to this problem because of its complex nature. The transition from the wellhead to the consumer involves regulated pipelines and distributors whose rates to consumers are fixed by state regulatory agencies. Throwing in the long-term contracts existing between producers and suppliers makes it impossible to easily estimate the conversion of wellhead prices to retail prices, except by a detailed analysis of rate structures for each distributor. Instead, we use a simple multiple to obtain a relationship as done below.

$$P_G^R = (7.0)(PW)$$

$$P_G^C = (5.5)(PW)$$

$$P_G^I = (1.68)(PW)$$

Higher multiples for residential customers and the lowest multiple for industrial are consistent with historical pricing strategies.

Four statistical equations and eight identities are needed to estimate the wellhead price of natural gas which equates natural gas demand and reserve additions. The model is used to find PW such that $GR(t) - GR(t-1) = Q_G^S(t) - Q_G^D(t) = 0$. The easiest way to conduct the analysis is to take an actual occurrence. At the start of 1974 gas recoverable reserves were estimated to be 249,950 TCF. By the end of 1974 reserves had shrunk to 237,132 TCF, a decline of 12,818 TCF. Gas reserve additions in 1974 was 8,678 TCF, making gas consumption 21,496. Gas inventory obviously shrunk. Our objective is to find a wellhead price that would have changed the demand and supply patterns such that gas reserves remained at 249,950, TCF.

In 1974 residential, commercial and industrial customers consumed 15,310 TCF of the 21,496 TCF produced. The remaining consumption went into a variety of other categories, which we assume are fixed and are not affected by price changes. If demand and supply were unrelated, our problem could be solved by raising the wellhead price of gas until enough new gas reserves were created to equal 21,496 TCF. Elasticities are used to calculate the required price input. The percentage increase in gas reserves needed is

$$(21,496/8,678)(100) = 247.7\%$$

and gives a price elasticity of reserve additions of 0.497 and an actual average wellhead price of $0.2616, the wellhead price needed is

$$= (247.7/0.497)(0.2616)$$
$$= (498.3\%)(0.2616)$$
$$= \$1.307$$

The average wellhead price must sextuple to $1.307 per MCF in order to increase reserve additions to 21,496 TCF. Note this is not the price of just new field and new pool gas, but the price for all gas.

Wellhead gas prices don't have to use 400% because any increase in gas prices also raises the price paid by consumers, which reduces demand. The quickest approach to find the market clearing price is to solve this system of equations simultaneously, as shown before, on a computer. The harder, but more illustrative technique is to select a wellhead gas price and then calculate gas demand and supply. This trial-and-error procedure is repeated until they are equal.

Finding reduction in demand requires the information summarized in Table 7.9. Let's raise the wellhead price of gas to $.90, a $0.90/0.2616 = 344% increase. Reserve additions will increase to

$$\frac{(344\%)(0.497)}{100} (8678) = 14,836 \text{ TCF}$$

Multiplying 344%/100 by 0.497 (the supply elasticity estimate) yields the percent response of gas reserves. This multiplied by actual gas reserves yields gas reserves accumulated under the higher price.

Table 7.9

Data for Gas Supply-Demand Example

Variables	Sector		
	Residential	Commercial	Industrial
Customers (000)	40,670	3,392	193.8
Price Elasticity	-0.103	-0.576	-0.016
Price in 1974	$1.22	0.95	0.58
Consumption/Customer	119.6	676.1	42,005

Prices facing consumers are calculated by multiplying $0.90 by the values 7, 5, and 1.68, or

$$P_G^R = (7)(0.90) = \$6.30$$

$$P_G^C = (5)(0.90) = \$4.50$$

$$P_G^I = (1.68)(0.90) = \$1.51$$

The percentage increase in price for each consuming sector equals these prices divided by the actual price in Table 7.9 and multiplying by 100.

$$\% \ P_G^R = 6.30/1.22 = 516\%$$

$$\% \ P_G^C = 4.50/0.95 = 473\%$$

$$\% \ P_G^I = 1.51/0.58 = 260\%$$

The reduction in gas demand is the percentage increase in price multiplied by the price elasticity, which is multiplied by actual consumption.

Residential: $\frac{516}{100}$ (-.103)(119.6) = 63.2 MM BTU/customer

Commercial:

Industrial: $\frac{260}{100}$ (0.16)(42,066) = 17,499 MM BTU/customer

Commercial demand is zero because with a price elasticity of -0.576 and a price increase of 473%, demand is reduced over 100% (=0). Gas demand obviously has a minimum value of 0.

Gas demand at the higher price is the sum of the demand after these reductions are subtracted from actual demand times the number of customers.

$$Q_G^R = (119.6 - 63.2)(40,670) = 2,293 \text{ TCF}$$

$$Q_G^C = = 0$$

$$Q_G^I = (42,005 - 17,499)(193.8) = 4,749 \text{ TCF}$$

Explained Demand = 7,042 TCF
Unexplained Demand = 6,186 TCF
13,299 TCF

Conversion to TCF from millions of BTU is obtained by multiplying each value by 0.001. Residential gas demand is 2,293 TCF instead of 4,964, industrial gas demand declines to 4,749 from 8,153 TCF, while commercial demand goes to zero, giving total gas demand explained with the equation of 7,042 TCF. Demand unexplained in the example of 6,186 TCF added to 7,042 TCF gives a total demand of 13,229 TCF when the average wellhead price of gas is $.90.

With expected reserve additions of 14,836 TCF, gas inventories are expected to rise by 1,607 TCF. Our goal was to keep gas reserves constant at the 1973 level, however. To find the right price equating supply and demand, wellhead price could be lowered until they are equal. Alternatively, we could just set commercial prices so that commercial gas demand was 1,607 TCF, which would equate gas demand and supply. Either approach would achieve our objective.

In the political environment mineral industries face such a decision. Any gas pricing policy which reduces demand in one sector to zero certainly faces hues and cries of complaint from that sector facing undue economic burdens. But from an allocative efficiency criterion. economic analysis maintains that this is the optimal result. Offsetting the lesson of allocative efficiency is the concept of distributive efficiency discussed in the introduction of the chapter. Distributive efficiency concerns how the burden of a change in mineral policy impacts different groups. It may be concluded by political bodies or mineral companies that the full impacts of the gas price increase must be allocated in a manner different from that considered here.

Moving into the arena of dividing the burden to be placed on each group is like opening Pandora's box again. There is no easy way out. Economics suggests putting the greatest burden on those consumers with the largest price or income elasticity, since this implies that they can substitute other fuels easier. Political reality seldom allows such easy solutions. Concrete

201

recommendations as to how to specifically allocate higher prices are impossible for this reason. Nevertheless, for whatever pricing strategy undertaken, supply-demand analysis can identify the likely consequences on the major consuming groups. Chapter 10 discusses in greater detail some ways to evaluate the alternatives.

Before concluding this chapter it is important to note that the discussion of mineral supply and demand here applies basically to free market economics where price is the primary vehicle employed to achieve specific goals. Obviously, methods other than price are also used in such economics. The process can be converted to analyze economic forces in countries where the economic system relies upon techniques other than price to allocate mineral resources. All that is required is the selection of the replacement for price that a country's planners can use to allocate to do this. Individual consultation with mineral and governmental representatives is necessary in each case since objectives selected for mineral industries and controlling economic forces vary from country to country. Regardless of the structure of a country's internal economic framework, international transactions are subject to the laws of supply and demand where prices and costs dominate. So, even though supply-demand analysis may be assigned a different role in each country and each mineral company in a country, it is still the underlying force that governs the workings of the mineral industry now and in the future. We can ignore the forces of supply-demand, and we can try to change them in the short-run, but eventually we must pay the price.

SUMMARY

One last point merits discussion. In the example of gas demand and supply, price and quantity changes ranging between 100 to 500% were dealt with, yielding some hard to accept conclusions, like commercial gas demand falling to zero. Supply-demand analysis has to be considered in two ways. First, it represents principles equivalent to physical laws in the physical sciences, with the exception that they may be bent for a while, but never broken. These principles will assert themselves at some point. Second, these principles cannot be observed nor tested for in the laboratory. Statistical techniques, used to make inferences about these principles, must therefore be used. Empirical estimates of supply and demand relationships with other economic forces often misrepresent the actual relationship, as is the case of zero commercial gas demand. In spite of empirical limitations, all that can be inferred is that the empirical estimates are off. Too often mineral professionals use misleading estimates to justify avoiding supply-demand analysis. This is an attitude similar to those who scoffed at Newton's concept of gravity because they couldn't see, feel, or smell it.

Another cause of the reluctance to embark on a road of supply-demand analysis by some professionals is the feeling that they already possess such information through experience. We are gratified to hear such statements because this confirms the usage of supply-demand analysis in mineral economic analysis, even though it is informal. Such informal analysis is often acceptable within specific groups of a mineral company for selecting sites for drilling or development. But in a broader sphere where mineral companies must convince skeptical audiences that a proposed mineral policy is deserved or unwarranted because "my experience" or "my information" says so is seldom accepted when there is a group of consumers facing significantly higher prices

202

for minerals. They want and deserve evidence that such experience is valid and is applicable to current and future problems. Growing skepticism over mineral programs that began in the seventies requires translating experience into a framework that educates the non-mineral professional.

REFERENCES

1. <u>An Analytical Framework for Evaluating the Oil Impact Program</u>. Charles Riecus Associates, Inc., Cambridge, Mass., 1969.
2. Ferguson C. E., <u>Microeconomic Theory</u>. Homewood, Ill.: Richard D. Irwin, Inc., 1966.
3. Hicks, J. R., <u>Value and Capital</u>. Oxford, England: Oxford University Press, 1968.
4. Henderson, J. E., and R. E. Quandt, <u>Microeconomic Theory: A Mathematical Approach</u>, New York: McGraw-Hill, Inc., 1971.
5. Kalter, R. J., et al., Alternative Energy Leasing Strategies and Schedules for the Outer Continental Shelf, Cornell University, 1975.
6. Debanne, J. G., A Continental Oil Supply and Distribution Model, SPE #2680.
7. Root, P. J., et al., The Impact of Changes in the Intangible Drilling Costs and Depletion Allowance Provisions on Independent Oil Producers in Oklahoma and the Oklahoma Economy, Bureau of Business and Economic Research, University of Oklahoma, 1969.
8. National Petroleum Council, U. S. Energy Outlook: Oil and Gas Availability, 1973.
9. McAvoy, P., <u>Price Formation in Natural Gas Fields</u>, New Haven, Conn.: Yale University Press, 1962.
10. <u>Studies in Energy Tax Policy</u>, Gerard Brannon, ed., Cambridge, Mass.: Ballinger Publishing Co., 1975.
11. Lovejoy, W., and P. Homan, <u>Economic Aspects of Oil Conservation and Regulation</u>, Baltimore, Md.: Johns Hopkins University Press, 1967.
12. Adelman, M., <u>The World Petroleum Market</u>, Baltimore, Md.: Johns Hopkins University Press, 1972.
13. National Petroleum Council, Impact of New Technology on the U.S. Petroleum Industry, 1946-65, Washington, D.C.: National Petroleum Council, 1967.
14. Chazeau, M., and Kahn, A., <u>Integration and Competition in the Petroleum Industry</u>, Port Washington, N.Y.: Kennikat Press, 1959.
15. Zareski, G. W., The Long-Term Prospects for Conventional U.S. Gas Supply, SPE #6342, 1977.
16. Schanze, J. J., Forecasting Energy Futures: Facility or Futility?, SPE #6344, 1977.
17. Campbell, J., and R. Campbell, Effects of Natural Gas Pure Controls: Past, Present and Future, SPE #6343, 1977.
18. Leontief, W., <u>Input-Output Economics</u>, New York: Oxford University Press, 1966.
19. Attanasi, E. E., Firm Behavior and Microeconomic Determinants of Petroleum Explanation, Southern Economic Association Annual Convention, 1977.
20. Scott, L., and J. Richardson, A Reclusive Model for Long-Range Regional Oil and Gas Production Forecasts, Southern Economic Association Convention, 1977.

8

ASSESSMENT OF RISK AND UNCERTAINTY

The professional never deals with <u>absolutely certain</u> numbers; some numbers are merely more certain than others. The logical first step in any analysis or decision using these numbers is to establish the effect of any uncertainty on the conclusion being made. The underlying cause for many serious decision errors is the failure to properly consider the quality of the input information to the decision process. In the worst case the effect of uncertainty is completely ignored and all numbers are arbitrarily assumed to be correct, or ... some <u>arbitrary</u> safety factor is used to hopefully compensate for the unknown uncertainty.

This latter, very simple, approach may prove satisfactory for a similar series of small decisions, no single one of which is critical, if:

1. The method of analysis is consistent

2. The variables affecting the system are few in number and can be measured in at least a semi-quantitative manner. (They are not purely judgemental.)

3. A finite length of time occurs between a series of similar decisions and the feed-back from early decisions is analyzed by one very knowledgable about the physical system and its performance characteristics for use in later decisions.

4. The variables affecting the performance of a system remain essentially unchanged during the experience period, i.e., the "rules of the game" do not change.

The above "ifs" limit this approach to those very rare persons with many years experience plus a proven track record of success. But, even for these rare persons this traditional, informal approach proves less valid as time goes on.

First of all, the number of decisions is decreasing and each is more critical. Some involve an amount of money that compromises a company's cash asset position (liquidity). Much of the input is more uncertain than in the past because of the over-whelming control over mineral exploration, production and price by political entities. Such political actions seldom are consistent in time, or in philosophy. All of this adds up to the fact that the rules governing mineral economics are changing rapidly. Whatever the quality of the traditional judgmental process it must be supplemented by some <u>formal</u> appraisal of the risks imposed by uncertain information.

This chapter is a summary of basic principles that govern the effect of uncertainty on value. Space does not permit a thorough development of practice in the area. There are many articles and books on the subject. Book references 1, 2 and 5 at the end of this chapter should prove particularly useful for mineral industry professionals.

Anytime one solves an equation all that is achieved is to find an answer that represents the results from the numbers used. There is an answer for each possible value of each quantity used in the equation. In effect, one has to solve the equation many times to find an array of answers. These may then be analyzed (formally of informally) to arrive at the answer for decision purposes. Today, this is simply done on a computer. If the results of the multiple calculations are reviewed by experienced, knowledgable persons familiar with the physical system better decisions should result.

Unless a person possesses detailed training in the areas of probability and statistics there is a natural resistance to this approach. There are several reasons for this:

1. Tradition

2. The classical method of presentation

3. The sometimes seeming conflict between the philosophies of engineering and science and those of this area

4. Lack of real understanding about what these areas contribute to the input.

By virtue of training and experience the professional develops a series of cause-effect relationships. The engineer attempts to relate these to equations and correlations which, although not exact, provide the relationship of dependent and independent variables necessary to implement the judgment process. The geologist uses his theories about petroleum formation, evolution of the earth's crust and the patterns observed as his basis for judgment decisions. The physician uses his own cause-effect relationships in his particular diagnostic area. Thus the principle is somewhat the same in all professional areas even though the details vary. The basic relationships establish the logic employed. Each observation is a sample of the system under study.

In preparing his "laws," concepts and theories the professional is sampling and then using what is called the induction process of thinking - developing a relationship that "fits" his interpretation of the samples obtained. The enlightened professional is not unaware that his sample is incomplete and imperfect. In fact, a hypothesis only becomes a law by successfully passing the test of time and usage. As more samples (data) are obtained, existing laws are modified or discarded.

A professional's training (and ego needs) dictate that he overcome any imperfection in his system model by the application of personal judgment. The necessity for exercising a significant amount of personal judgment in arriving at a meaningful answer is the primary element that separates the professional from the non-professional. The necessity for good judgment is obvious and cannot be compromised. When properly applied, statistics and probability do not compromise the judgment process; they merely change the character of it. It is this change that apparently frustrates.

This frustration is compounded by the manner in which statistics and probability are usually presented. A simple system (in concept) is obscured by vague language and an inordinant amount of trivia. Persons who use the tools but do not specialize in the area, find it difficult to separate the "wheat" from the "chaff."

Much of statistics in concerned with sampling and testing of the samples to determine if the results are significant. This is important in some types of planning, survey activities, and the like, where one can control the amount and type of data obtained. In research, for example, such sampling tests are important.

In the economic world of business operations, however, the amount and type of data are seldom controlled by statistical needs. We are required to get answers whether the data are adequate or inadequate, good, bad, medium or indifferent. The amount of raw judgment required simply varies in each circumstance. We must simply use these formal tools so far as they are significant and go on as before.

In spite of the above "flaws" and frustrations, statistics and probability may supplement the judgment process in a powerful manner. A professional who ignores these tools is naive or maybe downright incompetent.

With this as a preamble, let us look at some of the basic ideas involved.

SAMPLING AND STATISTICS

Statistics is nothing more than the name given methods for collecting, summarizing, presenting and analyzing data. Any data possess a certain randomness. This is due to the imprecision of the data gathering devices and the non-uniformity of the system being examined. In logging a well, for example, to determine porosity and water saturation, one obtains a range of values. A portion of the variation is due to the response of the logging tool and its interpretation. Also, the porosity and water saturation truly vary at different points. In using data one must first understand the character of the data before he may use it to determine the character of the system. This is particularly important. If one does not understand the character of the data the analysis of the process is sterile and the results may be invalid. The problems encountered in the area of sampling and statistics are more nearly a result of the people using the methods rather than a weakness in the methods themselves.

> Note: The examples in this chapter use petroleum terminology but the same principles
> apply for all mineral reserves. All minerals use comparable exploration sampling
> methods in principle. Involved is determination of gross reserve size which is
> corrected in some manner to give net yield. The analogy should be obvious if one
> remembers that all exploration and development activities represent a process for
> sampling the earth's crust.

What is the most realistic way to use data in a reservoir model? Take a simple volumetric equation (Chapter 16). It is realistic to obtain a simple "average" porosity and water saturation and merely plug these in to find "average" hydrocarbons-in-place. Or, are the data so random that in doing this one introduces too much potential uncertainty? Does one want the most probable answer or the average answer? They are likely different. How likely is any one answer compared to other possible (and reasonable) answers? This is what statistics is all about.

Language of Statistics

Like all areas of expertise statistics has a language of its own. There are a few key words used that are defined briefly below.

Population (Sample Space): the total amount of data available for analysis, expressed as a whole finite number or assumed to be infinite in size.

Sample: a single value or a group of values within the total population of numbers obtained by test or by interpretation of test results.

Raw Data: data collected which have not been organized in some numerical fashion.

Arrays: an arrangement of raw numerical data in some ascending or descending order of magnitude.

Element: a numerical value existing in the sample space. It may be convenient to classify sample space according to the number of elements contained therein.

Variable: any value of a given set of elements - permeability, porosity, etc.

The population (sample space) does not refer to an area or volume. It is a listing of all of the possible things that have been measured or can happen. Consider a standard 52 card deck. The population is 52, made up of 52 elements (cards). In rolling a standard pair of dice the sample space is composed of 36 elements (with both dice having 6 sides there are 6^2 or 36 ways numbers may be obtained). These are examples of situations where the size of the sample space is fixed by easy observance or analysis of a fixed physical system. In mineral economics the sample space size must be developed by some combination of existing data and subjective professional input. The absolute size may only be surmised in some manner, it is never known absolutely.

The sample space can be finite or infinite. In the case of our deck of cards or dice there are a finite number of elements in our sample space. In inspecting equipment for defects we have no way of knowing how many pieces must be inspected before a defective one is found. It could be the first, the second, ..., the thousandth, ..., or even the millionth. In not knowing how far to go, the number of elements in this sample space is countably infinite. In tossing a coin the results (heads or tails) comprise an infinite sample space because the number of tosses is infinite.

The elements comprising the sample space can be discrete or continuous. These elements are the data points collected when analyzing a given system. If one is measuring the number of wells per reservoir one can only obtain whole numbers like 1, 2, 3, ... but cannot obtain 3.5 or 4.68. In this case the elements are discrete.

In measuring a set of porosites for a given petroluem reservoir any porosity within the range of numbers may be obtained. For this case, the elements are continuous. As a general rule, measurements result in continuous data and counting results in discrete data. In the mineral economics area we are normally concerned with finite sample space composed of continuous variables (elements).

Sampling

Sampling is the process of obtaining data on a given system to develop an understanding of its behavior. The most common purpose is to predict future behavior based on the assumption that sufficient samples have been obtained to define the systems sample space. However, sampling is sometimes done to answer questions like:

1. How many samples must you take to determine the average composition of a given fluid within a specified narrow margin of error?

2. Several independent tests are made on a system with different results. Is the difference significant or merely chance fluctuation?

3. What is the likely future variation of a product from a system based on the sampling to date?

4. Was the change in the output of a system a result of an operating change or merely a chance fluctuation?

Statistical analysis is suitably applied only to <u>independent measurements</u>. If 30 tests are applied to reservoir cores, 10 each on three cores, then there are only three independent results (one for each sample). The multiple tests provide useful information about the precision of the test method. Average values for the three tests would be used for statistical analysis.

Statistical tests based on sampling may be used to determine if a valid relationship exists between variables, the variation to be expected in repeated measurements and the role that <u>chance</u> may play in the variation of data. One of the standard applications is to determine the <u>central tendency</u> of a group of data. A measure of central tendency tells how realistic it is to use some average to represent a variable whose value varies (like porosity and water saturation). One may also use statistical techniques on a variable like permeability to at least imply what the permeability distribution might be in a reservoir.

Some of these applications are discussed later in this chapter. References 1 and 2 discuss these applications in detail.

At this point we wish to concentrate on those aspects of statistics that lay the groundwork for the simple use of probability in the determination of value. In obtaining well data we are in effect establishing the <u>boundaries</u> within which the value of a variable must fall. When we measure initial reservoir pressure we presume that future values must be at or below this pressure. If a detailed appraisal of the wellbore via logs and cores is made, the sample yields the limiting values of porosity and water saturation as well as distribution of values within said boundaries. It is presumed that any value used to analyze this well will fall within these boundaries. A <u>sample space</u> is defined that hopefully includes the possible future value of said variable in the system being examined.

The basic problem is to define the sample space adequately so that any future outcome will fall within the numerical values defining that space. If we have too few samples the data must be extended subjectively. From a practical viewpoint the decision maker is faced with the question of how much data he needs before risking large amounts of capital funds.

In later sections we will discuss sampling <u>with or without replacement</u>. Concern for the difference between these two is an important aspect of assessing risk.

<u>Frequency Distribution</u>

The next step after obtaining data by sampling is to analyze it in some manner. If the number of data points is not too cumbersome the first step is to array the <u>raw data</u> in some manner. In many cases the easiest approach is group or classify the data. An array of numbers is set up in classes. Each <u>class</u> represents a range of values. If the proper number of classes is chosen this grouped data will give the same practical result as handling each data point separately.

To illustrate this we will use an example based on a series of porosity measurements in a reservoir rock. (Porosity is the fraction of total rock volume consisting of voids capable of holding fluid.) The table below shows 20 values of porosity obtained on 20 different cores from a reservoir.

0.17, 0.16, 0.10, 0.14, 0.12, 0.18, 0.14, 0.16, 0.19, 0.15
0.12, 0.16, 0.17, 0.19, 0.15, 0.16, 0.11, 0.21, 0.15, 0.19

These 20 values of the raw data can be divided into class intervals as shown. The first class interval is 0.10-0.12 while the second is 0.12-0.14, etc. The fact that 0.12 is the upper boundary of the first class and the lower boundary of the second class causes no difficulty so long as one is consistent. Arbitrarily we have used the convention that the upper class boundary is <u>exclusive</u> (only numbers below that number are included in the class).

Class Interval	Members	No. of Members
0.10-0.12	0.10, 0.11	2
0.12-0.14	0.12, 0.12	2
0.14-0.16	0.14, 0.14, 0.15, 0.15, 0.15	5
0.16-0.18	0.16, 0.16, 0.16, 0.16, 0.17, 0.17	6
0.18-0.20	0.18, 0.19, 0.19, 0.19	4
0.20-0.22	0.21	1

It is not necessary to have equal class sizes but this is normally more desirable. The mid-point of the class interval is called the <u>class mark</u>.

The number of class intervals chosen is a matter of accuracy and convenience. The equation

$$G = 1 + 3.3 \log n \tag{8.1}$$

may be used, where: G = no. of class intervals used and n = total number of data points.

The <u>number of members</u> shown above is the <u>frequency</u> of the occurrence of a data point in each class interval.

What has now been accomplished is to reduce 20 raw data points to 6 group data points the value of which is represented by the class mark. If enough classes have been chosen, the behavior of these six classes will be representative of all twenty data points.

<u>Relative frequency</u>. The relative frequency of a class is the frequency of that class interval divided by the total frequency of all classes (total number of points). It is often expressed as a percentage.

<u>Cumulative frequency</u>. This is the sum of all observations possessing values less than the maximum (upper) class boundary in a given class interval - the sum of the frequencies of that class and all lower classes. A plot showing cumulative frequency vs. the sample values is called an <u>ogive</u>.

Relative cumulative frequency. This value for each class interval is the cumulative frequency for the class divided by the total frequency.

The table below summarizes these quantities for the 20 porosity values.

(1) Class Interval	(2) Frequency	(3) Relative Frequency	(4) Cumulative Frequency	(5) Relative Cumulative Frequency
0.10-0.12	2	0.10	2	0.10
0.12-0.14	2	0.10	4	0.20
0.14-0.16	5	0.25	9	0.45
0.16-0.18	6	0.30	15	0.75
0.18-0.20	4	0.20	19	0.95
0.20-0.22	1	0.05	20	1.00
Total	20	1.00	--	----

What do the values as summarized above reveal? Column (3) tells what percentage of the time a given class interval of values occurred. For example, the measured porosity was 0.14-0.16, 25% of the time. Column (5) says that the porosity was less than 0.16 in 45% of the measurements.

Histograms and Frequency Polygons

Once continuous data has been arrayed as shown it is useful to prepare a graphical summary. There are two common forms:

1. Histogram - a series of rectangles with a base width on the abscissa equal to the class interval an area proportional to class frequency; the centerline of the base is at the class mark.

2. Frequency polygon - a line graph of class frequency on the ordinate versus class mark on the abscissa.

Figures 8.1 and 8.2 represent the corresponding graphical version of the preceding table for our porosity example. The frequency polygon is superimposed on the histogram for convenience. The same type of polygon would have been obtained by merely drawing the class mark lines.

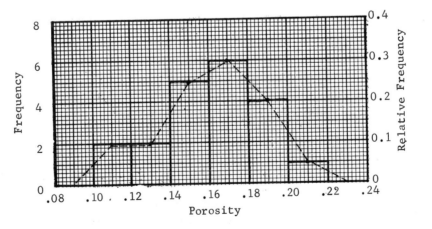

Fig. 8.1 Example of a Histogram and a
Frequency Polygon.

What may one read from these figures? In 8.1, for example, 30% (0.3) of the porosity values were between 0.16 and 0.18. From 6.2, 75% of the porosity values were <u>less than</u> 0.18. The relative cumulative frequency curve is a <u>less than</u> value. It would be possible to replot it as a <u>more than</u> value. In this case the lowest value in a class interval would be used instead of the highest value, as shown, in plotting the cumulative frequency curve.

Notice in Fig. 8.1 that the polygon has been extended outside of the histogram to porosity values of 0.09 and 0.23. This is customary. It merely assumes that the adjoining class marks possess a zero frequency.

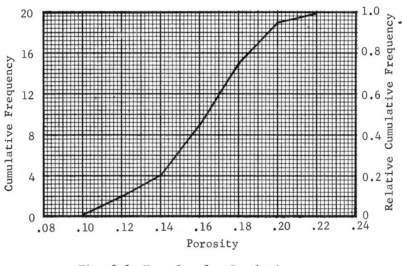

Fig. 8.2 Example of a Cumulative
Frequency Polygon

The polygon in Fig. 8.1 is a <u>distribution curve</u>. It does more than merely correlate data. It tells us the likelihood of the porosity that would be obtained from any future measurements on the system for which the curve was obtained, if the first 20 porosities constituted a <u>valid sample</u> and the factors affecting porosity did not vary with elapsed time. From the curve, the 21st porosity would have a 30% <u>chance</u> of being between 0.16 and 0.18; it would likewise have a 75% <u>chance</u> of being less than 0.18. So ... the distribution curve serves as the basis for the likelihood of occurrence of possible values of the random variable being considered. Stated another way, the variable being analyzed may <u>occur at random</u> within the boundaries of the distribution curve.

Notice that in Fig. 8.2 the curve is drawn through the upper value of the class interval. This was arbitrary. Some would prefer to draw the curve through the class mark. It makes no practical difference which is done.

This is <u>stochastic variation</u>; each observation possesses a random value. Successive observations are not related by a <u>known</u> fixed rule or relationship; only the sample space into which they should occur is known. This is the typical case for reservoir test variables, reservoir sizes, production performances, etc. Thus, the <u>frequency distribution</u> which tells us what fraction of the observations were in a given class interval furnishes the basis for a <u>probability distribution</u> that provides the chance (likelihood) that any future observation will fall in a given class interval, for the same system.

What has been accomplished is to develop a distribution curve which defines a sample space. The shape of this distribution curve will vary with the data. If insufficient data are available to develop a distribution curve one must use a standard curve or subjective curves discussed in later sections.

TYPES OF DISTRIBUTIONS

Figure 8.3 shows common types of frequency (probability) distribution curves.

Types (a), (b) and (c) are the most common encountered. A symmetrical curve is often encountered. It is normally assumed in determining random errors that occur in any given measurement.

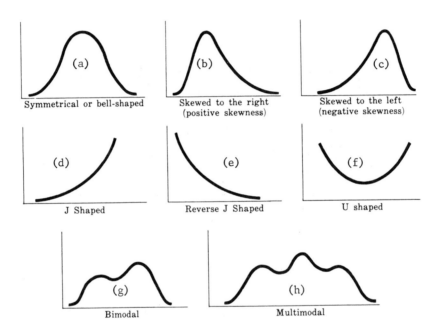

Fig. 8.3 Common Types of Frequency Curves (Ref. 3)

A skewed curve is common in measuring data of the type obtained from naturally occurring systems. The "tail" on one side of the maximum point is longer than on the other side. This is a natural shape when preparing a distribution for a commercial formation or field since there is a lower cut-off point that determines commerciality.

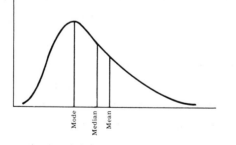

There are three types of "averages" that may be defined from a distribution curve - mean, median and mode. With a symmetrical curve these values are identical in value. For a skewed curve they are different. For a moderately skewed curve, like (b) or (c) in Fig. 8.3, a rule of thumb that may be used is

$$\text{Mean-Mode} = 3(\text{mean-median}) \qquad (8.2)$$

Fig. 8.4 Location of Mean, Median and Mode

on Skewed Frequency Curve (Ref. 3)

Mean

The word mean is a common contraction for the words arithmetic mean. Other central tendency measures such as geometric mean and harmonic mean will be discussed later.

The arithmetic mean is the expected value of the random variable in the distribution. It is simply the sum of the sample values in the population divided by the total number of samples.

$$\mu = \frac{\Sigma n_i x_i}{\Sigma n_i} \qquad (8.3)$$

where μ = arithmetic mean

n_i = the number of values (or the frequency of values in each class interval)

x_i = each individual value (or class mark when n_i is the frequency in each class interval)

Example: Calculate the arithmetic mean for the 20 porosity values shown previously using both the class values and the individual values.

$$\mu = \frac{1}{20} (0.17 + 0.16 +).10 + 0.14 + 0.12 + 0.18 + 0.14 + 0.16 +$$
$$0.19 + 0.15 + 0.12 + 0.16 + 0.17 + 0.19 + 0.15 + 0.16 +$$
$$0.11 + 0.21 + 0.15 + 0.19) = 3.12/20 = \underline{0.156}$$

(1) Class	(2) Class Mark	(3) Frequency	(4) Relative Frequency	(5) (2)x(3)	(6) (2)x(4)
0.10-0.12	0.11	2	0.10	0.22	0.011
0.12-0.14	0.13	2	0.10	0.26	0.013
0.14-0.16	0.15	5	0.25	0.75	0.0375
0.16-0.18	0.17	6	0.30	1.02	0.051
0.18-0.20	0.19	4	0.20	0.76	0.038
0.20-0.22	0.21	1	0.05	0.21	0.0105
Total	----	20	1	3.22	0.1610

$$\mu = 3.22/20 = \underline{0.161}$$

Column (6) in the above table shows that Equ. 8.3 may be used also with n_i being defined as relative frequency. The mean (μ) = Σ Col. (6)/Σ Col. (4) = 0.1610.

In most cases, when at least 8 class intervals are used, the different approaches will give results which are very close.

Median

The median is the value in a set of numbers which occurs at the mid-point of that set; the value of the median is exceeded by one-half of the values in the population of values. For our 20 porosities the mean is 0.160. If one arrays the numbers in order of size there are 9 numbers of 15 or less and 7 numbers of 17 or more. One-half of the numbers (10) occur in the four discrete values of 0.16 for porosity. Thus the median is about 0.16.

Mode

The value which occurs with the greatest frequency is the mode. On a continuous distribution curve it is the value of the peak. For grouped data it would be the class mark of the class interval with the highest frequency. For our porosity example the mode is 0.167 for the grouped data and 0.160 for the raw data. The value of 0.167 is the average of the 6 values in the most frequent class interval. If the class mark has been used, the mode would be 0.17.

Notice that in our example the mean, median and mode are very close to being equal. This is the mark of a fairly symmetrical distribution curve. Fig. 8.1 is not completely symmetrical even though it possesses general symmetry.

PROBABILITY

Probability is the word used to describe what the likelihood is of the occurrence of a future event. In numerical form it is expressed as a fraction from 0-1.0 or as a percentage from 0-100. If the probability is 0, the event cannot occur; if the probability is 1.0 the event is bound to occur. In real world systems, probabilities are never 0 or 1.0, they merely approach these as mathematical limits.

Before applying probability concepts it is important to define some of the rules and definitions governing the principles of behavior. First of all, there are three basic definitions of probability.

Relative Frequency (Empirical). - This form results from a statistical analysis of the type shown in preceding sections or is found from complete knowledge of the physical system. It is the ratio of the number of elements defined as an event divided by the total number of elements in the sample space.

What is an event? An event is any happening that can occur. We sometimes call this an outcome.

If you roll a fair, single die the number that comes up is the event (outcome). The total possible, equally likely, outcomes are six in number (one for each side of the die). So, the probability of any one number (event) occuring is 1/6 = 0.167, or 16.7%.

In tossing a fair coin there are only two possible outcomes (heads or tails). The probability of obtaining either is exactly 0.5 since the coin has only two sides. Likewise, in a deck of 52 cards, there are 52 outcomes. On the initial random draw of a card the probability of drawing a card from any one suit is 13/52 = 0.25; the probability of drawing any particular card is 1/52 = 0.019. The above are examples where the number of elements in the sample space are known absolutely. One can cite similar examples for all gambling games - roulette, slot machines, baccarat, keno, etc.

The previous porosity example is a practical minerals example of relative frequency probability. Figure 8.2 is an empirical curve from 20 measurements which we presume defines the sample space. If it indeed does, we can use it to predict the likelihood of a future event. Suppose the question is asked, "What is the probability of the porosity being less than 0.18 in the 21st sample? The answer, from Figure 8.2 is 0.75 (found by entering the abscissa at 0.18, proceeding vertically to the curve, and reading the result on the relative cumulative frequency scale). In order for this probability number to be valid we have assumed that 20 samples was sufficient to define the range of porosities that could feasibly occur in subject system. How does one determine how many samples are needed? This to some extent is subjective early in the process. However, at some point one finds that further samples do not change the position of the frequency distribution in any significant manner. At this point in time it is usually assumed that sufficient samples have been taken to define the system characteristics (sample space).

This is the same principle involved in exploring for minerals. Exploratory holes are drilled to find the extent and/or quality of the mineral reserve. How many such samples are needed? This is beyond the scope of this book but Reference 1 discusses it in detail.

Objective (Classical). - This form is obtained by logic. The word objective signifies that said logic is based on some professional basis and not on mere bias or unsupported opinion. This type of probability is a measure of the degree to which available information supports a given assumption.

This is the type of probability used often by the explorationist prior to developing hard data on a prospect. By mapping and other correlative techniques a conclusion can be drawn about the "odds" of a certain event being true.

Subjective. - In a sense it is impossible for a professional to be unbiased and purely objective. So objective probabilities are more of an ideal goal than a reality. A purely subjective probability is what the professional "thinks" it is without qualification of the source of that opinion. It is probably some combination of logic, biases, hunches and maybe a wee dab of wishful thinking.

In some mineral economics situations we do not have enough samples, so some degree of subjectivity must be used to develop decision criteria.

Sampling the Sample Space

Once a distribution has been established for a variable, regardless of how it is done, a sample space has been established. By a random draw from that space one can determine the probability of any further event occurring. Each draw is a sample of that space.

There are two ways one can sample - sampling with replacement and sampling without replacement. These are illustrated by the examples which follow, using the porosity example discussed previously.

> Example. 100 slips of paper are prepared. On each is written the class mark
> of each of the six classes used. The number of slips containing each
> class mark is in proportion to the relative frequency shown in Figure
> 8.1: 0.11-10 slips (10%), 0.13-10 slips (10%), 0.15-25 slips (25%),
> etc., making a total of 100 slips. These slips are then randomly
> mixed in a container before beginning to draw then one at a time.

Sampling With Replacement

> 1. On the first draw each of the 100 slips has an equal chance of being
>   drawn. The probability of any one number being drawn is the number
>   of slips of that number divided by the total slips (number of times
>   that event can occur ÷ total number of equally likely possible out-

comes. Even though each slip has an equal chance of being drawn there is only a 0.1 probability that 0.11 or 0.13 will be drawn.

2. Before making the second draw the first slip is replaced and all are randomly mixed again. For the second draw each slip has an equal chance of being drawn and the probability of obtaining any one number is the same as the first draw.

3. With replacement of the drawn slip before each subsequent draw, the "odds" remain the same whether it is the fourth, tenth, one thousandth draw, etc.

The probability (odds) of drawing say 0.11 remain at 1 in 10 (10%). Does this mean that one slip with the number 0.11 will absolutely be drawn when making ten draws? No! The number of such slips could be any number from 0 to 10 when making ten draws. We will expand on this later. For the time being let us simply say that there have to be a sufficient number of draws so that the probability number represents the relative occurrence of a specific event. As the number of draws increases the frequency of drawing a given number converges on (approaches) the frequency indicated by the distribution curve developed. With an infinite number of draws the results and the curve would coincide exactly.

Sampling Without Replacement

1. On the first draw each of the 100 slips has an equal chance of being drawn. Once again the probability of drawing any one number is the number of slips of that number divided by the total number of slips.

2. The first slip drawn is not replaced. Now there are only 99 slips left. Each of these remaining slips still has an equal chance of being drawn. But, the probability of drawing any one number has changed. Suppose the first number drawn was 0.13. There are now only 9 slips with that number left. What is the probability of drawing a 0.13 on the second draw? It is 9/99 or 0.091. What is the probability of drawing 0.11? It is 10/99 or 0.101.

3. Suppose 50 draws have been made without replacement and all ten slips for 0.11 have been drawn. What is the probability of drawing a 0.11 on the last 50 draws? It is zero for we know there are no slips left.

The important thing to remember is that the probability of any event changes when no replacement occurs. This is important. Most mineral evaluations are an exercise in sampling without replacement. Any distribution used must reflect this limitation in using a distribution. This is discussed briefly in later sections and rather thoroughly in References 1 and 5.

Regardless of whether you draw with or without replacement one thing is certain. You can predict the <u>chance</u> of an event occuring on any single draw but that does not fix the <u>real outcome</u> of that single event. The probabilities only govern overall performance on repeated events occurring in the same sample space (assuming of course that it has been properly defined).

It is impractical to apply probability principles using paraphernalia like slips of paper, marbles, etc. A <u>random number</u> table may be used to enable us to simulate random drawing, arithmetically, as in a later section.

Each of the numerical examples used to this point involved only <u>independent events</u>. Each porosity value does not depend on another in any known or presumed manner.

Types of Events

There are three types of events by definition.

<u>Independent Events</u>: All events (outcomes) are independent if the occurrence of any one event has no effect on the occurrence of other events. Suppose one is sampling a train load of coal to determine if it meets specifications. Each measurement of ash content is independent of all other measurements from different samples.

<u>Equally Likely Events</u>: Two or more events have an equal probability of occurring (0.50). An example of this is flipping a coin. On any one flip of a fair coin there is an equal chance of obtaining heads or tails. Each flip is also an independent event because a coin has no memory. The current flip is unaffected by the results of previous flips.

<u>Mutually Exclusive Events</u>: The occurrence of one event prevents the occurrence of all other possible events. A good example of this is drilling or coring. If no commercial reserves are found (dry hole) there is no chance for that location to have any economic value.

Since both mutually exclusive and independent events are important in mineral risk calculations, it is important to keep their difference clearly in mind.

Two variables in a system may be dependent to some degree on each other. If Y is dependent to some degree on X then the probability of an event in Y occurring depends on an event occurring with X. <u>Conditional probability</u> is the probability of a given event given that some other event has already occurred.

It is important to distinguish clearly between independent and dependent events because it affects risk analysis procedures. We will discuss this briefly in later sections but once again refer to References 1 and 5 for details.

A qualified professional usually knows whether the variables involved are dependent to some degree. There are statistical measures of correlation (see Correlation Coefficient section) to supplement this judgment. If in doubt one can make a cross-plot of the variables to see if they show any dependency. An example of this is shown in Figure 8.5.

Part (a) is for a plot of two random variables A and B. There is no apparent dependency. It would be safe to assume that variables A and B are independent.

Part (b) shows random variables C and D. They correlate quite well on a single curve. Variable D is an <u>explicit function</u> of C. The relationship shown could be represented quantitatively by a simple equation. It should be obvious that the value of D is dependent on the corresponding value of C. They therefore cannot be treated as independent variables.

217

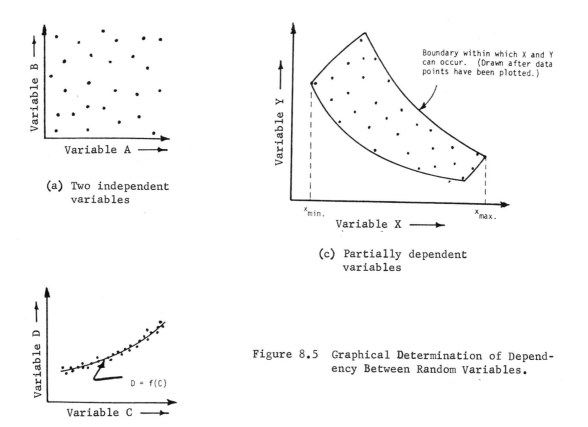

(a) Two independent variables

(c) Partially dependent variables

Boundary within which X and Y can occur. (Drawn after data points have been plotted.)

$x_{min.}$ $x_{max.}$

D = f(C)

(b) Completely dependent variables

Figure 8.5 Graphical Determination of Dependency Between Random Variables.

Part (c) is a plot of variables Y and X. There is no explicit relationship between them yet they are somewhat dependent on each other since there is a definite trend in values. Possible values of Y change with changing values of X. Thus Y is partially dependent on X. The important thing from a risk analysis viewpoint is that Y and X are not independent variables and cannot be treated as such. One method of handling such dependency is shown in References 1 and 18.

In the petroleum area there are many cases where at least partial dependency exists between common variables. A few examples include:

• reservoir porosity and permeability

• reservoir porosity and water saturation

• oil density and molecular weight

• oil density and viscosity

• formation thickness and productivity.

Similar dependent variables occur in all mineral areas. In a geothermal reservoir, for example, rate of heat deliverability per unit of time is dependent on rock permeability. Coal and oil shale would exhibit comparable dependency among variables.

Theoretically, it may be true that all of the variables affecting the characteristics of the earth's crust are dependent to some degree. As a practical matter, if we can discern no apparent dependency we assume they are independent for analysis purposes.

Basic Probability Theorems

There are two basic theorems governing probability operations. Each of these theorems can be expressed in terms of two events, A and B.

Addition Theorem. - This theorem is written as

$$P(A+B) = P(A) + P(B) - P(AB) \qquad (8.4)$$

where: P represents a probability

A is one event, B is another event

(A+B) signifies A and/or B can occur

(AB) signifies that both A and B occur

So, Equation 8.4 can be read as, "The probability of event A and/or B occurring equals the probability of A plus the probability of B minus the probability of both A and B occurring."

> Example. A box contains 100 balls each bearing a single number from 1 to 100. There is one ball for each number. What is the probability of drawing a 10, 20 or 30 (Event A) or a 70 or 80 (Event B)? From Equation 8.4,
>
> $$P(A+B) = (3/100) + 2/100 - 0 = \underline{0.05}$$
>
> The term P(AB) = 0 because a single ball cannot contain two numbers. In any draw like this P(AB) only comes into play if Event A and Event B contain one or more elements in common (share a common sample space.

Multiplication Theorem. - This theorem is written as

$$P(AB) = P(B|A) \, P(A) \qquad (8.5)$$

where: P(AB) is the probability of two events (A and B) occurring in sequence or simultaneously

P(B|A) is the conditional probability of B occurring given that A has already occurred

P(A) the probability of A occurring in the first place

> Example. Use our box of 100 numbered balls again. What is the probability of drawing a 10, 20 or 20 on the first draw and after replacing that ball, draw a 50 on the next draw? The chance of drawing a 10, 20, or 30 (Event A) is 3/100. The chance of then drawing a 50 (Event B) is 1/100. From Equation 8.5,
>
> $$P(AB) = \left(\frac{1}{100}\right)\left(\frac{3}{100}\right) = \frac{3}{10,000} = \underline{0.0003}$$

These two theorems are cited specifically because they are often used in mineral risk analysis.

Probability-Distribution Relationship

As noted previously, probability is always a positive number expressed as a fraction between 0 and 1.0 (or as a percentage between 0 and 100). Regardless of how the distribution curve is obtained, or its shape the area enclosed by it represents a probability of 1.0. This is the sample space. There is a 100% chance that any value of the variable will occur within that area (if the distribution is correct).

The sketches which follow illustrate the principle. These are frequency curves of the type shown in Figure 8.1. The probability that variable x is less than x_{min} is essentially zero. The probability that x is less than x_{max} is 1.0. The probability that x will be some intermediate value between 0 and 1.0 depends on relative areas.

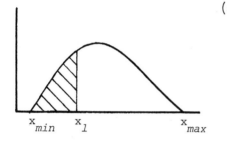

(a) The probability that the variable x will have a value equal to or less than x_1 is the cross-hatched area at left as a fraction of the total area under the curve.

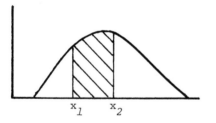

(b) The probability that variable x will have a value between x_1 and x_2 is the cross-hatched area shown at left, as a fraction of the total area under the curve.

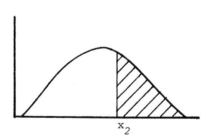

(c) The probability that the variable x will have a value equal to or greater than x_2 is equal to the cross-hatched area at left, as a fraction of the total area under the curve.

Figure 8.6 Schematic of Areas Which
            are Equivalent to Probability

If one knows the equation of the distribution curve these quantities may be found by integrating the equation of this curve between appropriate limits.

The probability that variable x will have exactly the value x_1 is zero, because a single line has no finite area.

Now that we have briefly summarized some of the basic rules and principles of probability let us look at some of the tools used.

STANDARD DISTRIBUTIONS

In addition to an empirical distribution curve illustrated with our porosity examples, there are some "standard" distributions. We will discuss several which are used commonly in the minerals industry.

1. Normal

2. Log Normal

3. Binomial (Bernoulli)

4. Hypergeometric

5. Subjective
   a. Triangular
   b. Rectangular

Each of these standard forms are represented by equations which facilitate probability calculations. Where possible, data are fitted to one of these standard forms. In some cases where data are limited one or more of these forms is chosen as a judgment decision.

NORMAL DISTRIBUTION

This is a symmetrical curve. The mean, median and mode shown in Figure 8.4 are coincident (at the mode).

The equation of this curve is

$$y = \frac{1}{\sigma\sqrt{2\pi}}\, e^{-\frac{1}{2}(x-\mu)^2/\sigma^2} \tag{8.6}$$

or

$$y = \frac{1}{\sigma\sqrt{2\pi}}\, e^{-\frac{1}{2}z^2} \tag{8.6a}$$

where: y = probability or frequency function

x = random variable

σ = standard deviation of x

μ = arithmetic mean value of x

π = 3.1416

z = $(x-\mu)/\sigma$, the standardized (dimensionless) variable

e = 2.71828

A normal curve as shown in the following figure possesses the same mean, median and mode.

An examination of Equ. 8.6 shows that the shape of the curve depends on the value of σ. The standard deviation σ is an indication of dispersion. If it is small the curve is thinner and taller. The larger the value, the broader and lower the curve. As the mean μ varies, the curve is merely shifted laterally; the shape of the curve does not change since μ is a measure of central tendency.

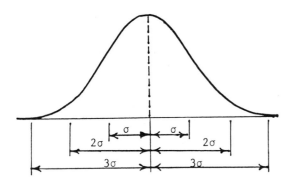

For a normal curve
 68.27% of the area lies within
  one σ limit of the mean.
 95.45% of the area lies within
  two σ limits of the mean.
 98.75% of the area lies within
  three σ limits of the mean.

Area possesses the same significance with a normal curve as any other distribution curve. The probability of occurrence of any value of x, within ±σ of the mean, is 68.27%.

The standardized variable z is convenient. One curve may be drawn to represent all normal curves. Variable x has been replaced by z.

Standard Deviation

The value is found from the equation

$$\sigma = \left[(1/n)\Sigma x^2 - \mu^2\right]^{0.5} \tag{8.7}$$

where: n = no. of values of x involved

x = values of the variable x

The value of σ found is used in Equ. 8.6. The value of μ for the same data is found from Equ. 8.3. These two values then fix the shape and position of the normal curve.

Example. The data shown below have been obtained from 42 successful wildcat wells drilled in a given area. Calculate μ and σ.

Bbl. oil/day	Bbl. oil/day	Bbl. oil/day
20	116	411
21	120	480
31	152	482
35	159	532
43	162	576
58	169	589
66	176	671
67	202	690
72	222	740
73	227	747
73	242	817
102	342	849
107	362	856
113	397	1,139

Class interval	Class mark	No. of wells Class frequency
0- 199	100	21
200- 399	300	7
400- 599	500	6
600- 799	700	4
800- 999	900	3
1,000-1,199	1,100	1
		42

For raw data

$$\sigma = \left[(1/42)(20^2 + 21^2 \ldots + 1{,}139^2) - (321.6)^2 \right]^{0.5}$$

$$= \underline{290.1 \text{ bbl. oil per day}}$$

For grouped data

$$\mu = \left[(100)(21) + \ldots + (1100)(1) \right] / (21 + \ldots + 1)$$

$$= \underline{328.6 \text{ bbl. oil per day}}$$

$$\sigma = \left[(1/42)(21 \times 100^2 + \ldots + 1 \times 1100^2) - (328.6)^2 \right]^{0.5}$$

$$= \underline{284.8 \text{ bbl. oil per day}}$$

Probability Calculation

The <u>cumulative probability</u> (frequency) curve is an s-shape on coordinate paper. It is a straight line on special <u>normal probability</u> paper. Remember ... cumulative probability is normally plotted as "percent less than." For any value of x or z the cumulative probability is found by determining the area from the left side of the normal curve to the value of x or z involved.

Table 8.1 is useful for this purpose. It contains values for relative area under the normal curve (total area = 1.0) for values of z from -3.5 to 3.509. This covers any practical area of concern.

The general procedure is as follows for a set of data for which the normal curve applies:

1. Compute a value of μ from Equ. 8.3.

2. Compute a value of σ from Equ. 8.7.

3. Compute a value of z for each value of x available using μ and σ from Steps (1) and (2)

4. Calculate the probability of any value of z or any range of values of z using Table 8.1

5. Plot the values of x vs. cumulative probability (% less than) on normal probability paper.

Example: A group of porosity data have been obtained on a well. The values of $\mu = 0.16$ and $\sigma = 0.02$, have been obtained by calculation. Determine: (a) the probability that porosity will be between 0.14 and 0.18, and (b) the probability that it will be less than 0.14.

(a) $z_1 = (0.14 - 0.16)/0.02 = \underline{-1.0}$

$z_2 = (0.18 - 0.16)/0.02 = \underline{+1.0}$

From Table 16.1, area for z_1 = <u>.1587</u>, area for z_2 = <u>.8413</u>. Area between z_1 and z_2 = <u>0.6826</u>. This is the probability for (a).

(b) The probability for porosities less than 0.14 is <u>0.1587</u>, the area to the left of z_1.

The probability that the porosity will be greater than 0.14 is equal to 1 - 0.1587 = 0.8413, since total area under the curve equals unity.

Table 8.1 Area Under Normal Curve From z = 0 to Any Value of z. (Ref. 17)

z	.00	.01	.02	.03	.04	.05	.06	.07	.08	.09
+0.0	.5000	.5040	.5080	.5120	.5160	.5199	.5239	.5279	.5319	.5359
+0.1	.5398	.5438	.5478	.5517	.5557	.5596	.5636	.5675	.5714	.5753
+0.2	.5793	.5832	.5871	.5910	.5948	.5987	.6026	.6064	.6103	.6141
+0.3	.6179	.6217	.6255	.6293	.6331	.6368	.6406	.6443	.6480	.6517
+0.4	.6554	.6591	.6628	.6664	.6700	.6736	.6772	.6808	.6844	.6879
+0.5	.6915	.6950	.6985	.7019	.7054	.7088	.7123	.7157	.7190	.7224
+0.6	.7257	.7291	.7324	.7357	.7389	.7422	.7454	.7486	.7517	.7549
+0.7	.7580	.7611	.7642	.7673	.7704	.7734	.7764	.7794	.7823	.7852
+0.8	.7881	.7910	.7939	.7967	.7995	.8023	.8051	.8078	.8106	.8133
+0.9	.8159	.8186	.8212	.8238	.8264	.8289	.8315	.8340	.8365	.8389
+1.0	.8413	.8438	.8461	.8485	.8508	.8531	.8554	.8577	.8599	.8621
+1.1	.8643	.8665	.8686	.8708	.8729	.8749	.8770	.8790	.8810	.8830
+1.2	.8849	.8869	.8888	.8907	.8925	.8944	.8962	.8980	.8997	.9015
+1.3	.9032	.9049	.9066	.9082	.9099	.9115	.9131	.9147	.9162	.9177
+1.4	.9192	.9207	.9222	.9236	.9251	.9265	.9279	.9292	.9306	.9319
+1.5	.9332	.9345	.9357	.9370	.9382	.9394	.9406	.9418	.9429	.9441
+1.6	.9452	.9463	.9474	.9484	.9495	.9505	.9515	.9525	.9535	.9545
+1.7	.9554	.9564	.9573	.9582	.9591	.9599	.9608	.9616	.9625	.9633
+1.8	.9641	.9649	.9656	.9664	.9671	.9678	.9686	.9693	.9699	.9706
+1.9	.9713	.9719	.9726	.9732	.9738	.9744	.9750	.9756	.9761	.9767
+2.0	.9772	.9778	.9783	.9788	.9793	.9798	.9803	.9808	.9812	.9817
+2.1	.9821	.9826	.9830	.9834	.9838	.9842	.9846	.9850	.9854	.9857
+2.2	.9861	.9864	.9868	.9871	.9875	.9878	.9881	.9884	.9887	.9890
+2.3	.9893	.9896	.9898	.9901	.9904	.9906	.9909	.9911	.9913	.9916
+2.4	.9918	.9920	.9922	.9925	.9927	.9929	.9931	.9932	.9934	.9936
+2.5	.9938	.9940	.9941	.9943	.9945	.9946	.9948	.9949	.9951	.9952
+2.6	.9953	.9955	.9956	.9957	.9959	.9960	.9961	.9962	.9963	.9964
+2.7	.9965	.9966	.9967	.9968	.9969	.9970	.9971	.9972	.9973	.9974
+2.8	.9974	.9975	.9976	.9977	.9977	.9978	.9979	.9979	.9980	.9981
+2.9	.9981	.9982	.9982	.9983	.9984	.9984	.9985	.9985	.9986	.9986
+3.0	.9987	.9987	.9987	.9988	.9988	.9989	.9989	.9989	.9990	.9990
+3.1	.9990	.9991	.9991	.9991	.9992	.9992	.9992	.9992	.9993	.9993
+3.2	.9993	.9993	.9994	.9994	.9994	.9994	.9994	.9995	.9995	.9995
+3.3	.9995	.9995	.9995	.9996	.9996	.9996	.9996	.9996	.9996	.9997
+3.4	.9997	.9997	.9997	.9997	.9997	.9997	.9997	.9997	.9997	.9998
+3.5	.9998	.9998	.9998	.9998	.9998	.9998	.9998	.9998	.9998	.9998

z	.00	.01	.02	.03	.04	.05	.06	.07	.08	.09
-3.5	.0002	.0002	.0002	.0002	.0002	.0002	.0002	.0002	.0002	.0002
-3.4	.0003	.0003	.0003	.0003	.0003	.0003	.0003	.0003	.0003	.0002
-3.3	.0005	.0005	.0005	.0004	.0004	.0004	.0004	.0004	.0004	.0003
-3.2	.0007	.0007	.0006	.0006	.0006	.0006	.0006	.0005	.0005	.0005
-3.1	.0010	.0009	.0009	.0009	.0008	.0008	.0008	.0008	.0007	.0007
-3.0	.0013	.0013	.0013	.0012	.0012	.0011	.0011	.0011	.0010	.0010
-2.9	.0019	.0018	.0018	.0017	.0016	.0016	.0015	.0015	.0014	.0014
-2.8	.0026	.0025	.0024	.0023	.0023	.0022	.0021	.0021	.0020	.0019
-2.7	.0035	.0034	.0033	.0032	.0031	.0030	.0029	.0028	.0027	.0026
-2.6	.0047	.0045	.0044	.0043	.0041	.0040	.0039	.0038	.0037	.0036
-2.5	.0062	.0060	.0059	.0057	.0055	.0054	.0052	.0051	.0049	.0048
-2.4	.0082	.0080	.0078	.0075	.0073	.0071	.0069	.0068	.0066	.0064
-2.3	.0107	.0104	.0102	.0099	.0096	.0094	.0091	.0089	.0087	.0084
-2.2	.0139	.0136	.0132	.0129	.0125	.0122	.0119	.0116	.0113	.0110
-2.1	.0179	.0174	.0170	.0166	.0162	.0158	.0154	.0150	.0146	.0143
-2.0	.0228	.0222	.0217	.0212	.0207	.0202	.0197	.0192	.0188	.0183
-1.9	.0287	.0281	.0274	.0268	.0262	.0256	.0250	.0244	.0239	.0233
-1.8	.0359	.0351	.0344	.0336	.0329	.0322	.0314	.0307	.0301	.0294
-1.7	.0446	.0436	.0427	.0418	.0409	.0401	.0392	.0384	.0375	.0367
-1.6	.0548	.0537	.0526	.0516	.0505	.0495	.0485	.0475	.0465	.0455
-1.5	.0668	.0655	.0643	.0630	.0618	.0606	.0594	.0582	.0571	.0559
-1.4	.0808	.0793	.0778	.0764	.0749	.0735	.0721	.0708	.0694	.0681
-1.3	.0968	.0951	.0934	.0918	.0901	.0885	.0869	.0853	.0838	.0823
-1.2	.1151	.1131	.1112	.1093	.1075	.1056	.1038	.1020	.1003	.0985
-1.1	.1357	.1335	.1314	.1292	.1271	.1251	.1230	.1210	.1190	.1170
-1.0	.1587	.1562	.1539	.1515	.1492	.1469	.1446	.1423	.1401	.1379
-0.9	.1841	.1814	.1788	.1762	.1736	.1711	.1685	.1660	.1635	.1611
-0.8	.2119	.2000	.2061	.2033	.2005	.1977	.1949	.1894	.1894	.1867
-0.7	.2420	.2389	.2358	.2327	.2296	.2266	.2236	.2206	.2177	.2148
-0.6	.2743	.2709	.2676	.2643	.2611	.2578	.2546	.2514	.2483	.2451
-0.5	.3085	.3050	.3015	.2981	.2946	.2912	.2877	.2843	.2810	.2776
-0.4	.3446	.3409	.3372	.3336	.3300	.3264	.3228	.3192	.3156	.3121
-0.3	.3821	.3783	.3745	.3707	.3669	.3632	.3594	.3557	.3520	.3483
-0.2	.4207	.4168	.4129	.4090	.4052	.4013	.3974	.3936	.3897	.3859
-0.1	.4602	.4562	.4522	.4483	.4443	.4404	.4364	.4325	.4286	.4247
-0.0	.5000	.4960	.4920	.4880	.4840	.4801	.4761	.4721	.4681	.4641

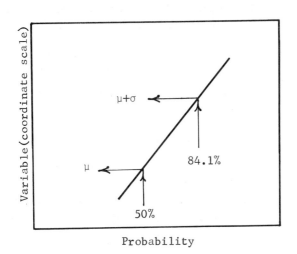

Probability

A plot of normal distribution on <u>normal probability</u> paper would appear as at left. The straight line may be obtained from the raw or classed data. From the curve both μ and σ may be found. Or, one can compute μ and σ from the raw or classed data. These two points then fix the location of the straight line on normal probability paper.

For classed data, for any distribution, the procedure shown in a previous section is used. The column labeled "Cumulative relative frequency" is plotted against the number representing the value of the variable in that class. The highest value in the class is used commonly but the use of the class mark is satisfactory.

How can you tell if a normal distribution describes the sample space for a given variable. The positive (and easy) way is to plot the data for the variable on normal probability paper. If you get a satisfactory straight line (even though all points do not fall on the line) assumption of normal distribution relationships should be satisfactory. Once you obtain this straight line it is easy to calculate μ and σ directly from this line as shown above.

As a good first guess, the distribution of a series of individual, independent measurements will tend to follow a normal distribution, if all measurements are taken in a consistent manner.

As an example of this consider the 20 porosities used earlier in this chapter. There are several clues that the distribution might be approximated by a normal distribution.

1. The mean, median and mode values are almost identical.

2. The cumulative frequency curve is s-shaped on coordinate paper (Figure 8.2).

3. The frequency curve (Figure 8.1) is not truly symmetrical but this might be due to the small number of data points available. Nevertheless it possesses general symmetry.

Replot the data on normal probability paper. Can you draw a satisfactory straight line through the points? If so, what are the values of μ and σ.

Hint: When too few data points are available or too few classes of data are used, the very low and very high points oftentimes do not fit a straight line very well even though the middle points do. It is often rational to ignore these extremity points and use the rest of them for constructing the straight line.

LOG NORMAL DISTRIBUTION

This distribution is very useful in the petroleum business for analyzing reservoir sizes, permeability, formation thickness, recovery, and the like.

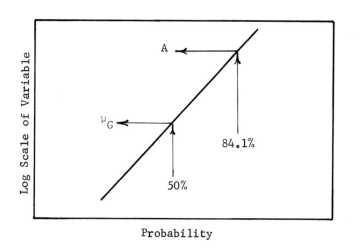

This distribution curve is skewed (like curves (b) and (c) of Fig. 8.3) when probability or frequency is plotted versus variable x or its standardized form z. It is a <u>normal curve</u> when the abscissa is plotted as log x or log z (instead of x or z).

This concept is usually applied by the use of <u>log probability paper</u>. The ordinate (for plotting the random variable) is a log scale. The abscissa is the regular probability scale applying for any normal distribution curve. As usually plotted, the probability is a "percentage less than" value.

The 50% "less than" point on the cumulative probability scale defines the <u>geometric mean</u> as shown on the sketch. This geometric mean (μ_G) is read from the ordinate at 50% probability. One cannot read arithmetic mean from this plot.

The log probability plot may be used to find standard deviation or the curve may be drawn knowing σ and μ_G. The value at Point A is read at 84.14% on the probability scale (50 plus one σ of 34.14). It can be shown that the following equation applies.

$$10^\sigma = A/\mu_G \qquad\qquad (8.8)$$

where: μ_G = geometric mean

If the line is obtained from the data directly one reads μ_G and A from the plot. Equ. 8.8 is then used to find σ. The same plot may be made very conveniently by determining σ and μ_G from the data, which fixes two points for constructing the straight line.

How can one guess if a log-normal distribution curve might apply? As a general rule this distribution might well apply for any variable obtained by multiplying or dividing uncertain numbers. A good example of this is permeability. It is not measured directly. It is calculated from an equation by measuring fluid flow rate, area of flow and pressure drop on a given sample. Each of these numbers will have measurement errors. The permeability distribution tends to be log-normal.

The log-normal distribution tends to apply reasonable well for all cases where the geometric mean is the most meaningful average. As a rule-of-thumb many variables in the earth's crust follow this trend because they are randomly occurring (at least the order is unknown).

<u>Geometric Mean and Standard Deviation</u>

When using classed data it is convenient to prepare a tabulation for calcuation of geometric mean and standard deviation. Proceeding from left to right the columns would be as follows:

(1) class interval; (2) log (base 10) of <u>class mark</u> of each class;
(3) frequency of each class (no. of numbers); (4) Col. 2 x Col. 3;
(5) Col. 2 squared; (6) Col. 3 x Col. 5; (7) cumulative relative
frequency from Col. 3 (by adding up numbers as in the porosity example).

$$\log \mu_G = \text{Sum of Col. 4 divided by Sum of Col. 3.}$$

$$\sigma = (\text{Sum of Col. 6/Sum of Col. 3}) - (\log \mu_G)^2$$

Of course, one could construct the line on log probability paper from Column (1) and Column (7).

It can be shown that the type of <u>mean</u> utilized <u>heterogeneous flow systems</u> depends on the geometry. For

Flow through parallel beds use arithmetic mean.

Flow in series use harmonic mean.

Flow through randomly occurring resistances use geometric mean.

Arithmetic mean was shown in Equ. 8.3.

<u>Geometric mean.</u> - This is found from the euqation

$$\mu_G = [(x_1)(x_2)(x_3)\ldots(x_n)]^{1/n} \tag{8.9}$$

where: μ_G = geometric mean

   x = value of each number

   n = no. of numbers

One simply finds the product of all the numbers and then takes the 1/n root of this product. The geometric mean may be found by taking the antilog of the arithmetic mean of the logarithms of the numbers in the set.

<u>Harmonic mean.</u> A harmonic mean is the <u>reciprocal</u> of the arithmetic mean of the recipro- cals of all numbers in the set. The equation expressing this is

$$\mu_H = \frac{n}{(1/x_1) + (1/x_2) + \ldots (1/x_n)} \tag{8.10}$$

One of the primary flow applications of a mean in a reservoir is the determination of an <u>average permeability</u> to use in a Darcy type equation. There is strong evidence that any single, average value used should be a <u>geometric mean,</u> regardless of how you may "layer cake" the reser- voir for model purposes. The arithmetic mean value of the type shown in core analysis summaries is seldom, if ever, compatible with the geometry. For this reason alone, the arithmetic average permeability should not give good results.

BINOMIAL DISTRIBUTION

The binomial distribution is a special case of the general case called Bernoulli processes. It is a special case in which only two outcomes can occur in a given trial. The equation for this distribution is

$$P(x) = (C_x^n)(p)^x(1-p)^{n-x} = \frac{n!}{x!(n-x)!}(p)^x(1-p)^{n-x} \qquad (8.11)$$

where: $P(x)$ is binomial probability of "x" successes in "n" trials.

        n = total number of trials

        p = probability of success in any single trial

        $q = (1-p)$ = probability of failure in any single trial

        x = number of successes in "n" trials, $(n-x)$ = number of failures

Expressions like n! and x! are called <u>factorials</u>. For example, $5! = (5)(4)(3)(2)(1) = 120$. The value of 25! would be found by writing all of the numbers from 1 to 25 and multiplying them together. This would be tedious manually.

What if we wish to find the probability of <u>at least</u> "c" successes instead of <u>exactly</u> "x" successes. This requires what is called cumulative binomial probability, defined by Equation 8.12

$$\sum_{x=c}^{x=n}(c_x^n)(p)^x(1-p)^{n-x} \qquad (8.12)$$

This is the same equation as 8.11 execpt for the summation. In effect one is solving 8.11 multiple times and adding the results.

For a large number of trials (n), it is convenient to use Ordnance Corps Pamphlet ORDP 20-1, Reference 19, to find cumulative binomial probability if you do not have a computer program available to do so. A sample of these tables is shown in Table 8.2 for n = 25. One can read cumulative binomial probabilities directly from this table. A separate table is needed for each value of n.

Example 1. In the toss of a fair coin (p = 0.50) what is the binomial probability of getting exactly 2 heads in 6 tosses?

$$n = 6, \ x = 2 \text{ and } p = 0.50$$

From Equation 8.11

$$P(x) = \frac{6!}{2!(4)!}(0.5)^2(0.5)^4 = \frac{720}{(2)(24)}(0.25)(0.0625) = \underline{0.234}$$

Example 2. What is the probability of obtaining atleast 8 heads in 25 tosses of a fair coin.

$$\text{From Table 8.2, } p(x) = 0.978$$

Example 3. What is the chance of obtaining 20 successful wells in a 25 well drilling program if the probability of success is 0.80. Notice the tables only go to p = 0.5. But p & q = 1.0. So let us solve Table 8.2 for q = 0.2. Then, x = 5.0 and n = 25. From this table, the probability of exactly 5 <u>or more</u> dry holes is 0.5793; for 6 <u>or more</u> dry holes is 0.196 (0.5793-0.3833). The probability of 5 dry holes is the same as 20 successful wells.

228

TABLE 8.2

Cumulative Binomial Probability of At Least x Successes For Various Values of p When n=25.

x \ p	.31	.32	.33	.34	.35	.36	.37	.38	.39	.40
1	.9999	.9999	1.0000	1.0000	1.0000	1.0000	1.0000	1.0000	1.0000	1.0000
2	.9989	.9992	.9994	.9996	.9997	.9998	.9998	.9999	.9999	.9999
3	.9932	.9949	.9961	.9971	.9979	.9984	.9989	.9992	.9994	.9996
4	.9737	.9793	.9838	.9874	.9903	.9926	.9944	.9958	.9968	.9976
5	.9254	.9390	.9504	.9600	.9680	.9745	.9799	.9842	.9877	.9905
6	.8344	.8593	.8813	.9006	.9174	.9318	.9441	.9546	.9633	.9706
7	.6981	.7343	.7679	.7987	.8266	.8517	.8742	.8940	.9114	.9264
8	.5319	.5747	.6163	.6561	.6939	.7295	.7626	.7932	.8211	.8464
9	.3639	.4057	.4482	.4908	.5332	.5748	.6152	.6542	.6914	.7265
10	.2213	.2555	.2919	.3300	.3697	.4104	.4517	.4933	.5347	.5754
11	.1188	.1424	.1686	.1975	.2288	.2624	.2981	.3355	.3743	.4142
12	.0560	.0698	.0859	.1044	.1254	.1490	.1751	.2036	.2346	.2677
13	.0230	.0299	.0383	.0485	.0604	.0745	.0907	.1093	.1303	.1538
14	.0083	.0112	.0149	.0196	.0255	.0326	.0412	.0515	.0637	.0778
15	.0026	.0036	.0050	.0069	.0093	.0124	.0163	.0212	.0271	.0344
16	.0007	.0010	.0015	.0021	.0029	.0041	.0056	.0075	.0100	.0132
17	.0002	.0002	.0004	.0005	.0008	.0011	.0016	.0023	.0032	.0043
18			.0001	.0001	.0002	.0003	.0004	.0006	.0008	.0012
19						.0001	.0001	.0001	.0002	.0003
20										.0001

x \ p	.41	.42	.43	.44	.45	.46	.47	.48	.49	.50
1	1.0000	1.0000	1.0000	1.0000	1.0000	1.0000	1.0000	1.0000	1.0000	1.0000
2	1.0000	1.0000	1.0000	1.0000	1.0000	1.0000	1.0000	1.0000	1.0000	1.0000
3	.9997	.9998	.9998	1.0000	1.0000	1.0000	1.0000	1.0000	1.0000	1.0000
4	.9983	.9987	.9991	.9999	.9999	.9997	.9998	.9998	.9999	.9999
5	.9927	.9945	.9958	.9969	.9977	.9983	.9988	.9991	.9994	.9995
6	.9767	.9816	.9856	.9888	.9914	.9934	.9950	.9963	.9972	.9980
7	.9394	.9505	.9599	.9677	.9742	.9796	.9840	.9876	.9904	.9927
8	.8692	.8894	.9071	.9227	.9361	.9477	.9575	.9658	.9727	.9784
9	.7593	.7897	.8177	.8431	.8660	.8865	.9046	.9205	.9343	.9461
10	.6151	.6535	.6902	.7250	.7576	.7880	.8160	.8415	.8646	.8852
11	.4548	.4956	.5363	.5765	.6157	.6538	.6902	.7249	.7574	.7878
12	.3029	.3397	.3780	.4174	.4574	.4978	.5382	.5780	.6171	.6550
13	.1797	.2080	.2387	.2715	.3063	.3429	.3808	.4199	.4598	.5000
14	.0941	.1127	.1336	.1569	.1827	.2109	.2413	.2740	.3086	.3450
15	.0431	.0535	.0656	.0797	.0960	.1145	.1353	.1585	.1841	.2122
16	.0171	.0220	.0280	.0353	.0440	.0543	.0663	.0803	.0964	.1148
17	.0058	.0078	.0103	.0134	.0174	.0222	.0281	.0352	.0438	.0539
18	.0017	.0023	.0032	.0044	.0058	.0077	.0102	.0132	.0170	.0216
19	.0004	.0006	.0008	.0012	.0016	.0023	.0031	.0041	.0055	.0073
20	.0001	.0001	.0002	.0003	.0004	.0005	.0008	.0011	.0015	.0020
21					.0001	.0001	.0002	.0002	.0003	.0005
22									.0001	.0001

x \ p	.01	.02	.03	.04	.05	.06	.07	.08	.09	.10
1	.2222	.3965	.5330	.6396	.7226	.7871	.8370	.8756	.9054	.9282
2	.0258	.0886	.1720	.2642	.3576	.4473	.5304	.6053	.6714	.7288
3	.0020	.0132	.0380	.0765	.1271	.1871	.2534	.3232	.3937	.4629
4	.0001	.0014	.0062	.0165	.0341	.0598	.0936	.1351	.1831	.2364
5		.0001	.0008	.0028	.0072	.0150	.0274	.0451	.0686	.0980
6			.0001	.0004	.0012	.0031	.0065	.0123	.0210	.0334
7					.0002	.0005	.0013	.0028	.0054	.0095
8						.0001	.0002	.0005	.0011	.0023
9								.0001	.0002	.0005
10										.0001

x \ p	.11	.12	.13	.14	.15	.16	.17	.18	.19	.20
1	.9457	.9591	.9692	.9770	.9828	.9872	.9905	.9930	.9948	.9962
2	.7779	.8195	.8543	.8832	.9069	.9263	.9420	.9546	.9646	.9726
3	.5291	.5912	.6483	.7000	.7463	.7870	.8226	.8533	.8796	.9018
4	.2934	.3525	.4123	.4714	.5289	.5837	.6352	.6829	.7266	.7660
5	.1331	.1734	.2183	.2668	.3179	.3707	.4241	.4772	.5292	.5793
6	.0499	.0709	.0965	.1268	.1615	.2002	.2425	.2875	.3347	.3833
7	.0156	.0243	.0359	.0509	.0695	.0920	.1185	.1488	.1827	.2200
8	.0041	.0070	.0113	.0173	.0255	.0361	.0495	.0661	.0859	.1091
9	.0009	.0017	.0030	.0050	.0080	.0121	.0178	.0252	.0348	.0468
10	.0002	.0004	.0007	.0013	.0021	.0035	.0055	.0083	.0122	.0173
11		.0001	.0001	.0003	.0005	.0009	.0015	.0024	.0037	.0056
12					.0001	.0002	.0003	.0006	.0010	.0015
13							.0001	.0001	.0002	.0004
14										.0001

x \ p	.21	.22	.23	.24	.25	.26	.27	.28	.29	.30
1	.9972	.9980	.9985	.9990	.9992	.9995	.9996	.9997	.9998	.9999
2	.9789	.9838	.9877	.9907	.9930	.9947	.9961	.9971	.9979	.9984
3	.9204	.9360	.9488	.9593	.9679	.9748	.9804	.9848	.9883	.9910
4	.8013	.8324	.8597	.8834	.9038	.9211	.9358	.9481	.9583	.9668
5	.6270	.6718	.7134	.7516	.7863	.8174	.8452	.8696	.8910	.9095
6	.4325	.4816	.5299	.5767	.6217	.6644	.7044	.7415	.7755	.8065
7	.2601	.3027	.3471	.3927	.4389	.4851	.5308	.5753	.6183	.6593
8	.1358	.1658	.1989	.2349	.2735	.3142	.3565	.3999	.4440	.4882
9	.0614	.0788	.0993	.1228	.1494	.1790	.2115	.2465	.2838	.3231
10	.0240	.0325	.0431	.0560	.0713	.0893	.1101	.1338	.1602	.1894
11	.0082	.0117	.0163	.0222	.0297	.0389	.0502	.0636	.0795	.0978
12	.0024	.0036	.0053	.0076	.0107	.0148	.0199	.0264	.0345	.0442
13	.0006	.0010	.0015	.0023	.0034	.0049	.0069	.0096	.0130	.0175
14	.0001	.0002	.0004	.0006	.0009	.0014	.0021	.0030	.0043	.0060
15			.0001	.0001	.0002	.0003	.0005	.0008	.0012	.0018
16						.0001	.0001	.0002	.0003	.0005
17									.0001	.0001

229

Notice in the third example that although the probability of obtaining a successful well on each attempt is 0.8, the odds of obtaining 80% successful wells in 25 attempts is rather poor. We continue to repeat the thesis that the chance of obtaining a certain outcome on a specific "draw" does not dictate the results for a relatively small number of finite draws.

For a binomial distribution,

Arithmetic mean, $\mu = np$

Standard deviation, $\sigma = (npq)^{0.5}$.

Equation 8.11 and 8.12 only apply for a process possessing all of the following characteristics:

1. Only two outcomes can occur.

2. Each trial is an independent event.

3. The probability of each outcome remains constant with repeated trials.

By virtue of these limitations it is useful in the quality control of manufacturing processes. There are other processes where more than two outcomes may be possible. In this case a multinomial distribution may be used, details of which are available in statistics texts.

Limitations (2) and (3) seriously affect the meaningful use of binomial distribution in mineral exploration. First of all, each successive trial is hardly an independent event. In a given exploration area the second attempt is certainly influenced by the results of the first attempt. We know of very few (if any) situations where people drill wells at random as truly independent events.

Limitation (3) implies sampling with replacement. The only way one can keep the probability constant is to sample this way. This is not how exploration works. Only a finite number of trials is possible. No reservoir is infinite. So, each time an outcome occurs it has to change the probability of future outcomes for our finite sample space. Those who use binomial probabilities for such applications are theoretically wrong. The degree of error will of course depend on the relative values of the variables involved.

The binomial distribution may be approximated by a normal distribution when n is large; $[np(1-p)] > 25$ is a good criterion for using this approximation.

HYPERGEOMETRIC DISTRIBUTION

This differs from binomial probability in that it is for sampling without replacement and a finite sample space. For determining the hypergeometric probability for two possible outcomes, Equation 8.13 applies.

$$P_h(x) = \frac{(G_x^R)(C_{n-x}^{N-R})}{C_n^N} = \frac{n!R!(N-R)!(N-n)!}{x!(n-x)!(R-x)![(N-R)-(n-x)]!N!} \qquad (8.13)$$

Where: n = size of sample

R = number of successes before n trials are made

N = population size (total elements in sample space before n trials are made

n = number of trials (sample size)

x = the number of successes in n trials

For many purposes it is convenient to define R by the equation R = pN, where p = a probability of success. In the minerals area p may be estimated by examination of analogous propsects based on subjective professional judgment.

Example: We wish to start testing 10 different prospects found by geophysics. It is desired to first test 5 of these structures by drilling one well on each. What is the probability of 2 discoveries in the program if 0.3 of the wells on similar prospects proved commercial? For these conditions, N = 10, x = 2, n = 5 and R = (0.3)(10) = 3. Substituting in Equation 8.13

$$P_h(x) = \frac{5!3!7!5!}{2!3!1!4!10!} = \underline{0.417}$$

Equation 8.13 can be expanded for more than two outcomes. References 1 and 5 discuss these and other applications.

For a hypergeometric distribution

Arithmetic mean, $\mu = \dfrac{nR}{N}$

Standard deviation, $\sigma = \left[\dfrac{nR(N-R)(N-n)}{N^2(N-1)} \right]^{0.5}$

Let us consider a comparable example to that above but use the drawing of balls to simplify the description. We have a container holding 10 colored balls, 3 black and 7 white. Each ball has an equal chance of being drawn but the chance of obtaining a black ball on the first draw is 0.3 (3/10). If the first ball is replaced, the odds remain the same in the second draw. Indeed, they remain the same for all draws so long as each draw is followed by replacement. An infinite number of draws could be made by this procedure.

If we call drawing a black ball a success, p = 0.3 in the applicable binomial distribution. Equations 8.11 and 8.12 apply for this situation. Suppose we wish to determine the probability of obtaining 2 black balls in 5 draws. From these equations the result is 0.3087.

Now suppose we do not replace the balls drawn. The first draw is black. On the second draw each ball still has an equal chance of being drawn but the total number of balls has changed as well as the relative number of black and white. We now have 2 black and 7 white. The conditional probability of drawing black on the second draw is now 2/9 given that black was drawn first. Without replacement we wish to consider the probability of drawing exactly 2 black in 5 draws total from our 10 ball sample space. This is exactly the same as the exploration example above. The answer is 0.417, about 33% higher than predicted by binomial probability.

For our 5 ball draw what is the probability that the balls are drawn in some particular order like BBWWW, WBWBW, etc? There are 10 possible orders in which two blacks can be drawn in a sample of 5. (You can prove this if you care to do so). Since these 10 orders are mutually

231

exclusive their probabilities can be added to obtain the probability of two blacks, regardless of the order.

Let us consider the order BBWWW. What is the probability of drawing 2 blacks on the first two draws followed by 3 consecutive whites on the next 3 draws (with no replacement)? For this case N = 10, x = 2, n = 5 and R = 0.3. The probability of drawing black first is 0.3. Given that this occurs, the probability of black on the second draw is 2/9; there are only 9 balls left, two which are black. The probability of white on the third draw is 7/8. On the fourth and fifth draws the probability of white is 6/7 and 5/6, respectively. Applying the multiplication rule for joint probabilities of the event BBWWW gives us (3/10)(2/9)(7/8)(6/7)(5/6) = 0.0417.

For the second order shown (WBWBW) the joint probability by the same procedure is (7/10)(3/9)(6/8)(2/7)(5/6) = 0.0417, the same as for the first order. In fact, all ten possible orders would give the same joint probability of 0.0417. For these ten mutually exclusive orders the probability of success, regardless of order, is (10)(0.0417) = 0.417, the number we calculated in the example using Equation 8.13.

This example has been shown to make a general point. The odds of any result really change on each draw (from a finite sample size) and are not constant as dictated by binomial probability. The unconditional probability of 0.3 that black will be drawn first is different from the conditional probability that black also will be drawn second, given that the first draw was black. The joint probability that the first two balls drawn are black is the product of these two probabilities (unconditional on first, conditional on second). Each ball remaining still has an equal chance of being drawn but the proportion of balls has been altered because no replacement occurs.

Do you see the differences between binomial and hypergeometric distributions? It is important to recognize these differences.

When N becomes infinitely large, the binomial equation gives the same result as the hypergeometric equation. When N is at least 10 times n the binomial will approximate the hypergeometric if p is found from the equation p = R/N.

SUBJECTIVE DISTRIBUTION CURVES

In many cases detailed data are so limited that no distribution curve may be developed from that data. But, on the basis of experience and general data, professional judgment may be exercised.

If a minimum, maximum and most probable value may be developed a triangular distribution is possible, like that shown. This distribution curve is handled the same as all others. The total area under it is unity. All of the general characteristics discussed earlier apply. It is merely simpler in origin.

In some instances it is not reasonable to predict a most probable value, only a probable minimum and maximum are possible. For this case a rectangular distribution may be drawn. It too possesses all of the same properties as any other distribution curve, except shape.

The following summarizes the properties of these two unique distributions.

Triangular

1. $\mu = (x_1 + x_m + x_2)/3$ (8.14)

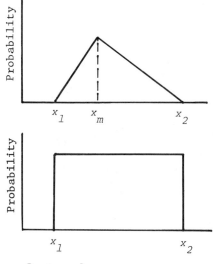

2. Cumulative probability

    a. For x less than or equal to x_m

$$= (x')^2/m \tag{8.15}$$

    b. For x more than or equal to x_m

$$= 1 - \left[\frac{(1 - x')^2}{1 - m} \right] \tag{8.16}$$

where: $x' = \dfrac{x - x_1}{x_2 - x_1}$ and $m = \dfrac{x_m - x_1}{x_2 - x_1}$

Rectangular

1. μ is arithmetic average of x_1 and x_2.

2. Cumulative probability is a straight line between probability = 0 at x_1 and probability = 1.0 at x_2.

Some find it difficult to accept the results of these two distributions based on so little hard information. What credence can you give to final output based on so little input? This is logical so far as it goes. The uncertainty of the input is obvious. But, with limited data you have to rely on professional judgment in some form. The proper question to ask is, "What is the advantage of using triangular and rectangular distributions compared to the other available alternatives?" About the only alternative is to guess some "magical" average number. Practically speaking, it is far more reasonable to determine a likely range of numbers than a single number in that range. In the face of uncertainty you are putting a boundary on values without the necessity of having to arbitrarily pick a single number within that boundary.

In a real sense you are "facing up" to the uncertainty rather than obscuring or ignoring it by using single numbers. The degree of uncertainty showing up in the output actually aids rather than hinders the inevitable judgment involved in the decision process.

A rectangular distribution normally will be assumed unless there are some positive data or reasons for picking a most probably value (mode) of the curve. A most probable value might be obtained from analysis of an existing reservoir in the same type of formation in the geographical area. If an average value is used for the middle value, this requires the triangle to be symmetrical (mode = mean).

If one has no positive reason for picking a most probable value, a rectangular curve is normally assumed.

RANDOM NUMBERS

Once a distribution curve has been developed for any variable, any number within that sample space can occur. It is thus a random variable within that space. The likelihood of the occurrence of any value within any range, is fixed by the shape of the curve.

The use of random numbers is more convenient for making random draws than the use of the slips of paper and marbles used previously to explain the principles. A series of numbers are

Table 8.3   Example of a Random Number Table.

53479	81115	98036	12217	59526	40238	40577	39351	43211	69255
97344	70328	58116	91964	26240	44643	83287	97391	92823	77578
66023	38277	74523	71118	84892	13956	98899	92315	65783	59640
99776	75723	03172	43112	83086	81982	14538	26162	24899	20551
30176	48979	92153	38416	42436	26636	83903	44722	69210	69117
81874	83339	14988	99937	13213	30177	47967	93793	86693	98854
19839	90630	71863	95053	55532	60908	84108	55342	48479	63799
09337	33435	53869	52769	18801	25820	96198	66518	78314	97013
31151	58295	40823	41330	21093	93882	49192	44876	47185	81425
67619	52515	03037	81699	17106	64982	60834	85319	47814	08075
61946	48790	11602	83043	22257	11832	04344	95541	20366	55937
04811	64892	96346	79065	26999	43967	63485	93572	80753	96582
05763	39601	56140	25513	86151	78657	02184	29715	04334	15678
73260	56877	40794	13948	96289	90185	47111	66807	61849	44686
54909	09976	76580	02645	35795	44537	64428	35441	28318	99001
42583	36335	60068	04044	29678	16342	48592	25547	63177	75225
27266	27403	97520	23334	36453	33699	23672	45884	41515	04756
49843	11442	66682	36055	32002	78600	36924	59962	68191	62580
29316	40460	27076	69232	51423	58515	49920	03901	26597	33068
30463	27856	67798	16837	74273	05793	02900	63498	00782	35097
28708	84088	65535	44258	33869	82530	98399	26387	02836	36838
13183	50652	94872	28257	78547	55286	33591	61965	51723	14211
60796	76639	30157	40295	99476	28334	15368	42481	60312	42770
13486	46918	64683	07411	77842	01908	47796	65796	44230	77230
34914	94502	39374	34185	57500	22514	04060	94511	44612	10485
28105	04814	85170	86490	35695	03483	57315	63174	71902	71182
59231	45028	01173	08848	81925	71494	95401	34049	04851	65914
87437	82758	71093	36833	53582	25986	46005	42840	81683	21459
29046	01301	55343	65732	78714	43644	46248	53205	94868	48711
62035	71886	94506	15263	61435	10369	42054	68257	14385	79436
38856	80048	59973	73368	52876	47673	41020	82295	26430	87377
40666	43328	87379	86418	95841	25590	54137	94182	42308	07361
40588	90087	37729	08667	37256	20317	53316	50982	32900	32097
78237	86556	50276	20431	00243	02303	71029	49932	23245	00862
98247	67474	71455	69540	01169	03320	67017	92543	97977	52728
69977	78558	65430	32627	28312	61815	14598	79728	55699	91348
39843	23074	40814	03713	21891	96353	96806	24595	26203	26009
62880	87277	99895	99965	34374	42556	11679	99605	98011	48867
56138	64927	29454	52967	86624	62422	30163	76181	95317	39264
90804	56026	48994	64569	67465	60180	12972	03848	62582	93855
09665	44672	74762	33357	67301	80546	97659	11348	78771	45011
34756	50403	76634	12767	32220	34545	18100	53513	14521	72120
12157	73327	74196	26668	78087	53636	52304	00007	05708	63538
69384	07734	94451	76428	16121	09300	67417	68587	87932	38840
93358	64565	43766	45041	44930	69970	16964	08277	67752	60292
38879	35544	99563	85404	04913	62547	78406	01017	86187	22072
58314	60298	72394	69668	12474	93059	02053	29807	63645	12792
83568	10227	99471	74729	22075	10233	21575	20325	21317	57124
28067	91152	40568	33705	64510	07067	64374	26336	79652	31140
05730	75557	93161	80921	55873	54103	34801	83157	04534	81368

random if they possess no discernible order and were developed in a manner to promote disorder. Even if theoretical randomness is impossible, a set of numbers may be developed that are random, practically.

A number of computer programs are available to develop random numbers and are normally used in machine computations. Table 8.3 is an example of a random number table useful for manual calculations.

To use such a table properly it is necessary to enter it at random. A simple way is to close your eyes, touch the point of a pencil to the table, and begin with the number touched. Or, one could use that number to denote which column and row to start at. Any random entering scheme that suits your fancy is adequate.

Suppose you entered the table at the number 61 (11th row down in the first two columns on the left), and you were going to use a series of two-digit random numbers. The successive numbers would be 04, 05, 73, 54, 42, 27, etc. Once you get to the bottom of the page you start at the top of the next two columns. If the numbers are truly random you are doing the same thing as drawing from a thoroughly mixed system.

The use of random numbers tends to bother people when they first use them. This is understandable. These numbers possess no discernible physical significance.

Suppose you obtain a number like 61 from a random number table. What does it represent? It represents one value of the cumulative probability between 0 and 100 (expressed as a percentage). This is the cumulative probability of some value of some variable being less than 0.61 (61%). The value of this variable will be fixed by its specific cumulative probability curve. This is shown in the next section on simulation.

If it is any help in understanding a random number table, it could be developed by placing 100 slips of paper in a container each containing one number from 0-100. Draw a number, record it, replace it in the container, mix the slips and then draw again. Repeat this process a large number of times. When you finish you have a random number table like 8.3. The numbers would be in a different order but they should be. There must be no order in a random number table.

When using the table we are merely approaching the random sampling process from a different direction.

SIMULATION APPLICATIONS

Simulation is the process of using equations that describe system performance together with formal consideration of the probability distribution of each variable occurring in the equation. This use in calculating value, reserves, interpretation of logs, cores, etc. is powerful.

One simple example is furnished by the equation for oil reserves

$$\text{Reserves} = (7758)(\text{Acre-feet})\,(\phi)\left[\frac{1 - S_w}{B_{oi}}\right](\text{Recovery Factor})$$

The key numbers affecting reserves are uncertain - acre-feet., ϕ, S_w, and R.F., assuming B_{oi} may be predicted fairly accurately. Question, "Are they all independent variables?" From available correlations likely not. Water saturation will depend on pore size distribution which usually varies with effective porosity. Recovery factor will vary with porosity and water saturation. Since these variables are not independent in "real life" they should not be treated as independent statistically. To be correct, some correlation of the dependent variable as a function of the independent variables would be needed. Or, one could prepare some summary showing the conditional probability of S_w for any given value of ϕ. The procedure could be that used in References 1 and 18, referred to in an earlier section.

In order to illustrate simulation simply we need an example where all variables are independent. To do this we can rewrite the above equation in the form

$$\text{Reserves} = (\text{Reservoir thickness})(\text{Reservoir area})(\text{Net oil recovery})$$

$$= (h)(A)(\text{Bbl/ac-ft})$$

where: h is in feet

A is in acres

Bbl/ac-ft is a net recovery factor based on experience and analogy.

Recovery factor probably is not completely independent but for early decision studies this may be a satisfactory assumption.

Example: We have an exploration prospect with geophysical data and the usual geographical maps. No exploratory drilling has been done as yet. The staff has been asked to develop a range of values for our three variables in as objective a manner as possible. In addition to a maximum and minimum value, a most probable value is requested if supported by evidence satisfactory to them.

Their conclusions follow.

h (ft.) Min. – 100, Max. – 200, Most probable – 130

Bbl/Ac-ft.: Min. – 300, Max – 600

Area (acres): Min. – 1500, Max. – 4000

A most probable value for thickness only was proposed based on isopachous maps prepared.

(These are considered valid numbers only if the reservoir contains oil (a fact as yet only presumed). Thus the results of this exercise when expressed as a probability are conditioned on the premise that the reservoir truly contains oil.

Recovery and area are rectangular distributions; h is triangular since three points are shown. The cumulative distribution curve for h is developed below. Values for x', m and cumulative probability for h are found from Equ. 8.14-8.16.

236

h	x'			Cum. Probability
100	0			0
120	.2	(0.2)(0.2)/0.3	=	13.33
140	.4	1 - (0.6)(0.6)/0.7	=	48.6
160	.6			77.15
180	.8			94.28
200	1.0			100

Figure 8.7 shows the plot of the above plot the straight lines describing the rectangular distributions.

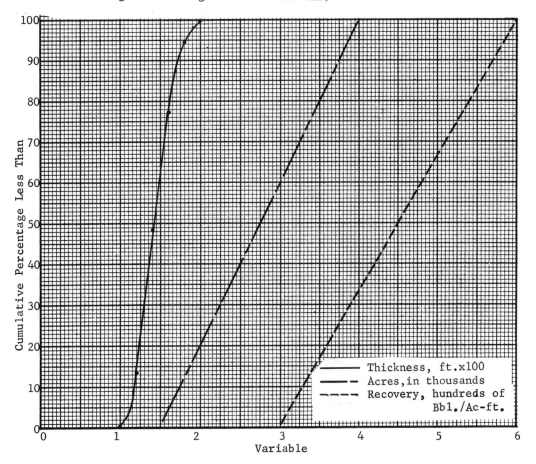

Figure 8.7 Cumulative "% less than" Curves for Example Problem

Now from these lines and a random number table it is possible to determine corresponding values for the dependent variable (reserves) in our example. For simplicity, just enter Table 8.3 at the upper left, using the first two columns.

Random No.	h	A	Rec. Factor	Reserves
53	145			
97		3930		
66			499	284×19^6
99	198			
30		2250		
81			545	243×10^6
19	123			
09		1720		
31			393	83×10^6

237

There are three passes shown in the preceeding table. We entered Fig. 8.7 to get a corresponding value for the three variables. For any one pass, multiplying the three results together gave reserves for that pass. If this calculation is made manually 100 passes should give reasonable input. In a computer solution 1000 or more passes might be made depending on the accuracy of the data and the need.

After making enough passes a distribution curve is made for reserves, the dependent variable. A cumulative probability curve will then be drawn. Any point on this curve shows the probability that the reserves will be a given value or less.

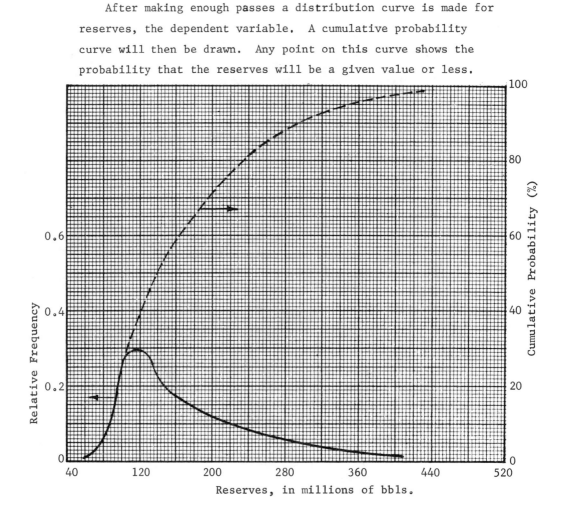

Figure 8.8   Results of Simulation Example.

Figure 8.8 was developed from a continuation of the above three passes. It is approximate since it is based only on 25 simulation passes but it serves to show the character of the output from this simulation example.

Based on the ranges of numbers provided by staff the minimum reserves are 45 million barrels [(100)(300)(1500)] and the maximum reserves are 480 million barrels [(200)(600)(4000)]. The arithmetic average of these two numbers is 263 million barrels. It could be found by using average values of thickness, area and recovery. Is this a good number? The results of our simulation will tell us.

From the distribution curve (solid line) the most probable value of the reserves is about 120 million bbls (the mode). The mean of the distribution curve is about 140 million bbls. This curve shows that the frequency of obtaining reserves of 260 million barrels is low.

The cumulative probability curve (dashed line) formalizes the "odds" involved. The probability of obtaining 260 million bbls. or less is 0.85. This means that the probability of obtaining more than 260 million is 0.15.

Any probability obtained from the example curve is the liklihood of said reserves, presuming any oil is present at all. The presence or absence of minerals can only be confirmed by drilling holes in the ground.

The important thing about this kind of presentation is that the decision maker has before him the chance of receiving a given level of rewards. He can compare these with the risk capital needs to see if the risk is acceptable. Are rewards for success compatible with the risk? If the ratio of reward to risk is satisfactory the amount of capital risk may be too large compared to available assets. If so, a joint venture may be indicated.

We did this calculation based on only one set of subjective numbers. Suppose the staff could not agree and several sets of numbers were advanced? Repeat the calculation for each set. From these a comparable set of decision parameters can be developed.

Whatever the weaknesses of this approach, it is superior to an arbitrary set of numbers developed in an informal manner, provided of course staff input is objective and not a sheer guess.

The above is just a simple example of the simulation procedure. Any distribution may be used. Time value of money and similar economic yardsticks may be incorporated into the simulation model. One may also introduce different investment strategies into considerations like farm outs, joint interests and the like. In essence, this approach has application in almost all reservoir performance-value calculations in this book. The references contain many examples that illustrate the procedure further.

EXPECTED MONETARY VALUE (EMV)

Oftentimes referred to as mathematical expectation this is an old proven concept that has many ramifications in the evaluation and decision process. For our purposes herein, it is only defined by using a simple example.

EMV involves the product of the probability of occurrence of a given outcome multiplied by the conditional value of that outcome (if it occurs).

Table 8.4 illustrates how EMV works. In making up this table it was assumed that a dry hole costs $50,000, a completed producing well costs $70,000 and that the investor will net $1.00/bbl. on the oil. The net revenue is the number of barrels of reserves times $1.00 minus $70,000. For a dry hole the net revenue is a constant -$50,000

Five possible strategies are considered to decide which one offers the best EMV. The probabilities used are obtained subjectively or from a simulated distribution of the type in the preceding example.

Table 8.4 Sample EMV Calculation (Ref. 7)

Possible Events	Prob. of Event Occurring	POSSIBLE ACTS				
		1	2	3	4	5
		Don't Drill	Drill with 100% Interest	Drill with 50% Partner	Farm-Out, Keep 1/8 Over-ride	Farm-Out, Come Back in for 50% Interest After Payout
Dry Hole	.60	$ 0	$ -30,000	$-15,000	$ 0	$ 0
50,000 bbls.	.10	0	- 2,000	- 1,000	- 625	0
100,000 bbls.	.15	0	4,500	2,250	1,875	2,250
500,000 bbls.	.10	0	43,000	21,500	6,250	21,500
1,000,000 bbls.	.05	0	46,500	23,250	6,250	23,250
Expected Monetary value		$ 0	$ 62,000	$ 31,000	$ 15,000	$ 47,000

Each number under each of the five columns represents the effect of five possible events for each strategy. The sum of each column is the EMV.

Consider Column 2, drill with 100% interest. The numbers are found as follows.

$$(0.60)(-50,000) = - \$30,000$$

$$(0.10)(50,000-70,000) = - \$2000$$

$$(0.15)(100,000-70,000) = \$4500 \quad etc.$$

All other columns are calculated in the same manner.

If one wishes to invest at all, and can afford to do so, drilling with 100% interest offers the best EMV. Is that good enough to risk $50,000? That is the decision of the investor. In this case one can risk nothing and have an EMV of $47,000 (Column 5). Is an extra $14,000 worth the $50,000 risk exposure? Whatever the answers, the EMV calculation provides one with a useful array of numbers for consideration.

The probability numbers must, of course, add up to 1.0 regardless of the number of outcomes considered. Each probability may be purely subjective, the result of a detailed study, or some combination thereof. Whatever the source, the EMV numbers proposed presume that the project involves a sufficient number of tries so that the probability numbers take on meaning.

The EMV approach applies to any investment strategy not just exploration. In fact, drilling is the most critical example of the principle. One may state (rather rationally) that a well is a one shot investment; there are no repeated trials on a single well. This is true so far as it goes. But, we believe that if EMV is used as a consistent strategy to determine drilling strategy it will optimize profit. Whatever the faults one may assign, EMV is at least the best and most reasonable consistent strategy among the alternative choices. Any inconsistent strategy is patently illogical.

ANALYSIS OF DATA

The principles covered earlier also may be used in a purely statistical manner to analyze data. Any data obtained will vary with time. How much of the variation is <u>physically meaningful</u> and how much is simple <u>random error</u>? As one example, consider the application of pressure measurements. If you attribute all pressure change to reservoir performance when in fact it was mostly random error, the error in the results may be astronomical. In preparing empirical correlations for reservoir characteristics and performance, what is the reliability of the correlation?

In this section we will be concerned only with:

1. The relationship, if any, between two chosen variables - correlation coefficients.

2. The variation to be expected in repeated measurements - variance.

3. The role of chance in the observed difference of two sets of data - significance.

Precision and Accuracy

Confusion often exists between the terms <u>precision</u> and <u>accuracy</u>. Precision merely refers to the degree with which a given set of numbers may be reproduced. An instrument may be very precise and woefully inaccurate. Accuracy refers to the correctness of the results.

There are two types of basic errors - systematic and accidental.

<u>Systematic errors</u>. This category includes personal errors - method of reading instruments that will vary from person to person. Instrument errors are systematic because of miscalibration. A systematic error may also occur by improperly handling raw data when reducing it to final form, failure to properly account for all factors affecting a measurement, and the like.

<u>Accidental errors</u>. A mistake is an accidental error. It is often of large magnitude. Chance errors also occur. The amount of these is never known exactly. They are caused by the interaction of all the factors that affect reproducivility. It is these chance errors that respond so well to analysis using statistical methods.

Correlation Coefficient

The usual procedure involves plotting two variables to see if they vary with each other in some logical manner. One may fit the best line through the data points using one of the techniques available. This still does not "prove" that the equation reliably describes the relationship between these variables.

The preparation of a figure showing the data points only is called a <u>scatter diagram</u>. If the two variables change systematically with each other, a correlation is presumed to exist (even though it might be a poor one). If the data points are all over the diagram (like a shotgun blast) the variables are presumed to be uncorrelatable.

The correlation coefficient is defined by the equation:

$$r = \frac{\Sigma xy - n\bar{x}\bar{y}}{\left[(\Sigma x^2 - n\bar{x}^2)(\Sigma y^2 - n\bar{y}^2)\right]^{0.5}} \qquad (8.17)$$

where: r = correlation coefficient

x,y = two variables

$\bar{x},\bar{y}$ = arithmetic mean (average) for each variable

$\bar{x} = \Sigma x/n$, $\bar{y} = \Sigma y/n$

n = no. of data points used.

Equation 8.17 may be derived by taking the slopes of the regression lines of y on x and x on y in a least squares fit. If perfect coincidence is obtained by this process, r = 1.0 (or minus one). If r = 0, no correlation exists. Therefore, the usual conclusion is that the closer r is to unity the better is the correlation.

Table 8.5 shows an example calculation for 25 data points relating oil gravity to depth. For this example r = 0.81 and r^2 = 0.66. Some interpret the square of the correlation coefficient as being equal to the fraction of the variations in y that is caused by the changes in x. For this example, 66 per cent of the change in oil gravity presumably results from change in sample depth.

Table 8.5   Example Calculation –
Correlation Coefficient

Produced Oil Gravity Versus
Formation Depth for 25 South
Arkansas Wells

Depth (ft.)	Gravity (°API)	Depth (ft.)	Gravity (°API)
1,356	16	4,100	36
1,994	26	4,800	36
2,100	31	4,816	31
2,100	33	6,112	41
2,100	34	6,330	29
2,106	15	6,530	35
2,150	22	7,136	40
2,235	24	7,255	44
2,245	27	7,350	38
2,480	31	7,600	38
2,650	19	7,848	44
2,785	26	8,260	53
2,800	30		

$\bar{x}$ = (1/25)(1,356 + 1,994 + ... + 8,260) = 4,289.5

$\bar{y}$ = (1/25)(16 + 26 + ... + 53) = 31.96

$n\bar{x}\bar{y}$ = 342.733 x 10^4

242

$$\sum xy = (1,356)(16) + (1,994)(26) + \ldots (8,260)(53) = 384.439 \times 10^4$$

$$\sum x^2 = (1,356^2 + 1,994^2 + \ldots + 8,260^2) = 595.175 \times 10^6$$

$$\sum y^2 = (16^2 + 26^2 + \ldots + 53^2) = 27,519$$

$$n\bar{x}^2 = 459.999 \times 10^6 \qquad n\bar{y}^2 = 25,536$$

$$r = \frac{(384.439-342,733) \times 10^4}{(595.175-459.999) \times 10^6 \times (27,519-25,536)^{1/2}} = \underline{0.81}$$

One must not use a correlation coefficient blindly. Two variables might correlate quite well but yet have no cause-effect relationship. A plot of birth rate versus the number of acres of timber might have a correlation coefficient of 0.95 even though the premise is sheer nonsense. The correlation must be physically reasonable from the cause-effect viewpoint, regardless of the correlation coefficient.

Also, be aware that a poor <u>linear</u> correlation coefficient might simply mean that a non-linear correlation exists, not that none exists.

Significance

We must always be concerned about whether the observed differences between our hypothesis and the results are significant or the result of pure chance. Statistical decision theory enables us to determine if the results are significant. A detailed study of these tests is beyond the scope of this study. However, some of the principles may be illustrated using the example in Table 8.5 and Fig. 8.9.

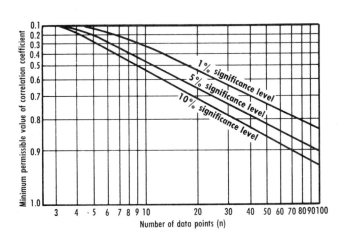

Figure 8.9 is based on one of the significance test tables. Note that it reflects the number of data points available. In practice, a 0.05 or 5 per cent significance level is usually chosen. What this means is that there are about 5 chances in 100 that we would reject the hypothesis represented by the correlation. We are 95 per cent confident that the variables are really related as shown.

Fig. 8.9 Minimum Value of Correlation Coefficient Needed at Various Significance Levels.

For the example, n = 25. From Fig. 8.9, a value of r greater than 0.6 shows a one per cent significance level. Since the value found is much higher the correlation appears to be highly significant. In addition, it is physically logical.

The range of depths was from 1000-9000 feet. If one repeated the calculation for the interval 3000-7000 feet, r = -0.04. In this range, no significant correlation of gravity with depth exists. The moral - a correlation coefficient only applies for the given range specified. Always specify this range.

SUMMARY

Through this book we will be solving equations and correlations. The numbers will be treated as certain ones to stress the technique involved. Remember ... in the real world all such numbers possess some degree of uncertainty. Assessment of the effect of this uncertainty on the calculation is necessary to obtain meaningful results.

REFERENCES

1. Newendorp, P.D., Decision Analysis for Petroleum Exploration. Tulsa: Petroleum Publishing Co., (1975).

2. Stanley, L.T., Practical Statistics for Petroleum Engineers. Tulsa: Petroleum Publishing Co., May (1973).

3. Spiegel, M.R., Statistics. New York: McGraw-Hill Book Co. (1961).

4. Kaufman, G.M., Statistical Decision and Related Methods in Oil and Gas Exploration. Englewood Cliffs: Prentice-Hall, Inc. (1963).

5. Schlaifer, R., Probability and Statistics for Business Decisions. New York: McGraw-Hill Book Co. (1959).

6. Miller, I. and J.E. Freund, Probability and Statistics for Engineers. Englewood Cliffs: Prentice-Hall (1965).

7. Grayson, C.J., Decisions Under Uncertainty, Boston: Harvard Business School (1960).

8. SPE Reprint No. 3. Dallas: Society of Petroleum Engrs. (1970).

9. Berry, V.J., Series from Oil, Gas and Petrochem Equip., June (1958) to Feb. (1959).

10. Campbell, J.M. and G.L. Farrar, Effective Communications for the Technical Man. Tulsa: Petroleum Publishing Co. (1972).

11. Lipschutz, S., Probability. New York: McGraw-Hill Book Co. (1965).

12. Chierici, G.L. et al., Jour. Pet. Tech., Feb. (1967), p. 237.

13. Capen, E.C. and R.V. Clapp, SPE Paper No. 3974, Oct. (1972).

14. Trudgen, P. and F. Hoffman, J. Pet. Tech., April (1967), p. 497.

15. Stout, J.L., SPE Paper No. 4126, Oct. (1972).

16. Root, P.J., Personal Communications.

17. Burr, I.W., Engineering Statistics for Quality Control. New York: McGraw-Hill Book Co. (1953).

18. Newendorp, P.D., J. Pet. Tech., Oct. (1976), p. 1145.

9

FORECASTING TECHNIQUES AND APPLICATIONS

Forecasting is an essential ingredient in the process of making mineral economic decisions. Decisions on the feasibility, profitability, and desirability of projects in the mineral industry depend in one way or another on forecasts of future events. The economics of a new reservoir, coal deposit, or shale lease depend directly on expectations about future prices of minerals, their availability, demand for the minerals, and the engineering economics of mineral removal and processing technologies. Forecasts of these factors are commonplace in industry and government.

Less common in both industry and government is a detailed understanding of the ramifications of their actions (or lack of action) on the economics of the mineral industry. Historically, this lack of understanding has taken the form of charges and countercharges that import quotas, depletion allowances, government leasing policies, and other policies were favoring one industry or making the U.S. dependent on unreliable foreign sources of energy. In the early seventies, new policies were unveiled - FEA control of crude oil prices and allocation programs, repeal of intangible drilling allowances, massive subsidies for new energy technologies and increased regulation of mineral industry working and environmental conditions, all in the name of energy independence and a better environment.

In every instance an essential input into the process leading to these decisions was a forecast of the economics of the mineral industry with and without implementation of these policies. Given the controversy usually surrounding announcement of these policies, a relevant question concerns the proper role of forecasting in making mineral policies and the proper procedures to employ in forecasting. These questions are the subject of this chapter.

The objective in developing quantative procedures for forecasting is to provide information to decision-makers about the likely consequences of following a particular policy on the mineral industry and the economy as a whole. Forecasts represent only one input into the decision-making process, and are the first step in a complicated, multiple step process for arriving at corporate and national policies.

At the corporate level, forecasts of mineral supplies and demands, prices, availability, profitability and government legislation are used to make new lease bids, producing decisions regarding existing wells, and new investments in other energy and synthetic fuels. Purchases of coal and nuclear deposits and development of coal gasification plants by petroleum companies, based on forecasts of petroleum demand continuing to grow faster than supply, illustrates one major mineral industry response to forecasts.

Within the political sphere, freeing the price of natural gas and revising mineral land leasing policies were touted by forecasters as enhancing mineral industry economics and society's wellbeing. Examples of policies based on forecasts are repeal of the oil depletion allowance, curbing the oil import quota, and modification of intangible drilling tax law.

Persistent and vocal rejection of mineral policies which are wrong is deserved. The practicality of many decisions affecting the mineral industry, such as control of natural gas prices, is certainly questionable, a point which history verifies continually. Inaccurate forecasts are to blame for some of these errors in mineral policy. Forecasts of higher domestic crude oil production by eliminating proration, reducing import quotas, and repealing the depletion allowance have certainly been in error. But, preparation of sound engineering and economic forecasts, which identify as best we can the effects of alternative mineral policies, provides a basis for developing the "best" policies for insuring the future wellbeing of society. Forecasts, which identify future conditions and are well documented, are essential inputs into the decision-making process, without which policies will vascillate according to the whims of decision makers. No mineral company would undertake any major new lease development or secondary recovery project without some forecast of future mineral demand and supply.

FORECASTING: ART AND SCIENCE

Everyone is a forecaster. In our personal lives we make shortrun forecasts of whether it is going to rain and what income and expenses we are going to have. Forecasting in this sense is an art since the actual forecasts are very _subjective_, with little reliance on sophisticated forecasting procedures. The minerals industry was completely dominated at one time by the art side of forecasting. Major petroleum and coal land developments were undertaken with little more than a feeling, and always the hope, that this particular venture would be profitable. Many East Texas and other early reservoirs were developed more on the belief that there was oil under the ground and that the finder of the oil could make a substantial profit than on the basis of precise analysis. Comparison of the crude nature of early reservoir evaluation, drilling, and refining technology characterizes the change which has occurred in forecasting.

In recent years the forecasting problems of reservoir volume and recoverable oil have become more sophisticated, in order to respond to the needs of engineering planning. Forecasting mineral economic conditions has changed in much the same way. Demands for better forecasts arising from the intervention of governments, growing complexity in the mineral business, and international disputes, all forced forecasting to develop more scientific procedures, consisting mainly of sophisticated, quantitative procedures made possible by the advent of the computer.

Growth in the science of forecasting has brought with it good and bad features. On the positive side, forecasters are now equipped with a stable of forecasting procedures allowing them to attack the problem of predicting the future, using scientific procedures whose logic and assumptions are well defined. Well specified approaches to forecasting permit persons receiving the forecasts to evaluate the reasonableness of the assumptions and methodology. Identifying erroneous or incorrect assumptions is a productive way to make recommendations for improving the accuracy of mineral forecasts. These positive benefits depend, however, on the decision-makers' ability to understand the scientific approach to forecasting.

But, the increasing use of sophisticated forecasting techniques has a negative side. One reason the scientific approach is desirable is the explicit detail introduced in the forecasting effort. Too much sophistication, which goes beyond the decision-maker's understanding of the assumptions and methodologies, prevents them from being able to thoroughly evaluate the forecasts. Too often forecasts are presented to decision makers in a form stating that a several hundred equation model or an autoregressive time series model were used to obtain the forecast of interest. Statisticians, econometricians, and operations researchers are all guilty of this, and in all too many cases, the decision maker is placed in a position of having to accept forecasts he does not agree with and/or does not understand. The result of this approach is often acceptance of invalid forecasts and rejection of valid forecasts. An example of this is the increased usage of complex reservoir simulation in petroleum engineering models to evaluate the feasibility of enhanced recovery projects. Executives are often swamped with discussions of random numbers, and probability distributions in presentations designed to indicate the economics of a reservoir. Statements like "I'm 75 percent sure that the mineral exists in commercially recoverable amounts" often times add more to the confusion than to the ability to make the decision to initiate such a project.

The function of scientific forecasting is to provide a basis to which the artistic element of forecasting can be added. The artistic side of forecasting cannot be avoided. This is because, even though anticipating the future on the basis of the past is valid to a point since the future is never entirely different from the past, future mineral economic conditions will never be exactly like past conditions. Scientific forecasting provides an understanding of what the past can tell us about the future. But the art or feel element of forecasting is necessary to anticipate future deviations from the past.

Combinations of art and science are common in technical analysis. In the case of downhole breaks or obstacles, "fishermen" begin with basic scientific principles and modify these principles according to their "feel" for specific problems. Rapid increases in the spot price of coal from about $20/ton after the OPEC embargo present a similar problem in forecasting. No scientific forecast can anticipate such drastic changes. Someone with a feel for the interrelationship between different mineral supplies, demands, and prices has to modify scientific forecasts.

Technically competent forecasters are not always the best judges of the artistic component of forecasting. Individuals with little or no training in technical forecasting may indeed do a better job. These individuals are often the ones with a better understanding of new technologies or eminent legal changes which may drastically alter the economics of past activity in the mineral industry. Perhaps the best way to describe the need for more interaction of science and art is to modify a common criticism of economists, which says, "economics is too important a subject to leave to economists." On the subject of forecasting the phrase becomes "forecasting is too important a subject to leave to forecasters." What we are saying is that the "art" and the "science" must be combined in an effective manner to produce meaningful forecasts.

TYPES OF FORECASTING METHODOLOGIES

Procedures for forecasting vary greatly in their complexity, basic assumptions, and usefulness. Some procedures are simple to use, but also provide little information; other procedures yield substantial amounts of information, but also take more time and effort to obtain. In between these two extremes are a variety of forecasting methodologies. Which methodology do we recommend is an often asked question that is impossible to answer easily. The importance of the forecast first must be decided upon. Obviously, if the forecast variable is critical to the decision-making process, detailed analysis may be justified. If the forecast is being·undertaken just for the sake of general conclusions, the time and money cost of engaging in a large-scale forecasting effort is hardly warranted.

Forecasts of future revenues from a petroleum reservoir or coal deposit for a present value or rate-of-return calculation of a $100,000 investment require less accuracy than a $1 billion investment. Crude oil price forecast accuracy is more important for a coal gasification or syncrude investment than for a small petroleum deposit. At $7.00/bbl. the coal gasification and syncrude investments may be undesirable. With an expected price of $15/bbl they may be desirable investments. In each case the importance of forecast accuracy dictates the forecasting effort to be undertaken.

Forecasting models are an abstraction from the real world, regardless of whether engineering or economic variables are being forecast. There are no exceptions. Even subjective forecasts which are made totally from the experience of knowledgable professionals are abstractions. Persons using the subjective forecasting approach have a model of the mineral industry in mind, without specifying explicitly what the model is. It is this feature - the degree to which a forecasting model is specified, that distinguishes among the forecasting methodologies.

Basic types of forecasting models, distinguished by the complexity of the forecasting procedures, are described below with regard to their strengths and weaknesses. The forecasting outline, of course, does not cover all forecasting approaches because of the number of such approaches and their high degree of interrelationship in some cases. But the outline does cover the approaches commonly used and is a convenient way of comparing them.

The basic forecasting models are:

1. Subjective

2. Deterministic

3. Statistical

Subjective Models

Subjective models characterize forecasts based on the experience of individuals which forms his model of the mineral industry. Subjective models are usually introduced by the preface "I think" or "from my experience" and proceed to "the (blank) is likely to occur." In certain cases this approach may be the most accurate forecasting procedure and should not be discounted simply because it is considered to be unscientific. It cannot be accepted unquestioningly either.

An example will illustrate the pitfalls of the subjective model. Several years ago at a meeting of the Association of Business Economists, attending members were asked to submit their estimate of Gross National Product in the U.S. for the coming year. At the next meeting the forecasts would be opened and the company forecast coming closest to actual GNP would be declared the winner. Representatives from a mineral oriented company, who had spent a considerable amount ot time and effort in forecasting GNP for their company decided that disclosure of the company forecast was improper. So they met, made up a number, and submitted the subjective forecast, which differed significantly from the scientific forecast used in company corporate planning. When the estimates of GNP were opened a year later, their estimate was not only the most accurate, but differed by less than one percent from actual GNP.

If the story stopped here, a strong case for subjective forecasts could be made. Impressed with their accuracy, these people used the same procedure for the next two years. In both years their forecast was the worst out of almost 500 forecasts submitted. The lesson is simple: forecasts based on subjective models are not always less accurate than scientific forecasts. Offsetting the disadvantages are the likelihood that the forecasts may be less accurate. The nature of subjective forecasts makes it difficult to identify the causes of any errors or to even judge the liklihood of potential error.

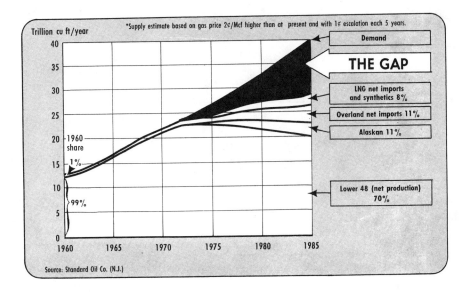

Figure 9.1 Estimate of Gas Demand and Supply. (Ref. 4)

In Figure 9.1 historical growth in gas demand is simply extended from 1970 to 1985. Gas supply, which parallels demand until approximately 1972 is assumed to become flat, producing an extremely large gap in supply and demand of 12 trillion scf. per year. These curves represent one subjective forecast relating to gas demand.

Subjective forecasts pertaining to energy consumption of all forms by Nelson (1) were published recently. Figure 9.2 summarizes historical energy consumption patterns from 1945 until 1975. In Figure 9.3 these curves are subjectively extrapolated to 1990. Nuclear and hydropower increase at a rate consistent with their growth in the 1965 to 1975 period. The share of petroleum and natural gas on the other hand diminishes in this period.

The forecasts in Figures 9.1 and 9.3 cannot be evaluated for accuracy because the basis for extrapolating past trends to 1990 is not known. Policymakers forced to make decisions based

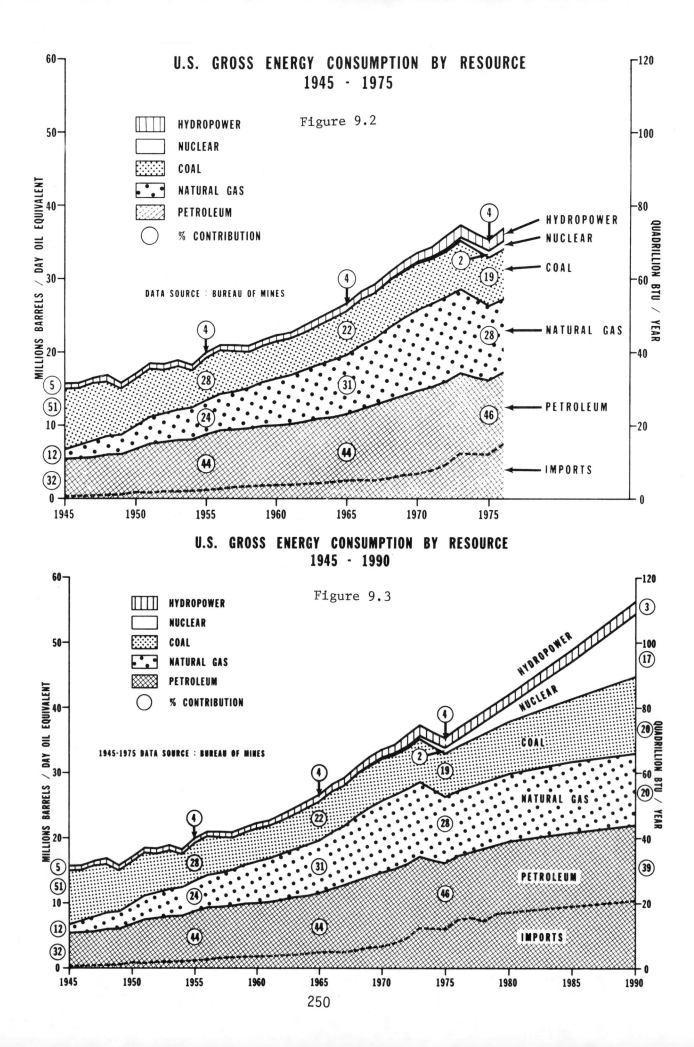

U.S. GROSS ENERGY CONSUMPTION BY RESOURCE
1945 - 1975

Figure 9.2

HYDROPOWER
NUCLEAR
COAL
NATURAL GAS
PETROLEUM
% CONTRIBUTION

DATA SOURCE : BUREAU OF MINES

HYDROPOWER
NUCLEAR
COAL
NATURAL GAS
PETROLEUM
IMPORTS

U.S. GROSS ENERGY CONSUMPTION BY RESOURCE
1945 - 1990

Figure 9.3

HYDROPOWER
NUCLEAR
COAL
NATURAL GAS
PETROLEUM
% CONTRIBUTION

1945-1975 DATA SOURCE : BUREAU OF MINES

HYDROPOWER
NUCLEAR
COAL
NATURAL GAS
PETROLEUM
IMPORTS

on these forecasts would have a hard time justifying energy decisions since they cannot really focus on concrete, defensible forecasts. Such forecasts may prove to be correct. But, unless people can understand why they might be correct, few will accept the forecasts.

Comparing the forecasts of Figures 9.1-3 illustrates the problem of evaluating subjective forecasts. Gas demand estimates in 1985 are about 40 trillion scf. from Figure 9.1 and approximately 24 trillion scf. from Figure 9.3. One reason for the difference is that the 9.3 forecast was made in 1976 and the 9.1 forecasts began in 1971. Starting date of the initial forecast does not account for all the difference because the 9.1 forecast of gas demand in 1975, the last year actual consumption used, is 27 trillion scf. Bureau of Mines data indicate actual gas consumption was approximately 23 trillion scf. for 1975.

Making a policy regarding gas demand for 1985 requires choosing the correct forecast - 40 trillion or 24 trillion scf. Without knowing the reasoning behind these forecasts the analyst has to make subjective judgments about which forecast is right.

A further factor affecting such curves is the price scenario used to predict supply. There is a question about whether or not the example forecasts are pessimistic. Price of gas will affect both supply and demand to some degree but it is difficult to gauge this since gas has always been underpriced compared to other fuel sources historically. There is strong (subjective) evidence that sharply escalating gas prices will increase supply substantially. This increased price will reflect itself in consumer price and should affect demand. We are already seeing a reflection of this with commercial customers who are going to alternate fuel sources. It is cheaper to burn coal, fuel oil or some such fuel rather than LNG which is not competively priced on a heating value basis. The homeowner cannot make this change quite so conveniently although high price tends to reduce demand via house design,, insulation, etc.

Subjective forecasts, using experience as a basis, are difficult to implement continuously for this reason. The dynamic, everchanging nature of the mineral industries makes it virtually impossible to accurately anticipate future events based on experience alone. Experience fails as a sole criterion for forecasting because few, if any, individuals have the sufficient breadth of experience to qualify as experts in anything but very narrow areas of mineral economics. Most readers are no doubt familiar with individuals whose experience rests in development of petroleum leases and who try to guess the future of oil shale development. Worst of all are the individuals who possess no real valid experience and who forecast events in both areas.

These criticisms do not mean that the opinions of such experts is to be completely discounted. Instead, the question has to be asked, "What is the specific experience which makes you believe that a certain event is going to occur?" Answers to this question will yield valuable insights in some cases and illustrate errors in judgment in others. The potential value of subjective forecasts warrants their serious consideration only as long as their worth is explicitly evaluated.

In engineering and science extrapolation of lines is very common. This is inherently more valid than subjective forecasting. First of all, there are fewer variables affecting the correlation. Second, most of these variables are relatively quantitative and values can be assigned with some degree of assurance, in a non-arbitrary manner. Third, the experienced person possesses at least an intuitive understanding about the proper relationship between variables. This is illustrated in Chapter 19 on decline curves. Too some degree they are <u>deterministic</u> models even though subjective in presentation.

Economic forecasts, on the contrary, involve many variables some of which have highly uncertain values. Arbitrarily using only a few and/or assuming fixed values leaves some real "holes" in the conclusions obtained. For this reason explanation of the logic and assumptions used is as important as the curve itself. Without this the reader cannot make the judgments necessary.

Deterministic Models

The word "deterministic" means that values are found from an explicit mathematical equation relating two variables. In analytical geometry there are a number of equations used to discuss various curve shapes when two variables are plotted versus each other. In general form these equations often are written in terms of "y" the dependent variable as a function of "x" the independent variable. There is quite an array of such equations.

We will review just a few to illustrate the most common forms. For our purpose "t" (time) will replace "x" as the independent variable. Each of these equations also contain constants to fit the equation to a set of data. We will use a, b, c, d, etc. to indicate such constants.

1. <u>Straight Line on Coordinate Paper</u>

   $$y = a + bt$$

   a is the value of "y" at t = 0 and b is the slope of the line. b is plus or minus depending on the slope of theline.

2. <u>Straight Line on Semi-Log Paper</u>

   $$y = 10^{a+bt}, \; y = ae^{bt}, \; \log y = a + bt, \; \ln y = a + bt$$

   ln is the natural logarithm; log is the logarithm to the base 10;

3. <u>Straight Line on Log-Log Paper</u>

   $$y = at^{b}, \; \log y = \log a + b \log t$$

4. <u>Parabolic</u>

   $$y = a + bt + ct^{2}$$

5. <u>Equilateral Hyperbola</u>

   $$y = t/(a + bt)$$

6. <u>Non-Equilateral Hyperbola</u>

   $$y = [(t-t_1)/(a+bt)] + y_1$$

   where: y_1 and t_1 are coordinates of any point on curve.

7. <u>Straight Line Followed by Curvature</u>

   $$y = a + bt + 10^{c+dt}$$

8. <u>Gompertz</u>

$$y = ab^{c^t} \text{ or } y = \alpha + ab^{cx}$$

9. <u>General Series</u>

$$y = a + bx + cx^2 + dx^3 + \ldots$$

The above are only the most common forms of such equations available. All are used to some extent in forecasting. Several forms, for example, will be used on decline curves in Chapter 19. Curves that flatten out on one end like (7) and (9) are often used in economic and production forecasting. Sometimes an "S" shaped curve like that shown for the Gompertz equation is obtained using an Error Function relationship.

Equations which correlate as straight lines should be used with caution over long periods of time. There are always factors which cause such curves to change slope. Consider Figure 9.1. Gas usage cannot keep rising forever. At some point, some combination of supply and price will cause a curtailment in usage and the curve will flatten out.

Other equations are often used. One, called the <u>compound growth</u> equation by economists, is written as

$$y = y_i(1+r)^t$$

where: y = production or consumption at time t

y_i = production or consumption at t = 0

t = time elapsed

r = growth rate, a fraction.

If we apply this equation in Figure 9.1 between 1970 and 1985 the following results:

$$40 = 23(1+r)^{14} \text{ or } \log 40 = 23 + 14 \log(1+r)$$

$$\log(1+r) = 0.017 \text{ and } r = \underline{0.04}$$

The growth rate in gas demand (r) is 4 percent per year. Applying the same equation to the Figure 9.3 forecast for the same time period yields a growth rate of about 1 percent per year. This example shows how subjective forecasts differ from deterministic models. We are now in a position to evaluate from judgment whether the forecasted growth rate is plausible.

These equations are common in engineering and economics. The characteristic common among all deterministic equations is that the variables being explained always depend <u>exactly</u> on the equation to the right side of the equality. There is nothing wrong with this in certain cases. But in forecasting, the relevant question is "Are future events likely to depend exactly on one or two variables and is the effect of these variables likely to remain the same over time?"

The answer to both questions is no. Most phases of activity in mineral industries - demand and supply of minerals, profitability, cash flow, and investment decisions, all depend upon a number of variables, whose impact on these parameters has never been clearly identified at any point in time. Furthermore, dramatic changes in the structure of the world in which mineral economic decisions are made suggests that the influence of variables often taken as explaining supply and demand is also changing over time. The issue is whether a deterministic model adequately represents the uncertainty involved in forecasts. If the appropriate numbers are inserted in $y = a + bt$, a forecast of y in a future year can be obtained. The criticisms of deterministic equations can be outlined as follows:

1. How well does $a + bt$ explain past values of y?

2. How are the coefficients a and b to be determined?

3. Do a and/or b change over time?

4. Do variables besides t explain y?

5. How accurate are the forecasts likely to be?

Answering these questions is difficult. This task is made more difficult by the failure of deterministic models to recognize the obvious uncertainty surrounding any attempt to explain demand and supply in the mineral industries.

A linear relationship between crude oil production (P) and price (R), $P = 80R + 1,000$, does not recognize that actual production depends on factors other than price and may differ from production perdicted by this equation. With the following historical data, the accuracy of the deterministic model can be tested.

Period	(1) Actual Production	(2) Crude Oil Price	(3) Estimated Production	(4) Difference (2-3)
1	1,200	$3.50	1,280	- 80
2	1,200	$4.00	1,320	-120
3	1,200	$4.25	1,340	-140
4	1,280	$5.50	1,440	-160

Column 1, Actual Production, lists crude oil production for four periods. Data on actual crude oil prices (Column 2) is inserted into the deterministic equation to yield estimates of production. Deterministic models of this type for predicting production in period 5 is likely to be even less accurate because the error continues to increase.

The difference between actual production and estimated production (Column 5), a measure of the uncertainty associated with the deterministic equations, is increasing over time and is consistently negative, indicating that the deterministic equation overestimates crude oil production as price changes. Production forecasts in Period 5 are therefore likely to be too large as well. Decisions based on overestimates of this type can have serious effects on a project's feasibility. This example illustrates the problems with forecasting dynamic systems using deterministic models which assume mineral economic systems are stable.

Stable physical systems, such as Kirschoff's Law, Newton's Second Law of Motion ($F = Ma$), Maxwell's equations for electromagnetism, and enthalpy can be approximated by deterministic equations without a great loss in the precision of the estimating equation. Accurate predictions in deterministic models of this type are due primarily to the stability of the physical system on one hand, and on another, because past experience has shown that small margins of error can be tolerated in the estimates of enthalpy. Two observations can be made. First, if the variable being forecast cannot be characterized as a stable system (which most mineral economic variables may not be), deterministic models can and often do lead to serious distortions in forecasts.

Second, even stable physical systems for which deterministic models have been constructed, are seldom 100% accurate. There is a small margin for error. If this margin of error is known and acceptable, a deterministic model is certainly satisfactory. But, unlike the properties of physical systems which can be studied in a laboratory in sufficient detail to determine the accuracy of a deterministic model, most major mineral industry decisions: price, supply, and demand, for example - cannot be studied in the laboratory. The past is the only guide available for identifying the determinants of supply, demand, and other factors. And, because the past is an imperfect guide, understanding the uncertainty of forecast errors in a model are necessary to obtain reasonable forecasts. Deterministic models are not the best tool for obtaining this information.

There is nothing wrong with deterministic models per se. In certain situations, usually engineering oriented, deterministic models are proper. For most non-engineering and many engineering situations, however, a model recognizing the uncertainty involved, called a statistical model, is more appropriate since it can do everything a deterministic model can do, plus providing more information about the relationship between variables.

Statistical Models

The term statistical describes deterministic models which have been expanded to consider uncertainty. The difference between a deterministic and a statistical model is illustrated by the addition of e_t in the following equations often used by economists.

Deterministic Model	Statistical Model
1. $Y_t = a + bt$	1. $Y_t = a + bt + e_t$
2. $Z_t = Ae^{rt}$	2. $Z_t = Ae^{rt} + e_t$
3. $V_t = Vo (1+r)^t$	3. $V_t = Vo (1+r)^t + e_t$
4. $W_t = Wo/e^{rt}$	4. $W_t = Wo/e^{rt} + e_t$

Distinguishing the two sets of equations is e_t, called an <u>error term</u>, which is calculated by using the portion of the equation on the right hand side of the equality to obtain a predicted value, usually indicated by an ($\hat{}$) above the variable being explained. For example, if $Y_t = a + bt + e_t$ was the equation being used, values for a, t, and b are used to predict Y_t, indicated by $\hat{Y}_t$. The difference between Y_t and $\hat{Y}_t$, or $e_t = Y_t - \hat{Y}_t$, provides a measure of how well the equation explains Y_t. Obtaining historical values for e_t and analyzing the pattern of errors of the statistical equation, the pattern of errors likely to occur in forecasts of the future is better understood.

Residuals in the crude oil production example were -80, -120, -140, and -160 were obtained by subtracting production predicted by P = 80R + 1000 from actual production. When R = \$4.00, P = 1,320 and the residual (e_2) = 1,200 - 1,320 = -120. From these residuals we know that predictions of future production are likely to be higher than actual production and the size of the error is increasing over time.

Statistical models, although generating important information about error terms for use in evaluating forecasts, also complicate the forecasting process because of the necessity of evaluating the error terms, and this evaluation can often be difficult. Evaluating error terms is made difficult by the multitude of statistical models available for such analysis. Procedures exist for performing this analysis and some will be discussed later.

Basic to all modelling, whether statistical or deterministic, are the issues of 1) what is the best model for explaining past behavior and 2) what is the best procedure for estimating the relationship implied by the model.

Attempts to answer these questions has led to the current view that two procedures have the highest probability of producing the most accurate forecasts. The first, called time series modelling, relies solely on past values of the forecast variable. Time series modelling is the most specific application of the philosophy that the past is the basis for the future. The second approach, econometric modelling, focuses on establishing the factors responsible for a certain event occurring. For example, if the demand for coal was being forecast, the econometric approach would relate coal demand in the past to coal price, price and availability of other fuels, level of industrial activity, and so on.

Time Series Forecasting

The primary assumption in the time series approach is that the future will continue to follow the past. This assumption is important because the forecaster must first believe that events in the future will follow the patterns established in the past. The emergence of the OPEC cartel in the early 1970's as a powerful economic unit, touching off a worldwide recession, is an example of an event which many people believe affected the accuracy of the past as a guide to the future. This is true only to a certain extent because historical mineral economic data include the effects of significant changes in economic conditions. Watergate, Vietnam, the Korean War, and recessions (57-59 and 68-69), natural gas price controls, tax changes, etc., have dotted the history of the mineral industry. On the international scene the OPEC cartel, changes in governments, expropriation of mineral resources, and other events have all affected past values of the variable being forecast.

Historical changes in mineral industry economics are desirable features to have in building a time series forecasting model. Why? Because the future will also experience similar events which we have no way of adequately anticipating. Forecasts based on historical data which include major shifts are more likely to occur in future mineral economic conditions. No one can forsee with certainty future recessions. But to the extent that recessions occur fairly regularly (which they do), they can be predicted by a time series model. Applications of time series models to the electric utility industry have shown that minor recessions occur every 12 or 13 years, the last one having taken place in 1973-74.

Significant changes in the structure of a mineral industry, such as that caused by OPEC in the petroleum industry, will not be picked up by time series models. But this is not a fundamental failing of time series modelling, since other forecasting procedures are not likely to pick up these upheavels either.

Even if time series forecasts fail to accurately predict the level of mineral industry activity over a year, these models often accurately predict the distribution of mineral activity over a year. Production planning in the mineral industries requires a lead time of several months. Refineries, coal mines, and other mineral procedures anticipate the demand for their product several months ahead in order to supply an adequate quantity of their product. Too small a supply means loss of sales and too large a production results in an increase in costly inventories.

Peak demand for gasoline occurs normally in the months of June, July, December, and January. In order to meet these demands refineries begin stockpiling gasoline several months prior to the anticipated peak. Overestimates of gasoline demand in these months lead to excess inventories and "price wars" to reduce these inventories. Underestimating peak demand often causes shortages. Petroleum companies lose income in both cases.

Monthly and even daily time series models performed adequately even during the economic turmoil in the early 1970's. Monthly gasoline and heating demand are examples of areas where time series models have, and are continuing to provide detailed information about the distribution of mineral demand between months. Even though major economic upheaveals, whatever the cause, cannot be always anticipated, time series models are valuable "aids" to assist in planning for the future. It should be emphasized again that forecasts are an aid to planning, not a substitute for judgment or experience.

Time series forecasting models are characterized by the following equation:

$$X_t = A_o + (A_1)(X_{t-1}) + (A_2)(X_{t-2}) + \ldots + (A_{t-n})(X_{t-n}) + e_t \tag{9.1}$$

The forecast variable, X_t, is explained by past values of the variable. The unit of time, t, can be a year, a month, a day, or any other unit of time for which data are available. The objective of such models is to determine the coefficients describing the relationship between X_t and past values, and the past values of $X_t (X_{t-1}$ to $X_{t-n})$ which minimize the unexplained variation in X_t historically. Minimizing the unexplained variation in X_t, represented by e_t, increases the liklihood that forecasts of X will be as accurate as possible.

Methods for estimating the time series equations range from the commonly used moving average approach, to the more complicated exponential smoothing procedure, and finally to the autoregressive integrated moving average approach (ARIMA). All of these procedures are based on the above equation. The advantage of the ARIMA approach is its flexibility. The ARIMA includes the moving average, the exponential smoothing, and other time series estimation procedures as special cases. Familiarity with the ARIMA procedure is desirable for this reason since the forecaster can compare the forecasting accuracy of all these approaches using a single estimation procedure.

An illustration of the ARIMA procedure is the forecast of coal demand. Monthly coal demand (CD) forecasts can be calculated from the equation

$$\text{Coal Demand} = 10 + 0.745 \, CD_{t-1} + 0.314 \, CD_{t-6} + 0.551 \, CD_{t-12} - 0.479 \, CD_{t-24}$$

Four past values for coal demand explain coal demand in the current period. Coal demand lagged one month, lagged six months, lagged one year, and lagged two years explain coal demand. The first three explanatory variables - t-1, t-6, and t-12 - are positive, indicating trends in coal demand. The negative coefficient for coal demand lagged two years indicates cyclical changes in coal demand.

257

Forecasts of coal demand in period t are found by inserting known past values for coal demand into the above equation. For

$$CD_{t-1} = 125$$

$$CD_{t-6} = 110$$

$$CD_{t-12} = 80$$

$$CD_{t-24} = 90$$

$$CD_t = 10. + (0.745)(125) + (0.314)(110) + (0.551)(80) - (0.479)(90)$$

$$= \underline{138.63 \text{ million tons}}$$

To forecast coal demand in period t+1, replace 125 with 138.6 for CD_{t-1}. Insert historical values for CD_{t-6}, CD_{t-12}, CD_{t-24} into the same equation. For $CD_{t-6} = 112$, $CD_{t-12} = 75$, $CD_{t-24} = 100$, the equation is

$$CD_{t+1} = 10. + (0.745)(138.6) + (0.314)(112) + (0.551)(75) - (0.479)(100)$$

$$= \underline{141.85 \text{ million tons}}$$

This process is repeated until the desired number of monthly forecasts is achieved. This procedure can be repeated for as many years as desired because we never run out of numbers to use in the equations. These values are produced by the forecasts and then inserted into the equation to calculate other forecasts.

The simplicity of this forecasting procedure is the ease with which it can be implemented. All that is required is historical data on the variable being forecasted, such as the coal demand example. Texts of the forecast accuracy of time series modelling indicate that it is at least as accurate as other forecasting procedures.

Econometric Forecasting

The econometric forecasting procedure differs primarily from time series modelling in identification of a causal relationship between the forecast variable and various explanatory variables. A simple equation for forecasting X_t might be

$$X_t = f(Z_t, Y_t, P_t, \ldots) + e_t \tag{9.2}$$

where X_t is the variable being forecast. The explanatory variables represented by the function sign f are those which cause event X_t to occur. The forecasting equation differs from the time series modelling equation as the result of the variables Z_t, Y_t, P_t, instead of past values of X_t.

The oil production forecasting, $P = 80 (R) + 1000$, is an example of an econometric model. An alternative oil production forecasting model is $P = 80 (R) + 60 BI + 1000$, where BI is a measure of the demand for crude oil. The number of variables explaining X_t in Equation 9.2 depend on the complexity of the variable being forecasted. The more complex the variable, the greater the number of explanatory variables needed to explain it.

Econometric forecasting models estimate future economic conditions in a mineral industry, given forecasts of the explanatory variables. For example, if X_t was the demand for natural gas, Z_t was the level of business activity, Y_t was consumer income, and P_t was natural gas price, the demand for X_t in year (t+n) can be predicted once values for Z_t, Y_t, and P_t in (t+n) are known. If the econometric model explaining the demand for natural gas is X_t = 3000 + 20 Z_t + 60 Y_t - 19 P_t, forecasts of natural gas demand are found by inserting values for Z_{t+n}, Y_{t+n}, P_{t+n} into the equation. Values for Z_{t+n} = \$600 million, Y_{t+n} = \$200 million, and P_{t+n} = 75 cents give X_{t+n} = 3000 + (20)(600) + (60)(200) - (19)(75) = 3000 + 12,000 + 12,000 - 1,425 = 25,575 million cubic feet in the year (t+n).

Econometric models also address questions regarding appropriate policies to follow. For example, a relevant question associated with natural gas demand concerns selecting the price of natural gas at the wellhead which reduces natural gas demand to available supplies. If natural gas supply is estimated to be 20,000 million cubic feet in year (t+n), the natural gas price reducing demand to 20,000 is \$3.68. This assumes governmental units do not control either business activity or consumer income. Natural gas price is found by setting 20,000 = 3000 + 12,000 + 12,000 - 19 P_{t+n} or P_{t+n} = (20,000 - 27,000)/ - 19 = \$3.68. Alternatively, the natural gas price required to increase demand to 28,000 million cubic feet in year (t+n_ is (28,000 - 27,000)/ 19 = 52 cents.

This is an example of the "what if" questions econometric models are designed to answer. Other types of "what if" questions are: what if coal prices are controlled; what if the petroleum industry is subjected to utility type regulation; what if government mineral leasing policies, both onshore and offshore, are revised; what if OPEC ceases to be an effective economic force; and what if environmental and occupational safety regulations are enforced without regard for industry economics.

Econometric models have enormous potential for both public agencies and private corporations involved with minerals. But it is an understatement to say that econometric modelling has not been highly accurate in past forecasts and has never been widely accepted as a valid analytical procedure. The reason is a lack of understanding about the econometric model by decisionmakers on one hand, and an often flagrant disregard by many econometric modellers for the "realities" of the mineral industry on the other. The result: an almost complete misunderstanding of forecasts which were in error to begin with.

Econometric forecasting is a useful tool for planning for the future. But, like any complicated piece of machinery, it must be understood before it can be of any use to anyone. Practitioners of the science of econometric modelling must first learn something about the artistic elements of the mineral industry, and experienced mineral industry personnel must learn something about the science before econometric forecasting can realize its full potential.

GOLDEN RULES OF FORECASTING

Forecasting future mineral industry conditions is a complicated process. Three rules apply to all forecasting efforts. Selection of a forecasting procedure is Rule 1.

Golden Rule 1. Use several types of forecasting procedures as checks on the reasonableness of the forecasts.

259

Using several forecasting procedures also involves determining the basis on deciding which forecasting procedures to use. This is Rule 2.

> Golden Rule 2. Match the time and monetary cost of forecasting with the benefits of an accurate forecast; building expensive forecasting models is only worthwhile if they can make the company or nation prosper through better planning.

Implemention of Rules 1 and 2 still leaves perhaps the fundamental issue plaguing the introduction of reasoned forecasts: namely, what to do with the forecasts themselves.

> Golden Rule 3. Forecasts from models are not a substitute for judgment; they complement judgment. This is the art and science of forecasting.

Golden Rule 3 is the single most important factor in forecasting. Forecasts from a model are never sacred; neither are forecasts based wholly on judgment. Both approaches have produced horrendous errors in the past, ranging from the Edsel to national econometric models. Individually, the art and science of forecasting face many obstacles in producing reasonably accurate forecasts. But, the art and science together is the only hope any nation has of consistently generating forecasts which are accurate guides in planning for the future.

ECONOMETRIC FORECASTING

Overview

Building an econometric forecasting model is composed of nine steps. The sequence of steps is:

1. Development of a model which includes all of the variables suspected of having an effect on the variables being forecast.

2. Collection and verification of historical data.

3. Modification of models to reflect data availability.

4. Estimate the relationship between the forecast variable and the explanatory variables.

5. Construct forecasts for the desired time period.

6. Subject the forecasts to experienced mineral industry personnel for review.

7. Repeat steps 4 and 5 until the review process fails to turn up major weaknesses in the forecasts.

8. Update the forecasts.

9. Use the final model to ask "what if" questions.

These nine steps provide a simple framework for econometric forecasting. Each step, however, involves judgment at every step. The effort involved in each step and procedures for completing Steps 1-9 are discussed individually below.

As each step is discussed, an example is presented to illustrate the application of each procedure. Forecasting models of monthly gasoline demand and natural gas supply will be used throughout as examples. The procedures and steps are also applicable to most of the forecasting problems arising in other segments of the mineral industry.

MODEL DEVELOPMENT

Econometric forecasting is the application of statistical principles to the study of mineral economics. Development of a mineral industry econometric forecasting model is guided by economic principles for this reason. Basic to mineral economic analysis are the concepts of supply and demand, which are the cornerstone for all forecasting. Increasing mineral supplies and controlling mineral demand are the objectives of many of the proposed changes in governmental operating rules and regulations, especially during the period of the seventies. Mineral industries, in addition, plan new ventures on the basis of profitability, which depends on supply and demand. Demand and supply relationships are described to illustrate the procedures for mineral industry model development.

Demand Models

The demand for minerals is determined by a number of factors, including the price of the mineral, the income of consumers, prices of goods used in conjunction with the minerals, prices of goods which can be substituted for the mineral, and preferences of consumers. In the case of gasoline demand, the factors determining demand expressed in equation form are:

$$G_g = f(P_g, Y_t, P_s, P_c, T) \qquad (9.3)$$

where G_d represents the demand for gasoline, P_g the price of gasoline, Y the income of consumers, P_s the price of substitutes for gasoline (which there are none commonly used), P_c the price of complements such as cars, and T is the preferences of consumers regarding gasoline usage.

Variables on the right hand side of the equality in Equation 9.3 form the basis for beginning most demand forecasting models. As indicated in Steps 2 and 3 above, the list of arguments explaining gasoline demand varies depending on data availability and other factors. But Equation 9.3 captures the factors most likely to affect demand, mainly the price of the good, consumers ability to purchase the good, and the price of other items.

In the case of gasoline demand and demand for other minerals, other explanatory variables affect demand. Gasoline demand depends on the number of miles cars are driven. So add car mileage to Equation 9.3. Home heating oil or natural gas demand which depend partly on temperature is another example of the type of variable affecting mineral demand. Add any variable to the right side of Equation 9.3 you feel affects mineral demand.

Supply Models

Mineral supply depends on the profitability of the producing project. Factors affecting profitability include development costs, costs of labor, cost of materials, the cost associated with building an adequate transportation network, and, of course, these costs are traded off against the price received for the mineral. A simplified expression for some of these factors is:

$$Q_s = g(K,L) \qquad (9.4)$$

261

where Q_s is the quantity of the mineral supplied, K and L are the capital and labor used to produce a unit of mineral output, respectively, and g is the function describing the relationship between the capital and labor inputs and the quantity of the mineral produced. Mineral producers are faced with the task of taking information about this production function, mineral demand, and the cost of capital and labor, and then supplying the mineral in quantities maximizing company profitability.

Natural gas supply, however, depends on several, very important factors. Among these factors are the number of wildcat wells drilled, the success of the drilling projects, and the size of the reservoir. Each of these factors is affected by economic conditions. Lack of drilling capital reduces the number of wildcat wells drilled. Faced with a capital shortage, drilling is likely to occur on lower risk acreages rather than unknown lease sites; thereby raising the success rate. Drilling projects emphasizing the more certain domestic, onshore acreages are also likely to yield smaller reservoirs. Equation 9.5, consisting of four equations, illustrates the interaction of these factors:

$$Q_s = W \times S \times D \qquad\qquad (9.5)$$
$$W = F_1(P_g, X)$$
$$S = F_2(P_g, X)$$
$$D = F_3(P_g, X)$$

where the quantity of new natural gas supply is the product of wildcat wells drilled (W), the success ratio (S), and the acreage size of new reservoirs (D), P_g is the price of natural gas, and X represents all other variables affecting W, S, and D.

In Equation 9.5 forecasts of Q_s are made after the effect of natural gas price and other variables (X) is computed for W, S, and D separately. Analyzing W, S, D individually provides a better picture of how natural gas prices affect natural gas supply than an equation such as $Q = F_4(P_g, X)$.

Misconceptions About Supply and Demand

Mineral supply and demand as represented in Equation 9.3-9.5 mean something entirely different than the demand and supply concepts in Figures 9.1-9.3. Historical production of X gallons are often plotted over time as measures of mineral supply and demand. Figure 9.4 illustrates a typical plot of historical production and sales of gasoline.

In economics, supply and demand refer to the quantity of gasoline supplied and demanded in a year given certain prices, incomes, and other factors. Mineral economic supply and demand is a schedule of points representing the quantity of gasoline (or natural gas) supplied or demanded at different prices only. Figure 9.5 illustrates the economic concept of supply and demand. The supply curve, sloping upward and to the right, indicates an increase in supply at higher prices. Negatively sloped demand curves indicate consumers reduce consumption as prices increase.

At $2.00/Mcf producers find and produce more natural gas than at $0.50/Mcf. Consumers likewise curtail natural gas consumption following a price increase from $0.50/Mcf to $2.00/Mcf. The slope of the supply and demand lines measures the change in mineral supply (or demand) for

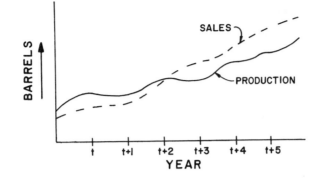

Figure 9.4   Historical Production of and
Sales of Gasoline

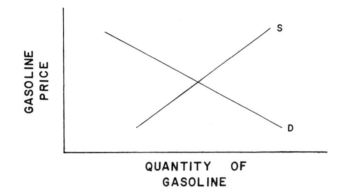

Figure 9.5   Economic Picture of Gasoline
Supply and Demand

each change in price. If, for example, natural gas price goes from $0.50/Mcf to $0.75/Mcf is associated with a reduction in natural gas demand of 1,000 Mcf to 850 Mcf, the 50% increase in price (50-75)/50 reduces natural gas demand by 15 percent [= (1000-850)/1000].

Proponents of freeing natural gas prices argue implicitly that higher natural gas prices also stimulate producers to increase the supply of natural gas. If every increase in natural gas price increases supplies by the same percentage, tripling gas prices from $0.50 to $2.00/Mcf would increase supply by 300% [= (2.00-.50)/0.50]. Estimating the precise relationship between natural gas price and supply, demand is one objective of an econometric model.

Fundamental differences exist between Figures 9.4 and 9.5. Figure 9.5 suggests that demand and supply in a given year can vary depending on the price of the mineral, holding other things like income prices of other goods constant. Figure 9.4, which traces the pattern of demand and supply over time, does not differentiate among the many factors likely to cause a particular demand or supply pattern to emerge.

Differentiating the relative influence of price and other factors on demand or supply is an important reason for econometric forecasting, and is the reason why demand, supply model lines proceed on the basis depicted in Figure 9.5. Why is this differentiation important? Because ac-

263

curate forecasts require at least a general understanding of why the lines such as those in Figure 9.4 occurred before the effect of future changes in mineral policies can be anticipated.

Dramatic changes in mineral industry supply and demand in the early seventies, caused by a series of significant movements in the world mineral picture, affected every nation's economy. Drawing in a sharp decline in Figure 9.4, such as in Figure 9.3 for natural gas, does not measure the relative importance of the rapid rise in crude oil prices, the reduction in consumer incomes, and the increased costs of new cars. Without an understanding of mineral supply and demand responses to such factors, it is difficult, if not impossible, to plan for the future.

A second reason for developing models around the idea expressed in Figure 9.5 concerns the "What if" questions being asked about the mineral industry. A simple example illustrates the importance of concentrating on Figure 9.5. Assume gasoline demand (G) is determined only by the price of gasoline (P_g) and consumer income Y, and has the following relationship:

$$G_g = 0.4 \ P_g + 0.5 \ Y \tag{9.6}$$

This equation is interpreted to mean that a one-unit increase in P_g reduces gasoline demand by -0.4 units, and a one-unit increase in I increases G by 0.5 units.

Gasoline demand observed in the past (Figure 9.4) represents the net result of changes in income and gasoline price. If income and gas price both rise by 1 unit (a $ in both cases), then gasoline demand will actually increase by 0.1 gallons [(0.5)(1) - (0.4)(1) = 0.1]. If income had not changed when gasoline price increased, gasoline demand would be reduced by 0.4 gallons [(0.5)(0) - (0.4)(1) = -0.4]. Further, a one-dollar decline in income associated with rising gasoline prices, such as occurred in 1973-74, reduces gasoline demand 0.9 gallons [(0.5)(1.0) - (0.4(1) - 0.9].

Historical data therefore represent the <u>net</u> effect of a variety of factors acting on mineral supply and demand. Forecasting models which ignore effects of price, income, and other factors will only be accurate forecasting models as long as the explanatory variables change at the same rate over time. When a relative change does occur, future gasoline demand will deviate greatly from historical patterns.

With an econometric model of the type illustrated in Figure 9.5 the forecaster can evaluate the importance of relative changes of price and income on gasoline demand. In Equation 9.6 equal changes in gasoline price and income affected demand equally because the relative magnitudes of the coefficients (0.4 and 0.5) are about the same. But, if the income coefficient was 2.0, small changes in income would have an impact on demand substantially larger than gasoline price.

DATA COLLECTION

Models, such as those developed in the first step, provide the basis for collecting data. Two problems are encountered in every mineral industry study. First, data for many of the variables selected as being major determinants will not be available at all or are not available in a form that is useful for analysis. An obvious example of data unavailability is the "taste" variable in the mineral demand equation. Taste is unmeasurable, so it cannot be collected. But in spite of this problem, it is better to start with a complete forecasting model and discard variables later than to ignore potentially important variables.

Models should be as completely specified as possible because, even if data are not available, a suitable proxy (substitute) variable may be found. For taste, suitable proxies are unlikely to be found, even though some inadequate proxies are occasionally used. Another example is represented by FEA requests for data on the actual price paid for minerals, not just the posted price. Posted prices for crude oil sold in the international market overstate the actual price received by the producer. If the posted price is $14.25/bbl and the actual price is $13.25/bbl, due to various quantity discounts, producers receive $1.00/bbl less than the posted price. Profits on a barrel of crude oil are less than profits based on posted price in this case. Given that actual price is not always available, should crude oil price be dropped from forecasting equations. It depends on how well posted prices proxy actual prices. If posted prices are always a set percentage higher than actual prices, there is no problem. If the percentage varies, more detailed evaluation is required.

Proxy variables are not the same as the desired variable, and as such, contains errors. The issue of errors in the measurement of a variable is the second real problem facing the data collection task. It is safe to say that along with taxes and death, the only sure thing is that all data will contain some errors, and this is a fact which has to be lived with in developing forecasting models. There is nothing that can be done about most of these errors. What is required is an understanding of the type and magnitude of the errors. Are the errors always in the direction of making the observed data larger or smaller than the desired data (such as international crude oil prices for example)? How large is the error? These are but two of the questions which must be answered. By answering these questions the forecaster has a hint of the meaningfulness of the forecasting model.

Everyone who has used mineral statistics published by companies, trade associations, lobby groups, or government agencies quickly realizes that the data are not always precise. The lack of precision does not imply incompetence or devious motives on the part of any group; rather it merely indicates the high cost and extreme difficulty in obtaining the data in the first place. The statistical data should be understood for what it is - much needed information which is less than 100 percent accurate. But, in essence, data for which the errors are understood are better than no data at all.

MODEL MODIFICATION

Reworking the original model in light of recognized data deficiencies is perhaps the most important, and least understood element of econometric forecasting. Economic theory has developed a rich repository of material to draw upon in the course of model development. But when available data does not match that material, little if any guidance can be given. Choice of substitute variables is part of the art of econometric forecasting, and these choices must be based on a thorough understanding of the mineral industry under study. There is no adequate substitute for this knowledge.

Model modification is not an exact science. The science that does exist suggests that the substitute variables should be closely correlated with the desired variable. Posted crude oil prices may be correlated with actual prices paid, and, as such, are an adequate substitute for the unknown actual crude oil price. In the absence of any data on the desired variable, the scientific approach reverts quickly to the art of forecasting.

One simple rule should be followed, however - always attempt to understand why a particular proxy variable was chosen and what the expected error is. Misunderstandings among those evaluating the forecasts and the reason for including a particular variable can be avoided in this way.

MODEL ESTIMATION

Estimating the relationship among variables - 0.4 for price oil and 0.5 for income in Equation 9.6 is the easiest forecasting step in the sense that a number of well-documented, accepted procedures are available for selecting the best econometric forecasting model, and is, thus, the area which is the most scientific in forecasting. Scientific procedures discussed as a part of model estimation hinge on judgmental factors used in developing the model and selecting the data.

Going from model development to estimation involves three distinct phases:

1) Selection of equation functional form

2) Selection and application of estimation procedure

3) Evlauation of model adequacy.

Selection of Functional Form

Mineral demand and supply models detailed in Equation 9.3 and 9.4 were given an implicit functional form in that "f" and "g" did not specify a precise relationship between the variable being explained, often called the dependent variable, and the explanatory variables. In selecting the functional form, "f" and "g" have to be expressed explicitly. Some common functional forms and examples of demand models using the functional forms for $G = F(P_g, Y)$ are:

Function Name	Function Form	Example
Linear	$Y = mx + b$	$G = A_o + a_1 P_g + A_2 Y$
Logarithmic	$Y = bx^m$	$\log G = \log A_o + A_1 \log P_g$
Semi-Log	$Y = be^{mx}$	$\log G = A_o + A_1 P_g + A_2 Y$

Almost any functional form felt to be valid can be estimated. But the most commonly estimated equations are the linear and logarithmic. For this reason, the following discussions work with these functional forms.

There is no easy basis for selecting the functional form. But the functional form should be logical. An example of one illogical form encountered while advising an energy company on forecasting was the model presented by the staff, which was:

$$Q_t = b_1 + b_2 X_t + b(Q_t - Q_{t-1}) + e_t$$

The dependent variable being forecast, Q_t, was made a function of X_t and the change in Q_t between the last two years. By all the standard procedures for evaluating forecast accuracy, this equation was very good. What is wrong? Suppose the year is 1980 and we want to forecast Q_t in 1981.

It can't be done because the value of Q_t in 1981 has to be known to compute Q1981 - Q1980, which is not known. Logical _consistency_ is obviously an essential ingredient in model specification.

Graphical analysis of changes in variables is a convenient first approximation of the proper functional form to choose. If after plotting natural gas supply against gas price and line A in Figure 9.6 is obtained, an exponential or double log functional form might be appropriate. A line like B indicates a linear functional relationship would be more appropriate. Each variable in a mineral industry econometric model can be analyzed in this way as a first step in selecting the proper functional form.

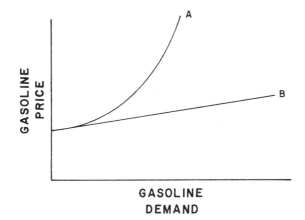

Figure 9.6 Various Relationships of Supply and Demand Curves

Oftentimes the dependent variables in econometric models display an exponential relationship with one variable (A) and a semi-log relationship with another variable (B). For the gasoline demand model relating demand as a function of price and income, gives an equation like:

$$Q_t = (A_o)(P_g)^{A_1}(e^{A_2Y})(e^t)$$

$$\ln Q_t = A_o + A_1 \ln P_g + A_2Y + t \qquad (9.7)$$

after taking logarithms of both sides. Other equation specifications are possible and must be determined by the model builder.

Selection of Estimation Procedure

An estimation procedure exists for almost every functional form. Rather than present and discuss these estimation procedures, the estimation procedure which is the most commonly used and has the most desirable statistical properties for many functional forms is emphasized. The term ordinary least squares, Gauss-Markov, and curve fitting are all names given to the procedure which minimizes the unexplained variance in the forecasting model.

The importance of unexplained variation in forecasting models warrants further discussion. The distinguishing feature of statistical models is the presence of an error term, e_t, in the forecasting model. In a linear gasoline demand equation, we have $G_t = A_o + (A_1)(P_g) + (A_g)(Y_t) + e_t$. The variation in gasoline demand which is not explained by $A_o + (A_1P_g) + (A_2Y)$ is included

267

in e_t. The term e_t is called a <u>residual</u> for this reason. What is required is an estimation procedure which computes A_o, A_1, A_2 in such a way that the sum of all the e_t's is the smallest. Why? Because only by obtaining the best statistical fit on past data are the forecasts likely to be most accurate. Larger errors in computing A_o, A_1, A_2 create the potential for large forecast errors.

The objective of the model estimation process is to select a model and estimation proceedure which minimizes e_t. Estimation procedures are judged by comparing the residuals, and selecting the estimation procedure with the smallest sum of error terms.

Suppose two econometric forecasting equations are estimated for a mineral decline curve using data for five years. To determine the accuracy of the equations, compare the residuals of the equations. These might look like the numbers under column heading e_t.

Year	Model 1		Model 2	
	e_t	e_t^2	e_t	e_t^2
1	10.	100.	-14.	164.
2	-5.	25.	-8.	64.
3	1.	1.	-20.	400.
4	-14.	164.	-12.	144.
5	8.	64.	-4.	16.
Sum	0	364	-56.	788.

In Model 1 the errors vascillate between positive and negative, indicating overestimates of production in one year and underestimates in other years. Model 2, where all residuals are negative, is a poor forecasting model because the signs are consistently negative, suggesting that forecasts of the future may understimate mineral production rates. Residuals covering both positive and negative values, such as Model 1, are preferred since the forecasts from the model are more likely to accurately anticipate future production. The forecasts will be too large in some years and too small in others, but will also be closer to the mineral production rates on balance.

A second way to evaluate models 1 and 2 is to compute the sum of the squared residuals. Square the residuals as in columns e_t^2 and then add the squared terms. Model 1 has a sum of squared term of 346 and Model 2, 788. Model 1 is better because its value is smaller. Remember, the smaller the value of e_t^2, the better the model is likely to be for forecasting.

Squared residuals are used to compare different models instead of the residuals themselves, since the ordinary least squares estimation procedure is one designed to produce residuals which sum to zero. This is the situation with Model 1. When the sum of the residuals does not equal zero, as in Model 2, there is something wrong with the mineral forecasting model and the model should be re-evaluated.

The objective of the estimation procedure is to minimize e_t. Suppose that in the gasoline demand model e_t equalled 80 gallons and the average gasoline demand in the sample was 400 gallons per month. The average error or (unexplained gasoline demand) in the model is 20% = 80/ 400. Forecasts of future gasoline demand can be expected to be <u>at least</u> 20% different than actual demand. With a forecast of 500 gallons in a future period, for example, actual gasoline demand could fall within a range of 420 (= 500-80) to 580 (= 500+80) gallons.

If, after re-estimating the model for the same data, e_t is reduced to 20 gallons, the forecasting accuracy of the model is improved. A 5% forecast error (5 = 20/400) implies that in a forecast of 500 gallons in a month actual demand is likely to fall in a range of 480 (= 500-20) to 520 (= 500+20) gallons, which is obviously a more accurate forecasting model. A natural gas forecasting model similarly should minimize e_t.

The lesson is simple: if a forecasting model cannot explain past changes in gasoline demand or natural gas supply, it is not likely to explain future demand and supply.

Forecasting Models for Monthly Gasoline Demand

In order to illustrate the application of the econometric forecasting modelling procedure, six econometric equations are presented as examples. These models are intended as illustrations and are not intended to actually forecast monthly gasoline demand. The six monthly gasoline demand equations are:

$$1) \qquad G_t = C_o + C_1 P_t + C_2 Y_t + e_t$$

$$2) \quad \log G_t = d_o + d_1 \log P_t + d_2 \log Y_t + \log e_t$$

$$3) \quad \log G_t = F_o + F_1 P_t + F_2 Y_t + e_t$$

$$4) \qquad G_t = g_o + g_1 P_t + g_2 Y_t + g_3 G_{t-1} + e_t$$

$$5) \qquad G_t = h_o + h_1 P_t + h_2 Y_t + h_3 G_{t-12} + e_t$$

$$6) \qquad G_t = i_o + i_1 P_t + i_2 Y_t + i_3 Y_{t-1}, + i_4 C_{t-1} + e_t$$

where G_t is demand for gasoline in month t, P is average price for gasoline in month t, Y is consumers income in month t, G_{t-1} is gasoline demand in the previous month, and G_{t-12} is gasoline demand in the same month one year ago.

Each variable is designed to measure different factors determining gasoline demand, although it is obvious that gasoline demand depends on other factors as well, such as a number of fuel consumption characteristics of automobiles.

Gasoline price captures the higher cost of driving as prices increase, which should tend to lower the demand for gasoline. Offsetting the expected negative effect of higher gasoline prices is a positive coefficient associated with income. If income increases at a faster rate than gasoline price, gasoline consumption may increase even though gasoline prices are increasing, because consumers may feel that they can afford to pay the higher price. The lagged variables, G_{t-1}, G_{t-12}, and Y_{t-1}, capture the dynamics of adjustment in gasoline demand over time. In a period of rapidly changing prices and incomes, the price and income coefficients may not be able to adequately measure consumer responses. Lagged variables, which represent adjustments in gasoline demand over time, are intended to incorporate the effects of dramatic changes in the economy.

Lagged income (Equation 6) also focuses on the problem that gasoline demand is often the greatest in the summer months when consumers travel more in taking vacations and weekend outings. Because the cost of these trips is often paid for from family savings, income in the previous month is a proxy for the resources available for financing these trips.

The price and income variables are divided by the consumers price index to convert these monetary variables to a measure of the relative cost of gasoline versus the cost of other goods and services in the economy. Deflating monetary variables in this way accounts for periods when gasoline price or income are rising (or falling) relatively faster than other products. If, for example, gasoline price rises at an annual rate of 10 percent and the inflation rate is 15 percent, gasoline is still cheaper relative to other products even though the normal price is rising. A similar situation exists for natural gas supply. Holding natural gas supply at a fixed level of $0.52 per Mscf as the FPC did in the mid-1970's lowers the profit of producers. Inflation of 10% in a year lowers the purchasing power of this $0.52 to $0.468 because the cost of goods used to produce natural gas has risen by 10%. Producers making 5.2¢ (= $0.52 x .10) exactly offsets his profits.

Table 9.1 presents the ordinary least square (OLS) estimates of the six monthly gasoline demand models. The estimated equations are generally good for monthly data as explained by the criteria outlined below.

Evaluation of Gasoline Demand Equations

Estimated equations are evaluated by:

1) Comparing the sign and magnitude of the coefficients with the sign expected from economic theory and the magnitude of the coefficients obtained in other studies of mineral supply and demand.

2) Determining the statistical significance of the coefficients.

3) Analyzing the summary statistics indicating the overall statistical properties of the forecasting model.

There are no definitive rules for identifying the best mineral forecasting model since estimated models seldom meet all of the desired criteria. Science helps in making these judgments, but the artistic element in forecasting is almost always the basis for selecting the forecasting model to be used.

Sign and Significance of Coefficients

The coefficients presented in Table 9.1 should conform to the theory of economic demand outlined earlier in the chapter: gasoline demand is expected to be negatively related to gasoline price and positively related to income. The coefficient signs are important because the sign of the coefficient determines the effect of future price changes on gasoline demand. Should a positive coefficient be used, increased gasoline demand as gasoline price rises is implied. Negative coefficients on the other hand suggest that future price increases will lower the demand for gasoline. Policy-makers would suggest programs for lowering gasoline prices to reduce gasoline demand with a positive sign, and raise prices for a negative demand. Differences in coefficient signs obviously can have significant effects on future mineral industry supply and demand.

Under column heading "Gasoline Price" in Table 9.1, the six gasoline price coefficients are negative as expected. In the income column, the signs of the income coefficients are all positive. The signs of the remaining variables also coincide with what was expected, except for income lagged one month in the last column of the independent variables.

Table 9.1

ESTIMATED EQUATIONS

DEPENDENT VARIABLE	INDEPENDENT VARIABLES*						SUMMARY STATISTICS		
Monthly Gasoline Demand (G)	Intercept	Gasoline Price (P)	Income (Y)	Gasoline Demand Lagged 1 Month (G_{-1})	Gasoline Demand Lagged 12 Months (G_{-12})	Income Lagged 1 Month (Y_{-1})	R^2	SE	DW
1. Linear Specification	255.10	-4.82 (6.59)	0.38 (13.81)				0.73	30	1.87
2. Log-log Specification	2.37	-0.264 (5.48)	0.683 (14.0)				0.75	0.066	1.98
3. Semi-log Specification	5.65	-0.11 (6.9)	0.00086 (14.4)				0.74	0.067	1.93
4. Linear Specification	134.09	-2.53 (3.2)	0.207 (5.01)	0.468 (5.5)			0.788	27.	1.98
5. Linear Speciation	160.39	-3.22 (5.7)	0.131 (4.2)		0.747 (10.7)		0.84	21.9	1.21
6. Linear Specification	134.42	-2.47 (3.1)	0.52 (1.09)	0.469 (4.4)		-0.32 (0.67)	0.78	27.4	1.99

* Numbers in Parenthesis are t-Statistics

271

With the exception of income lagged one month, the correctness of the coefficient signs does not indicate the presence of any major statistical problems. Occurrence of an incorrect sign is the first flag that the equation specification may be wrong or the data contains errors previously unnoticed. Re-examination of the data and a re-evaluation of the model specification should automatically follow in these cases. For example, if the logarithmic equation specifications produced positive price coefficients and the linear equations the expected negative sign, it would be deduced that the logarithmic equation specification is inappropriate as a forecasting model of monthly gasoline demand.

A more difficult problem occurs when one coefficient has the correct sign and another does not. In the gasoline demand example the logarithmic equations might have produced a positive gasoline price and a positive income coefficient, while the linear equations yielded negative signs for both coefficients. Both equation specifications are revised in such instances.

For econometric forecasting in the mineral industry the signs are important primarily because the value of econometric forecasting models is their detailed recognition of cause-effect relationships. Is the supply of coal likely to increase as coal prices decrease? What about petroleum and other mineral supplies? Most mineral industry experts would definitely argue that a positive coefficient is expected and would not believe an equation suggesting that coal and other mineral supplies increase as prices decrease.

Statistical significance of the coefficients is the second tool for evaluating model adequacy. Many people mistakenly believe that the larger the coefficient, the more important it is in explaining the dependent variable. This is wrong. The size of the coefficient depends only on the units in which the dependent and independent variables are measured. If the dependent variable is measured as millions of barrels or tons and price is defined as cents per barrel or ton, the coefficient of price will be larger than another coefficient measured in millions of dollars.

To illustrate this situation, in the following equation

$$Y = 10.0 + 1,011 \ X_1 + 10.3 \ X_2$$
$$(2,000) \qquad (1.1)$$

the coefficient for X_1 is 1,011 and 10.3 for X_2. The former coefficient is obviously larger. Underneath each variable in parenthesis is the standard error of each coefficient. Standard errors of regression coefficients measure the dispersion of the coefficient values around the most likely estimates of the coefficients, the values 1,011 and 10.3. That is, the coefficient is like an average measure of the relationship between the independent variables X_1 and X_2 and the dependent variable Y, while the standard error indicates the dispersion of possible coefficient values around this average. In the above example, the range the coefficients might reasonably be expected to take on is 1011-2000 = - 989 to 1011 + 2000 = <u>3011</u>, which is quite a large range. On the other hand, the coefficient for X_2 has a range of 10.31-1.1 = 9.2 to 10 X 3+1.1 = 11.4, an obviously smaller range. Coefficient X_2 is preferred because of its smaller dispersion.

If X_1 was gasoline price and X_2 was income, a one-dollar increase in the price of gasoline could reduce gasoline consumption by as low as 3011 gallons or, with equal probability, increase gasoline consumption by 989 gallons, giving a possible range in gasoline demand of 4,000 gallons. Such ranges are too large to produce accurate forecasts of future gasoline demand to be of practical value.

The range of the X_2 coefficient, 9.2 to 11.2, is useful in practical forecasts of gasoline demand. For every dollar increase in income, gasoline demand will increase by 9.2 gallons to 11.4 gallons, a fairly small range. Smaller ranges, such as exhibited by X_2, raise the probability that forecasts of gasoline will be more accurate than those derived from X_1.

The relevant test for coefficients is whether they differ significantly from zero. If the coefficient is statistically equal to zero, changes in the coefficient do not affect the dependent variable. The t-test is used to determine statistical significance. The t-test is defined as

$$t = \frac{\text{value of coefficient - zero}}{\text{standard error of the coefficient}}$$

Compute t in the above example by t_{X_1} = (1011-0)/2000 = 0.50 and t_{X_2} = (10.3-0)/1.1 = 9.26. These are meaningless except when compared to a table indicating significance levels. Table 9.2 presents numbers for the t-distribution used in computing the t-statistic. With 25 observations on Y and 2 independent variables (X_1 and X_2), the equation has 23 degrees of freedom (25-2 = 23). Look down the column headed df, the degrees of freedom column, until 23 is reached. Next, determine the test statistic, say a 95% confidence interval, listed under heading $t_{0.95}$. Go down this column until row 23 is reached. The number at the intersection of row 23 and column $t_{0.95}$ is 1.714. This number, 1.714, is the basis for determining the statistical significance of the coefficients. Coefficients with t-values greater than 1.714 are significantly different than zero, and those less than 1.714 are considered to be equal to zero.

In the above example the t-value of X_1 (0.50) is less than 1.714, indicating that the coefficient 1011 is not statistically different from zero. Variable X_2's t-value (9.36) exceeds 1.714 and is significantly different than zero. Of the two variables, X_2 thus is the most important in explaining Y.

Previously the sign of the coefficients was discussed. It is important to realize that when a coefficient is statistically insignificant, a positive or negative sign ceases to be a measure of model adequacy. For example, in the above example if the expected sign of X_1 was negative instead of the positive, but insignificant value observed, we would not be alarmed. Insignificant values are taken as zero, and the sign of insignificant values is therefore irrelevant.

Ranges for each coefficient in Table 9.1 are calculated by dividing each coefficient by the t-statistic in parenthesis. In Equation 1, the gasoline price standard error is -4.82/6.59 = 0.72. A one-cent increase in the real price of gasoline reduces gasoline demand by 4.82 million barrels each month. The range of this reduction is 4.82 - 0.72 = 4.10 to 4.82 + 0.72 = 5.54 million barrels. The range of the income coefficient is 0.38/13.8 = 0.027. Increasing real consumer income by one dollar increases monthly gasoline demand by 380,000 barrels, with a range of 0.38 - 0.027 = 353,000 to 0.38 + 0.027 = 407,000 barrels.

These ranges should be considered in relation to total monthly gasoline demand. With monthly gasoline demands averaging 400 million barrels, the effect of a cent increase in gasoline price or a dollar increase in income does not affect total gasoline consumption greatly. Gasoline

Table 9.2

SIGNIFICANCE LEVELS FOR t-TESTS

df	$t_{0.60}$	$t_{0.70}$	$t_{0.80}$	$t_{0.90}$	$t_{0.95}$	$t_{0.975}$	$t_{0.99}$	$t_{0.995}$
1	0.325	0.727	1.376	3.078	6.314	12.706	31.821	63.657
2	0.289	0.617	1.061	1.886	2.920	4.303	6.965	9.925
3	0.277	0.584	0.978	1.638	2.353	3.182	4.541	5.841
4	0.271	0.569	0.941	1.533	2.132	2.776	3.474	4.604
5	0.267	0.559	0.920	1.476	2.015	2.571	3.365	4.032
6	0.265	0.553	0.906	1.440	1.943	2.447	3.143	3.707
7	0.263	0.549	0.896	1.415	1.895	2.365	2.998	3.499
8	0.262	0.546	0.889	1.397	1.860	2.306	2.896	3.355
9	0.261	0.543	0.883	1.383	1.833	2.262	2.821	3.250
10	0.260	0.542	0.879	1.372	1.812	2.228	2.764	3.169
11	0.260	0.540	0.876	1.363	1.796	2.201	2.718	3.106
12	0.259	0.539	0.873	1.356	1.782	2.179	2.681	3.055
13	0.259	0.538	0.870	1.350	1.771	2.160	2.650	3.012
14	0.258	0.537	0.868	1.345	1.761	2.145	2.624	2.977
15	0.258	0.536	0.866	1.341	1.753	2.131	2.602	2.947
16	0.258	0.535	0.865	1.337	1.746	2.120	2.583	2.921
17	0.257	0.534	0.863	1.333	1.740	2.110	2.567	2.898
18	0.257	0.534	0.862	1.330	1.734	2.101	2.552	2.878
19	0.257	0.533	0.861	1.328	1.729	2.093	2.539	2.861
20	0.257	0.533	0.860	1.325	1.725	2.086	2.528	2.845
21	0.257	0.532	0.859	1.323	1.721	2.080	2.518	2.831
22	0.256	0.532	0.858	1.321	1.717	2.074	2.508	2.819
23	0.256	0.532	0.858	1.319	1.714	2.069	2.500	2.807
24	0.256	0.531	0.857	1.318	1.711	2.064	2.492	2.797
25	0.256	0.531	0.856	1.316	1.708	2.060	2.485	2.787
26	0.256	0.531	0.856	1.315	1.706	2.056	2.479	2.779
27	0.256	0.531	0.855	1.314	1.703	2.052	2.473	2.771
28	0.256	0.530	0.855	1.313	1.701	2.048	2.467	2.763
29	0.256	0.530	0.854	1.311	1.699	2.045	2.462	2.756
30	0.256	0.530	0.854	1.310	1.697	2.042	2.457	2.750
40	0.255	0.529	0.851	1.303	1.684	2.021	2.423	2.704
60	0.254	0.527	0.848	1.296	1.671	2.000	2.390	2.660
120	0.254	0.526	0.845	1.289	1.658	1.980	2.358	2.617
00	0.253	0.524	0.842	1.282	1.645	1.960	2.326	2.576

price affects total consumption by about 1.2 percent (= 4.82/400) and income alters consumption by a tenth of one percent (0.1 = /480/400,000).

Of the six monthly gasoline demand equations reported in Table 9.1, the price and income coefficients are statistically significant in every equation except 6, where the income t-value is 1.09. The other income t-values generally exceed 3.2. The table t-value for 120 observations (January 1966 to December 1975) and roughly 4 explanatory variables, which we take as df = 120, is 1.658. Judged against 1.658, the only insignificant coefficients are income and income lagged one month in the sixth equation. The explanatory power of income, which is significant, in all other equations where lagged income is not present, is obviously affected adversely by lagged income. Equation 6 should be discarded for this reason.

Gasoline demand lagged one month and twelve months are positive and statistically significant. Lagged gasoline demand performs two functions: first, it tells us the length of time required for gasoline users to <u>completely</u> respond to changing gas prices. The adjustment period is calculated by dividing the coefficient of the lagged dependent variable by 1 minus the coefficient. In Equation 4 (Table 9.1), the mean adjustment lag is 0.468/(1-0.468) = 0.87, or approximately 10 months (= 0.87x12 mo.). Consumers adjust to higher gasoline prices, but this adjustment takes some time because of the time required to adjust investments in larger cars and other gasoline-using capital equipment.

Second, lagged dependent variables, such as G-1, are used to calculate the percentage reduction in gasoline demand associated with a percentage gasoline price or an income change over the 10-month period.

Measuring the effect of percentage changes in the independent variables, termed <u>elasticities</u>, is an effective way to evaluate the magnitude of the coefficients relative to other empirical studies. Comparing coefficients of one model with those found in other studies is a convenient way of testing the consistency of the results. Major differences from studying elasticities indicates potential problems which must be explained before either finding can be accepted.

The measure of the percentage reduction is called <u>elasticity</u> - <u>price elasticity</u> when dealing with gasoline price and <u>income elasticity</u> when related to income. Elasticities are calculated differently depending on the equation specification. If the equation is linear, $G = A_1 + A_2P + A_3Y$, the price elasticity is $E_p = A_2(\overline{P}/\overline{G})$ and the income elasticity is $E_Y = A_3(\overline{Y}/\overline{G})$ where the $^-$ indicates the mean value of each variable. For log-equations the elasticities are just the coefficients, or, given $\log G = \log b_1 + b_2 \log P + b_3 \log Y$, $E_p = b_2$ and $E_Y = b_3$. For semi-log equations, such as Equation 9.3 in Table 9.1, $E_p = c_2P$ and $E_Y = c_3Y$. A convenient summary for 14 linear equations is:

Type of Equation	Example	Elasticity
Linear	$Y = d_0 + d_1x$	$E_x = d_1(\overline{X}/\overline{Y})$
Log-log	$\log Y = \log d_0 + d_1 \log x$	$E_x = d_1$
Semi-log	$\log Y = d_0 + d_1x$	$E_x = d_1\overline{x}$
Semi-log	$Y = \log d_0 + d_1 \log x$	$E_x = d_1(1/\overline{Y})$

For equation specification 4 in Table 9.1 the mean historical monthly gasoline demand is 509 million/bbl, the mean gasoline price is 27.94 cents, and mean income is $866 million. The price and income elasticities are E_p = 2.53 (27.94/509) = - 0.138 and E_y = 0.207 (866/509) = 0.352, respectfully. The corresponding elasticities for the log-log equation (Equation 9.8, Table 9.1), read directly from the table), are E_p = 0.264 and E_y = 0.683. The logarithmic elasticities are larger because of the exponential relationship assumed by a log-log equation specification.

Price elasticities of - 0.138 and 0.264 are consistent with elasticities reported in other studies of gasoline demand, and, as such, do not indicate any major inconsistencies.

Elasticities are convenient measures of how gasoline demand changes as price and income vary. Previously a cent increase in gasoline price was shown to reduce demand by about one percent. Using calculated elasticities, the percentage change in gasoline demand is calculated directly. A 10 percent increase in gasoline price, regardless of the actual level of gasoline price ($.50 per gallon or $1.00 per gallon) reduces gasoline demand by 1.38 percent (Equation 9.1). A ten percent increase in real income increases gasoline demand by 2.64 percent.

Practical application of these elasticities says that if gasoline demand was 1,000 million barrels and gasoline price increases 10 percent from $1.00 per gallon to $1.10 per gallon, gasoline demand will fall by 13.8 million barrels (= 100 x 0.138). Gasoline demand after the ten percent price increase is 986.2 million barrels. The increase in gasoline demand associated with a ten percent rise in income is 26.4 million barrels.

This example illustrates the problem of observing movements in gasoline price and demand over time. Casual observation often suggests that price does not affect demand. Instead what may be happening is that every increase in price is associated with an increase in income. In the above example a ten percent increase in price reduced gasoline consumption by 13.8 million barrels, but, as the result of a 26.4 million barrel increase in consumption resulting from a ten percent increase in income, actual gasoline consumption rose 12.6 million barrels. Price had an effect which was not apparent.

These elasticities are short-run elasticities in that they refer to the adjustment in gasoline price and 0.392/0.468 = 0.837 for income. A ten percent increase in gasoline price and income changes gasoline demand by - 2.94 percent and 8.37 percent respectively over a ten-month period.

Long-run elasticities give specific information about how changes in future prices and incomes will affect gasoline demand. If gasoline price and income both increase ten percent at the same time, the long-run elasticities tell us that gasoline demand will increase in the future by 5.43% (= 8.37 - 2.94). Again the increase in income more than offsets the increase in price.

These elasticities partially explain why demand for some minerals continued to grow in 1974 and 1975 even though mineral prices rose rapidly. The fourfold increase in mineral prices in this time period dominated the income elasticity. But after a short time, rising income began to offset the price effect until consumers could afford to return to their former levels of consumptions. For forecasting the future, the question is whether income will rise faster in the future than gasoline price. If this happens, gasoline demand will continue to rise even with higher gasoline prices.

Equation Statistics

The equation statistics in Table 9.1 are measures of the forecasting models overall ability to explain gasoline demand. The three summary statistics - the coefficient of multiple determination (R^2), the standard error of the regression (SE), and the Durbin-Watson statistic (DW) - all identify specific properties of the model which are useful in determining the accuracy of the model for forecasting.

The R^2 statistic measures the variation in gasoline demand explained by the equation. An R^2 equal to 1 means that all of the variation in gasoline demand has been explained by the model. An R^2 of 0, the lower bound of the statistic, means that none of the variation in the dependent variable is explained by the model. Obviously, the closer R^2 is to 1, the more desirable the model. But, in most practical cases, the R^2 will never reach 1.

The coefficient of multiple determination is defined as the variation in the dependent variable (V_g) divided into the variation of the values predicted from the estimating equation (V_g'), subtracted from 1, or $R^2 = 1.0 - V_g/V_g'$. V_g is just the variance of the original observations on the dependent variable, while V_g' is calculated by plugging the actual values of P and Y into the estimating equations $G = 255.10 - 4.82P + 0.38Y$ and then calculating the variance of the estimated values for G.

Comparison of the R^2 values in Table 9.1 shows that there is not a great difference in the explanatory ability of equations 9.1-4 and 9.6, all having R^2 values around 0.75. About 75 percent of the variation in G is explained by these equations and approximately 84 percent is accounted for by equation 5. The absolute level of these equations is not as large as they could be nor as large as a forecaster would like them to be since roughly 25 percent of the variation in gasoline demand remains unexplained. These equations are altered later to illustrate ways of improving forecast accuracy.

The second summary statistic, the standard error of the regression, is another indicator of the explanatory power of the estimating equation. The standard error is useful in conjunction with the R^2 because it provides some insights into the accuracy of an R^2 of 0.75. Standard errors are calculated by summing the absolute value of the difference between the actual dependent variable and estimates of the dependent variable, or $SE = SQRT [\frac{1}{n} (G-G)^2]$. In Table 9.1, the estimated monthly demand in each month differs from the actual monthly demand by an average of 30 barrels. Mean monthly demand is about 509 barrels, so the average model error is about 5.8 percent of actual monthly demand.

An error of 30 barrels or 5.8 percent does not mean that forecasts of the future will have an error of 5.8 percent. In some months the forecast value will be less than actual monthly demand by more than 5.8 percent and less than 5.8 percent in other months. In still other months the forecasts will exceed actual monthly demand in the same way. Equation 1 only produces a 5.8 percent forecast on average.

The concept of average forecast error is the empirical counterpart to the story related earlier about the forecasters submitting GNP forecasts at the convention. Their forecasting procedure was bad because their average forecast error was large, not because the error was large in one or two years.

Is 5.8 percent good or bad? It depends on how accurate the forecasts have to be. If the costs to a mineral company of their magnitude of error are small, 5.8 forecast error is acceptable. If the costs are significant, further work is obviously required. Equation specification 5 with an average error of 21.9 bbl. (4.3 percent forecast error) would be better in such cases, or entirely different models might have to be developed. Again, this is an instance where the art aspect of forecasting becomes important.

Care must be exercised in applying the standard error approach. Comparison of equations 2 and 3 with the other equation specification standard errors shows considerably smaller figures. Why - because the dependent variable is measured as the log of monthly demand. Standard errors, the difference between the actual value of the dependent variable and the estimated value, depend on the units in which the dependent variable is measured. Standard errors of dependent variables measured in barrels, such as 21.9 in Equation 9.5 are larger than standard errors when logarithms of barrels is used, such as 0.066.

Differences in the size of the standard errors are compared by transforming all standard errors to similar units. In the case of the log-log equations take the anti-log. If the standard error in log form is 0.06, the anti-log is 1.16.

Another approach divides the standard error by the mean of the dependent variable. For the equation with a standard error of 0.066, the mean of the dependent variable is 6.23, giving 0.066/6.23 = 0.01. The _average forecast error_ is about one percent of the mean value of the dependent variable.

Durbin-Watson statistics measure the correlation between the residual in month t and the residual in the month t-1. A DW statistic between 1.5 and 2.5, as those reported in Table 9.1, is not correlated with the previous month's gasoline demand. High correlations between successive months indicates further work on the model is necessary to correct that misspecification.

Constructing Initial Forecasts

Forecasting is more difficult using statistical models because we presume that future levels of mineral activity depend upon other economic forces. To forecast, values for the _explanatory variables_ have to be used. Dramatic changes in the explanatory variables, either up or down, allows the forecasts to follow patterns other than just a straight line. This is what differentiates statistical forecasting from trending. In _trending_ the assumption must be made that all the explanatory variables follow their historical patterns. Significant deviations from these patterns are ruled out. Mineral industries facing dynamic movements in price, technology, costs, and other forces, trending in any form is highly inaccurate. Trending is only accurate when explanatory variables follow systematic, identifiable patterns.

Trending is used by many forecasters since it is easier than trying to anticipate further movements in the explanatory variables. Several procedures are available to obtain these values. One thing is certain, however - as much art as science will be involved, and these forecasts contain errors. One approach is to develop separate forecasting models for each explanatory variable. Such an approach is too complex, and time consuming for most applications. The alternative is to pick-up forecasts from professional forecasters. Chase Econometrics, Data Resources Institute, the Wharton School of Business, Georgia State, the Federal Reserve Board, and others publish or sell forecasts of many economic forces commonly employed in statistical models.

Another procedure allows the forecaster to specify any values for forecasts of the explanatory variables they desire. Concensus opinion might suggest that expected values for explanatory variables will follow the trend of past values, or increase at some other specified values. Table 9.3 illustrates this process using the gasoline demand example. Both gasoline price and income are assumed to increase by one percent each month. Income in the first forecast month is 1300 x 1.01 = 1313, in month two it is 1313 x 1.01 = 1326. Gasoline prices are forecast in the same manner.

Plugging the forecast values of income and price into any one of the estimating equations listed in Table 9.1 yields forecasts of monthly gasoline demand, contingent on the specified values of income and price occurring in the future. For forecast period 1 the fourth equation gives

$$GD = 134.09 - 2.53P_G + 0.207Y + 0.468GD_{-1}$$
$$= 134.09 - (2.53)(48.0) + (0.207)(1313) + (0.468)(557)$$
$$= 545$$

Forecasts of gasoline demand are calculated in the same manner. The other estimating equations are also useable, and each will give slightly different forecasts. (See Table 9.3)

Statistical forecasts of this kind vary in accuracy according to how well the anticipated explanatory variables match the values that actually occur. Errors in forecasting the independent variables do not always produce bad forecasts. Suppose the price of gasoline actually was 49¢ instead of 48¢, an underestimate of one cent. A forecast of 545 gallons is still obtained if actual income in period 1 is $1324 instead of $1313.

$$GD = 134.09 - (2.53)(49) + (0.207)(1324.6) + (0.468)(557)$$
$$= 545$$

Differences between the actual and expected values for the independent variables "washed-out" in this example. Forecast errors seldom exactly cancel each other in reality, but a certain portion of such errors will wash-out.

FORECAST REVIEW

Subjecting forecasts to a review by knowledgeable mineral industry personnel is perhaps the most essential ingredient in forecasting accurately. Ideally, the forecaster and the mineral expert (who may be the same person) should work together in a spirit of harmony to match the scientific procedures of forecasting with the knowledge of the mineral industry gained through many years of experience. But the ideal seldom occurs. Forecasters, regardless of their actual training, often look upon management reviewing the forecasts as not having the necessary experience to judge the adequacy of the forecasts. Mineral experts, on the other hand, reciprocate by arguing that forecasters are so engrossed in methodology that they lose sight of the objectives of the forecast and their results are basically unintelligible to non-forecasters.

Alas, both groups are correct. Without communication between the forecasters and the people using the forecasts, the time and monetary cost incurred in the forecasting endeavor is

TABLE 9.3

ACTUAL AND FORECAST VALUES OF
GASOLINE PRICE, INCOME, AND GASOLINE DEMAND

	EXPLANATORY VARIABLE		DEPENDENT VARIABLE
PERIOD	Gasoline Price (P)	Income (Y)	Monthly Gasoline Demand Equation 4 (linear)
A. Actual			
1	47.4¢	1,244	488
2	48.0	1,262	488
3	48.0	1,278	473
4	48.1	1,287	506
5	47.6	1,295	471
6	47.6	1,300	557
B. Forecast			
1	48.0	1,313	545
2	48.5	1,326	541
3	49.0	1,339	540
4	49.5	1,352	541
5	50.0	1,365	543
6	50.4	1,378	546
7	50.9	1,391	548
8	51.5	1,409	551
9	51.9	1,417	554
10	52.4	1,430	557
11	52.8	1,444	560
12	53.3	1,457	563

Source: Equation 9.4 $G = 134.09 - 2.53P_G + 0.207Y + 0.468G_{-1}$

wasted. Too often, management takes forecasts without understanding them and changes them to fit their subjective ideas. This is wrong. If you don't want to accept the professional guidance of the forecaster all the way through, don't waste his time and your money. Also, management is just as likely to make errors in judgment as the forecaster. The forecaster has insights to offer, so use them.

If the forecasts are not acceptable, why not? There are two possible reasons to check:

1. reevaluate the model itself

2. evaluate the adequacy of the forecasts of the independent variable.

In an atmosphere of mutual respect for the respective skills of each group, what might be said about the forecasts of gasoline demand in Table 9.3? Consensus would easily be reached that these forecasts don't make much sense and probably will not resemble the demand for gasoline which actually occurs in the forecast period. The reason: monthly gasoline demand clearly possesses variations by season, with January, December, and August typically being peak demand months. Failure of the forecasts to follow a similar pattern should be questioned unless events in the future are expected to destroy the historical seasonal variation.

The assumption of a 1 percent growth per month in both the gasoline price and income data also does not reflect events which are likely to happen. Income and gasoline price vascillate between months and this movement is what causes the variation in gasoline demand. By using gasoline price and income data increasing linearly over time, the historical pattern of gasoline demand has been altered. Restoring the forecasts to a form similar to the historical can partly be accomplished by using more realistic gasoline price and income data.

There is still the issue of how good is a model which does not account for seasonal variations. This can be accomplished by adding dummy variables to the estimating equation. Dummy variables are explanatory variables which have a value of 1 for the period which is being analyzed and is zero otherwise.

REVISION OF ESTIMATING EQUATIONS

Review of forecasting equations points out areas where the model is deficient. Incorporating identified deficiencies into the model, hopefully, increases the forecasting accuracy by making the model a better measure of reality. In the gasoline demand example two problems must be corrected - the inability of the model to reflect seasonal variations, and doubts about the forecast of gasoline price and income. The first problem, seasonal variability, is often handled by adding a dummy variable for those months when demand differs significantly from most months.

Introduction of Dummy Variables

Inspection of historical gasoline demands might suggest that demand is highest in certain months, usually August, December and January. In order to account for these deviations, the typical functional relationship is modified. Line A in Figure 9.7 shows the functional relationship between gasoline demand and income. Adding a dummy variable for the summer (DVS) and winter (DVW) in the following way

$$G = A_0 + (A_1)(P) + (A_2)(Y) + (A_3)(G_{-1}) + (A_4)(DVW) + (A_5)(DVS)$$

281

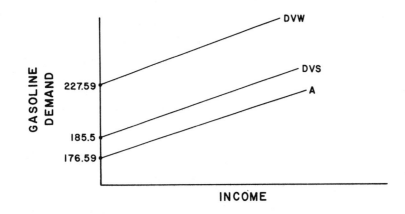

Figure 9.7 Using Dummy Variables to Explain Seasonal Variation

changes the intercept terms. This dummy variable approach assumes that gasoline demand responds to price and income changes in equal magnitudes in every month. The primary difference is in the average level of demand in each month. Equal responses are illustrated by the same slopes for the three curves in Figure 9.7. Different rates of basic demand are represented by different intercepts.

The intercept values in Figure 9.7 are obtained from the regression results

$$G = 176.59 - (3.33)(P) + (0.269)(Y) + (0.276)(G_{-1}) + (51)(DVW) + (8.9)(DVS)$$
$$(5.5)(8.3)(16.2)\phantom{(G_{-1}) + }(9.1)(1.2)$$

$$R^2 = 0.876, \quad SE = 20.9, \quad DW = 1.87$$

Adding the dummy variables increases the R^2 from 0.78 to 0.87 and reduces the standard error of the regression to 20.9 from 27.0. The coefficients still possess the right sign and are statistically significant, except for DVS. A t-value of 1.2 suggests that the intercept in summer months does not differ significantly from non-summer months. Incorporating intercept dummy variables improves the model's statistical accuracy without changing the acceptability of coefficient signs and magnitudes.

Intercept dummy variables are evaluated in the following way. The intercept value for months other than January, August, and December is 176.59 million barrels of gasoline.

In January and December the intercept term increases by 51.0 million bbls. or 227.59 = 176.59 + 51.0 million bbls. The new intercept for August is 185.49 = 176.59 + 8.9. For the other nine months the intercept is 176.59.

Forecasts Using Different Explanatory Variables

Recalculating the 12-month forecast using the same gasoline price and income data and the new estimating equation gives

282

Forecast Period	New Gasoline Forecast	Old Gasoline Forecast
1	574	545
2	530	541
3	520	540
4	518	541
5	520	543
6	523	546
7	526	548
8	536	551
9	532	554
10	533	557
11	536	560
12	589	563

The gasoline forecast under the "new gasoline forecast" heading displays a significantly different pattern than the old gasoline forecast. The new gasoline forecasts are also closer to the expectations discussed in Step 6. These forecasts, however, are still based on the questionable forecasts of monthly gasoline price and income. Further corrections are therefore required.

Table 9.4 presents an alternative set of forecasts of gasoline price and income which follow a varying pattern over time instead of a constant rate of growth as assumed in Table 9.3. These figures are again aribitrary in the sense that they are not based on a specific forecasting procedure; instead they are intended to be a more adequate reflection of monthly movements in gasoline price and income.

Comparison of the three sets of forecasts shows a substantial difference between forecasts. Table 9.4 forecasts are significantly higher than the earlier forecasts in period 12 and period 1, at a level of demand of 602 million bbl compared to 563 million bbl for the other forecasts. The best forecast is chosen from the three alternatives. The forecasts in Table 9.4 are preferred since they capture the seasonal gasoline demand pattern and because the forecasts are based upon more realistic assumptions about future movements in gasoline price and income.

More than three forecasts are considered in reality. But the principles for choosing among forecasts of mineral industry activity as presented here are to test the soundness of the forecasting equation and the quality of forecasts of the independent variables.

UPDATING FORECASTS

Tracking the accuracy of a model's forecast over time is an essential ingredient to successful model building. Models change to reflect new information generated by comparing forecast with actual values that occur. This learning process is similar to how professionals develop expertise. By making decisions and then observing the actual successes and failures they identify the important forces that they must concentrate on in future professional decisions. Statistical forecasting evaluation follows the same pattern.

Evaluating forecast accuracy involves comparison of the forecast of mineral activity with actual mineral activity. Obviously since actual mineral values are unknown at the time of the forecast, one has to wait until that data becomes available. The objective is to find significantly different patterns which consistently emerge. For example, if forecasts consistently overstate or understate actual mineral activity, further work on the model is needed. Should

TABLE 9.4

REVISED FORECASTS

Forecast Period	EXPLANATORY VARIABLES		DEPENDENT VARIABLE*
	Gasoline Price (P)	Income (Y)	Monthly Gasoline Demand (million bbl's)
1	46.8	1,313	578 (579)
2	46.2	1,325	538 (509)
3	45.6	1,336	532 (530)
4	45.2	1,346	535
5	46.1	1,357	535
6	46.4	1,374	539
7	47.3	1,382	540
8	48.5	1,389	546
9	46.9	1,405	549
10	46.1	1,411	551
11	47.4	1,425	555
12	49.4	1,434	602

*Forecasting Equation $G = 176.59 - 3.33P + 0.269Y + 0.276G_{-1}$

$$+ 51.0 \ DVW + 8.9DVS$$

the forecasts vary randomly around the actual values, some being too high and some too low, the model is acceptable.

In Table 9.4 the numbers in parenthesis indicate actual monthly gasoline demand occuring in the first three forecasting months. The forecasts are accurate in each instance, deviating from actual demand by less than 2 million bbl. This is an average forecast error of 2.0 percent [= (579-578)/579 + (540-509)/509 + (532-530)/530]/3.

The forecast accuracy for the first three months can be deceiving. The estimating equation indicated that the average forecast error in the sample period was about 20 million bbl. Continuation of this average error in the future suggests that somewhere in the future the forecasts in Table 9.4 will deviate from actual demand, perhaps by as much as 40 million barrels. At this point there is no way of knowing when these larger forecast errors are likely to occur. This is the uncertainty of forecasting. Future forecasts in general have a larger forecasting error than estimates based on historical data. Larger errors are the result of increasing errors in the forecasts of the independent variables for future periods because they are also unknown.

SUMMARY

Econometric models of mineral supply and demand forecasts the same as other procedures. But in addition, the econometric procedure also gives users the capability of evaluating the effects of policy decisions eminating from corporate management or government planners. Understanding the consequences of major mineral policy changes is an important element in deciding whether to implement a particular mineral-oriented strategy and determining the precise strategy to follow, as explained in Chapter 10.

These advantages of econometric forecasting depend almost entirely on first developing an adequate model and second, providing accurate forecasts of the independent variables. Bad models and data certainly produce bad policies. Following the nine steps outlined in the section protects against the development of bad policies. But as the discussion has emphasized, this protection hinges on the judicial use of the art, as well as the science, of forecasting. Scientific procedure does not protect against bad judgment. Combining bad judgment with poor scientific procedure is even worse.

One obvious source of errors in econometric forecasts occurs because econometric models fail to explain 100 percent of the variation in the dependent variable. Explaining 86 percent of the variation in historical gasoline demand, as the model presented previously does, means that future gasoline demands will not be predicted exactly. But the imprecision in these forecasts can be accounted for by including a range measuring the likely value future gasoline demand is expected to fall. This range is calculated by adding and subtracting the standard error of the estimating equation to the forecast variable. For the forecast values given in Table 9.4, add and subtract the standard error of the estimating equation (20.9). This gives:

Lower Range (-)	Mean Forecast	Upper Range (+)
557	578	599
517	538	559
512	532	553
514	535	556
518	539	560
519	540	561
525	546	567
528	549	570
530	551	572
534	555	576
581	602	623

The upper and lower ranges indicate the boundary of the most likely values in gasoline demand. Actual gasoline demand will fall outside this range occasionally, as gasoline demand of 502 million bbl's in Table 9.4 does. This is expected. But if too many actual values fall outside the specified range, revision of the econometric model is required.

A more difficult problem pertains to forecasting future values of the independent variables since future values of these variables are also unknown. Forecasts obtained from econometric models are only as good as the forecasts of the independent models. Some independent variable forecasts are available from national forecasting models, such as monthly income. Others, such as mineral prices, often are reported by governmental agencies. In the absence of published forecasts, separate forecasting models for the specific independent variables are necessary. One approach is often to build an econometric forecasting model of mineral prices and other variables. Linking several econometric models together creates a set of forecasting equations, all requiring implementation of the nine steps presented in this chapter which can become a time consuming and costly process.

An alternative approach to developing additional econometric models is application of the time series monthly procedure to the independent variables. Time series modeling is easier and less expensive. Forecasts generated by a time series model are particularly useful in the policy evaluation phase of econometric forecasting because the independent variable forecasts represent continuation of past trends and cyclical variation in these variables. As such forecasts of the dependent variable from the econometric forecasting model reflect likely levels of mineral supply and demand if major policy changes are not undertaken. Altering the forecasts of mineral supply or demand by inserting new values for the independent variables resulting from the policy change, such as lower profits arising from rescinding the depletion allowance, provides detailed estimates of the impact of such policies.

Errors in econometric forecasts, thus, arise from the forecasting equation used and from incorrect values assigned to the independent variables. In each instance these errors are positive, not negative, which are features of this type of forecasting for two reasons. First, these errors explicitly recognize that forecasting is surrounded by an aura of uncertainty. Recognizing and understanding uncertainty is essential in generating accurate forecasts. Second, a well-conceived, well-developed forecasting model can be modified as its deficiencies are recognized since the pattern of errors in the forecasts can be analyzed. In deterministic or objective models such modifications are more difficult. Most importantly, evaluation and discussion of the forecast errors emphasizes that forecasting is not pure science, but is by necessity a place where sound judgment is combined with available scientific procedures to anticipate an area where nothing is known for sure, the future of mineral supply and demand.

REFERENCES

1. Nelson, T.W., "Prospects for Energy Self-Sufficiency," SPE #6111, 1976.

2. Campbell, J.M. Jr. and R.A. Campbell, "Effects of Alternative Gas Pricing Stragegies on Natural Gas Reserve Additions," SPE #6318, 1977.

3. Theil, H., Theory of Econometrics, New York: North Holland Publishing Co., 1970.

4. Kinney, G., "Higher Gas Prices Coming. Question is How Much?," OGJ, June 1970.

5. Johnston, J., Econometric Theory, New York: McGraw Hill Co., 1967.

6. Theil, H., Applied Economic Forecasting, New York: North Holland Publishing Co., 1968.

7. National Petroleum Council, United States Energy Outlook Oil and Gas Availability, 1972.

8. Box, H. and L. Jenkings, Applied Statistical Forecasting, London: Cambridge Press, 1970.

9. Merrill, W.C. and K.A. Fox, Introduction to Economic Statistics, New York: John Wiley and Sons, 1970.

10. Zapp, A.D., "Future Petroleum Producing Capacity of the United States," USGS Bulletin 1142-H, 1962.

11. Adelman, M., World Petroleum Market, Cambridge: MIT Press, 1973.

10

EFFECT OF POLICY ON MINERAL SUPPLY AND DEMAND

Professionals in every discipline involved with minerals are constantly being asked questions like, "How much of a mineral actually exists in the world (or a country)?", "Will raising prices really help solve the energy crisis?", or "What is the value of the minerals located in a certain spot?" Answering these questions usually involves the statement, "It depends." It depends upon a number of technical and economic forces that have to be identified before concrete answers can be formulated.

Unfortunately, professionals often hedge when answering these questions. Hedging is correct when no definitive ideas about the forces affecting mineral supply and demand, and economic value are available. Occasions arise, unfortunately, in the real world where the professional employed in both the private and public sectors is asked to supply such answers.

In public institutions, professionals must evaluate the impact of environmental regulations, leasing agreements, pricing controls, tax changes, new technologies, and many other factors which affect minerals. Professionals working for private institutions also must respond to similar questions. To most professionals it usually is obvious that any conclusions which are reached depend upon a variety of unseen and unknown conditions.

Recognition of "it depends" factors is small consolation when some conclusions must be reached. Development of dynamic procedures for incorporating as many of these forces as possible into such analyses is the objective of this chapter. One single approach for resolving all policy issues does not exist. But by following the procedures outlined, most of the major "it depends" issues can be evaluated with reasonable accuracy.

POLICY ANALYSIS - WHAT IS IT?

Arriving at plausible solutions to mineral problems is a difficult problem at best, since the answer depends upon future economic and technological conditions. Policy analysis is a methodology designed to measure the impact of every plausible set of economic and technological forces on the production and consumption of minerals. Policy analysis does not necessarily identify the optimal solution to a mineral problem. But it does illuminate the likely outcome of public or private institutional actions on mineral economics.

In essence, policy analysis is designed to ask "what if" questions -

"What if OPEC ceases to control crude oil prices?"

"What if petroleum companies are forced to divest refining and marketing?"

"What if environmental regulations on coal are made more restrictive?"

"What if natural gas price controls are changed?"

"What if policies for leasing federal lands are changed?"

"What if investment credit and other tax laws are changed?"

Questions such as these could be generated covering several pages. But the essence of policy analysis is clear. Sound planning within both public and private institutions requires the study of the impact of new policies on mineral consumption and production. Asking "what if" questions is the first step in finding these answers.

An important consideration in policy analysis concerns its relationship to the political decision-making process. Policy analysis, as presented in this chapter, is an input into the political arena where decisions are made; it does not replace or compete with political decision making.

Professionals employed by public institutions are often reluctant to voice their professional opinions on mineral policies, even when asked directly, because they feel that it is a "political" situation. Several conflicting, but self-evident, forces are at work. One is that almost all of the decisions or recommendations made by publicly employed professionals impact political decisions, directly or indirectly. To forego making technical decisions when there is a chance of political ramifications means that no truly technical decisions will be made.

Second, it is inconsistent to hire a professional engineer or geologist on the basis of their knowledge, and then tell them that their technical opinions are not wanted because they overlap the political decision-making process, or, even worse, give them the answer and ask them to justify it with pseudo technology.

Even though the discussion has centered on public sector professionals, a similar situation exists for their counter-parts in private industry. Management in private companies formulates its "political" decisions apart from recommendations by technical personnel. Oftentimes technical personnel feel reluctant to interfere with the management prerogative for the same reason that public employees hedge in relating to political decisions.

Fundamental to resolving the reluctance of technical personnel to provide inputs in policy analysis is to recognize that the two events--technical and management (or political)--overlap. Technical professionals hold the key to sound political and management decision making since they provide the basic inputs essential to planning. Poor inputs result in poor decisions. The objective of policy analysis is to provide a framework for the professional to supply the best available technical inputs.

Given these technical inputs, politicians or management can evaluate the impact of their decisions on social welfare or corporate goals, respectively. Policy analysis does not and cannot make these decisions. All policy analysis does is quantify the likely impact of one decision versus another on mineral production and consumption. The final decision must be made based on the goals of management or the politicians' concept of social welfare.

As an example, suppose that a mineral company is considering two alternative investments. Investment A is in the traditional line of business--coal production. Investment B represents a venture into a new line of business, geothermal development. The technical and economic staff recommends Investment A because of its higher return and less risk. Management, however, selects Investment B. Does this reflect a rejection of the inputs? Not necessarily. Management may feel that technical factors are not the most important in determining the future direction of the company.

But technical personnel still have an obligation to supply the best possible information on technical matters. Supplying that information is not the same as interfering with the political or managerial process; for, without these inputs, sound planning is difficult, if not impossible.

RELATIONSHIP TO ECONOMICS AND FORECASTING

Previous chapters have laid the basic framework for incorporating engineering and geological data into a format suitable for decision making. Economics and the forecast of economic conditions derived from technical information provide the key elements in policy analysis. Given an economic framework describing the relationship between mineral supply and demand and economic forces, policy analysis becomes simply the description of the change in mineral supply, demand, or both under each policy proposed.

Among mineral policies, leasing procedure stands as an important element in determining the quanity of mineral supply. Should that leasing policy be changed? The pertinent question is how will mineral supply be altered. To measure the impact of a change in leasing policy, an economic model is employed. Suppose the following model was given:

$$MS_t = 10 - 0.35 \ LC_t$$

where: MS_t = new addition to mineral supply in year t
LC_t = lease cost in year t ($)

Mineral supply could be in tons, barrels, Mscf, Btu's, whatever units are appropriate for the mineral being studied.

With a forecast of expected lease costs under the present lease bidding system, estimates of the reduction in mineral supply attributed solely to lease costs are obtained. Table 10.1 shows a hypothetical example of lease costs in total dollar expenditure for five years.

The forecast of new mineral supply (Column B) is based on an economic model (MS = 10 - $0.35 \ LC_t$) and a forecast of lease costs under the current system (Column A).

The policy change proposed is expected to reduce lease costs by 1/2. The impact of such a policy change is found by taking 1/2 of Column A (Column C). Plugging Column C values into the economic model yields the estimates of mineral supply reported in Column D. The total increase in mineral supply is the difference between Columns D and B: 3.5, 5.25, and so on in the various years. The actual magnitude of the values in Column E depends on the units in which MS is measured, such as million Mscf, thousand tons, and so on.

Table 10.1

Hypothetical Illustration of Policy Analysis

	(A)	(B)	(C)	(D)	(E)
Year	Expected Lease Costs (Million $)	Expected Mineral Supply Given (A)	New Lease Costs	Mineral Supply	Impact of Leasing Policy
1	20	3.0	10.0	6.5	3.5
2	30	−0.5	15.0	4.75	5.25
3	35	−2.25	17.5	3.9	6.15
4	40	−4.00	20.0	3.0	7.0
5	50	−7.5	25.0	1.25	8.25

Column B equals 10 − 0.35 Col. A

Note that the figures representing the actual impact of the leasing policy will never actually be observed. That is, 1.25 units of mineral supply are expected to occur in year 5, not 8.25 units. The estimated figure of 8.25 units, nevertheless, is important since it measures the net increase in mineral supply which is generated by the alternative leasing policy.

Knowing the estimated change in mineral supply resulting from such a policy revision allows public decision makers and management to quantify the benefits (or losses). With this information they possess more accurate inputs for selecting the best course of action.

Policy evaluation of the type illustrated in Table 10.1 may also be extended to consider the time value of money (Chapter 5) by calculating the rate of production of the new mineral supplies, mineral price, and cost of production to derive cash flow projections. Discounting the cash flows yields net present values. By comparing the net present values, rates of return, or the time value of money indicator selected, the optimal leasing cost policy can be determined.

Policy analysis, thus, is the end result of economic analysis and forecasting. And, since both the economic models and forecasts hinge upon technical inputs, policy analysis requires detailed interaction among all three of these areas. Input errors usually compound as they flow through the various stages of analysis. Constant review of all inputs by experienced technical personnel, as suggested in the forecasting chapter, obviously is a necessary condition for accurate policy analysis.

GOALS OF POLICY EVALUATION

Too often evaluation of proposed policy changes attempts to establish preconceived notions or experiences, with little regard for determining the actual impact of the proposed policy. In congressional testimony, reports to management, and responses in position papers, the overriding theme is how do we demonstrate the harmful (or beneficial) aspects of the policy. Every professional knows of reports prepared in this way. "If policy X is implemented, the world faces shortages of the mineral in the future," is a common statement found in many evaluations of proposed policies.

Policy analysis may be used to make such statements. But policy analysis is much more than this. Every element inherent in policy evaluation--engineering evaluation of producible mineral reserves, rate of production, mineral demand, mineral prices, and costs of production--involves varying degrees of uncertainty. The actual shortage or abundance of a mineral is a function of the combination of these elements which occur in the future.

The goal of policy analysis is to provide a consistent framework by which different combinations of forces and their impact on mineral supply and demand are observed. Final selection of the set of events most likely to occur again eventually rests with politicians and management.

Selection of events leading to an impact often appears arbitrary in policy analysis. It is! But by formalizing the analysis using a consistent framework, some basis exists for challenging what appear to be arbitrary conclusions. Specifying an alternative set of economic forces may lead to a different, less arbitrary conclusion. Note that the primary source of disagreement in policy analysis is the specification of future economic and technological forces. In Table 10.1 a new leasing policy was assumed to cut leasing costs in half. The alternative specifications are reductions in costs of 1/4, 3/4, and so on.

The cause of conflicting opinion comes from the fact that we have never observed the impact of the proposed leasing policy, and therefore must make educated guesses. Just as two geologists will desagree about a mineral prospect, two engineers on the best operating procedure, and two economists giving 3 answers, policy analysis also includes the opportunity for disagreement.

Even with the lack of agreement, policy evaluations are definitely worthwhile. Compare the results of policy analysis based on a sound engineering and economic model of a specific mineral industry to reports circulated in the public press regarding Mr. X's or company Y's position on a policy. Such reports usually contain phrases like "in our opinion" or "according to our research" followed by a conclusion. How does a professional evaluate these results other than simply accept or reject the conclusions. He cannot. Imagine two geologists arguing about the mineral prospects of a lease without ever bothering to find out specific details of the lease or well. Such conversations quickly degenerate to statements like those above.

Anyone who has followed mineral policy discussions in recent years knows that there have been too many _arbitrary_ answers. The decision from among such answers must likewise be arbitrary.

The political process is basically one of compromise. There are two choices--intelligent compromise or arbitrary compromise. Intelligent compromise is only possible when one can examine the data and assumptions that lead to a given answer to establish the _relative_ merit of that answer. If nothing else, this educates the decision makers about what factors govern the problem. They can appraise input from parties on both sides of a confrontation, examine each input factor and come up with an intelligent compromise. It is surprising how often one can read a report whose results you agree with and yet on examination find the conclusion is based on a lot of "gibberish."

In the pages which follow we will present an overview of quantitative policy evaluation concepts. This is not for the purpose of providing magical procedures but to outline techniques that enable one to evaluate _alternative_ policy procedures.

ISSUES IN POLICY EVALUATION

Differences in perspective perhaps represents the major source of disagreement in policy evaluation. Experienced participants in reviewing reports commenting on proposed policy are often amazed at how two reports, taking basically the same data, arrive at totally opposite conclusions. Similar methodologies may even be used in some cases. Perspective, selfish interests, by whatever name one chooses, precipitates most of these differences. Company A wants a policy that is "best" from their view, consumer group B desires a policy that supplies all of the mineral they desire at a low cost, and public agency C pushes for a policy they can administer effectively.

Allocative Efficiency

From the viewpoint of a professional economist the two main considerations are the allocative efficiency and distributive efficiency of a mineral policy. A mineral policy allocates efficiently when the maximum quantity of a mineral is supplied at the lowest possible cost. Included in this price is a fair return to companies producing the mineral.

Both consumers and producers benefit when minerals are allocated efficiently--producers realize a fair return on their investment and consumers pay the lowest possible price for the mineral and mineral-derived products. Examples of distortions in allocative efficiency include natural gas price controls, oil entitlement programs, prorationing, import quotas, and any other program where producers and consumers cannot bargain freely over the price of a mineral.

Experienced mineral professionals will observe that the first two examples were severely criticized by the mineral industry, while the latter were supported. A reverse position, however, was taken historically by consumer groups, governmental agencies, and politicians. Which group was correct? Neither one! Allocative efficiency is the closest concept to a physical law in economics. There is one and only one combination of mineral price and quantity that yields the smallest price and desired quantity of production. Any other quantity-price combination is inefficient, and therefore less desirable.

One hundred percent allocative efficiency is never achieved in practice. It is a standard. Its counterparts in the physical sciences are absolute zero and isentropic processes, which are used only as a basis for relative measurements. In policy evaluation deviation from allocative efficiency measures the trade-off in mineral price and quantity made necessary by other goals specified by a society or the world community. Allocative efficiency, once found, represents a single point against which the cost and benefits of a proposed policy can be measured.

Controls on natural gas prices, for example, exemplify one society's evaluation that lower gas prices are more desirable than those which allocate gas efficiently. In this instance society through its elected officials has chosen to deplete its warehouse of a natural resource at a faster rate and a lower price than allocative efficiency would dictate. Depletion of a scarce resource is the cost of controlled gas prices. But, lower gas prices for a period of years is a benefit to consumers. Actual magnitudes of these costs and benefits are the difference between the allocative efficient levels of mineral price and quantity and the actual gas prices and consumption under controlled prices.

Distributive Efficiency

In societies where overt attention is given to special interest group needs, such as low income families, the unemployed, the aged, regional location, inflation, and so on, consideration of efficient allocation is inadequate politically. Distributive efficiency concerns the impact of policies on specific groups within society.

It is the distributive efficiency aspect of policies which has justified many mineral policies. Import quotas were felt to aid national security, prorationing to protect a scarce resource from rapid depletion, and controlled natural gas prices provided low cost energy. Other popular phrases indicating an interest in distributive efficiency are full employment, controlled inflation, life-time employment, a job with meaning, health insurance, retirement protection, etc. These items generally represent the interests of general sub-groups of the population.

Mineral companies also have an interest in distributive efficiency. This is the reason for the existence of lobbying groups representing small mineral companies. The Independent Petroleum Association of America (IPAA) is such an agency in petroleum. Coal has similar groups, as do other mineral companies. On many policy questions before public agencies the IPAA and the American Petroleum Institute (API) take different sides because one group would benefit at the expense of another. So distributive efficiency does affect the relationship among mineral companies, and between these companies and the users of the mineral.

Techniques exist within economics for determining how minerals might be distributed efficiently. Public policy formulations seldom rely on these procedures, however, preferring instead to distribute the "pieces of the pie" based primarily on political perceptions of constituent needs. Regardless of the precise reasons for deviating from allocative efficiency, costs to society will be incurred and, as will be shown later, these costs will hopefully be offset by benefits to "deserving" members of society.

Up to this point we have discussed distributive efficiency in the context of mineral companies versus public agencies. Similar trade-offs also occur within a firm. As mineral companies integrate horizontally via acquistion of other mineral product lines, management is constantly faced with the task of allocating resources among the product lines efficiently. Suppose a company has a budget of $100 million to allocate among its petroleum, coal, and synthetic divisions, and that the technical staff has recommended an even 1/3 split as being the most efficient allocation. Management, like public employees, may distribute the funds differently for reasons that are never fully explained.

What the mineral professional can provide is a measure of the cost, usually in dollars, of management's budget allocation by comparing the cash flows associated with the allocatively efficient budget with the cash flows from the budget actually implemented. This difference which is usually negative, representing foregone income, has to be offset by the goodwill, favorable public image, or other benefits that are hard to quantify.

Distributive efficiency in its general sense is designed to capture the real world phenomenon of decision makers choosing inefficient allocation of resources, both public and private, for a variety of reasons. Mineral professionals often are perplexed by such deviations from their recommendations (which they feel are efficient). They should not be. As long as human beings are subject to different motives, either as members of public agencies or as executives, they will formulate and implement policies that do not allocate mineral resources efficiently. Mineral professionals play an important role in specifying the cost of all such deviations.

TECHNIQUE OF POLICY EVALUATION

The critical phase in policy evaluation is determination of the cause-effect relationship between a policy and mineral supply and demand. Implicit in such evaluations is the need to prepare forecasts of mineral supply and demand. Like any forecasts, actual predictions of future events will contain errors. Nevertheless, policies cannot be evaluated unless forecasts exist. The key to the accurate evaluation of policy impacts is having "reasonable" forecasts of the future. Forecasts are important for the simple reason that no policy by itself is likely to destroy a mineral company. To say that a tax law or a set of price controls will affect the mineral supply and company profits is not enough. A quantitative estimate of the amount of mineral supply reduction due to lower profits from the policy is necessary; otherwise, policymakers are incapable of comparing the other benefits and costs of their policies.

Forecasts are used to answer the basic questions: What will the mineral picture look like if the policy is implemented, and what characteristics will mineral supply-demand have without the policy. An important element of these "with-without" alternatives often ignored in most policy analyses is that these are not equivalent to a before-and-after forecast. In the before-after approach, current production and consumption patterns, and economic conditions are inventoried before a policy begins. Any expected change in forecast for these conditions after the policy is started is attributed to the policy. This is the after case and it is an erroneous analysis.

Figure 10.1 illustrates the fallacy behind the before-after policy evaluation. In Figure 10.1a total production of coal is traced from period t-n to the current year, labelled t. In the before-after approach, the assumption is made that the production of coal would not change after year t, represented by the straight-line segment AB. The line segment AC signifies a forecast of coal production after the proposed policy is started. The area between ABC, represented by the lines, is the increase in coal production attributed to the policy. Concluding that the policy produces a net welfare increase to the population in the form of greater coal production given by the areas ABC is erroneous.

The cause of the error in a before-after policy analysis is indicated in Figure 10.1b. Here, a forecast of future coal production without the policy is given by line segment AE. Here, the coal production is forecast to be increasing, not remaining constant. Comparing the without increase in coal production to the with-policy increase in coal, still line AC, shows a reduction in coal production for the forecast period equivalent to the area ACE. In the before-after approach the policy is credited with increasing coal production by ABC, while the more representative with-without evaluation shows that, even though coal production is increasing, it is increasing at a slower rate. This is a social welfare loss if increased coal production is considered a desirable goal.

Similar analyses have to be performed for every element that might be identified as being impacted by a policy. The with-without evaluation is the appropriate procedure in each instance.

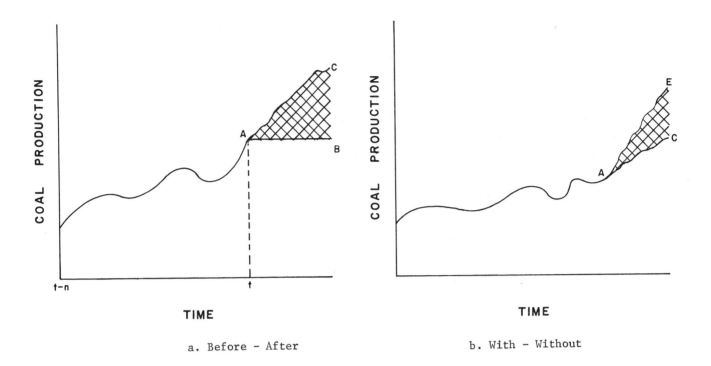

a. Before – After   b. With – Without

Figure 10.1 Hypothetical Illustration of Evaluation of
Impacts of a Coal Policy

Before-after procedures will yield reasonable policy impact estimates only when the environmental element being studied is not expected to change, as line AB indicates. The before-after estimates are accurate only because they are equivalent to the with-without estimates in these instances. Because the before-after approach is conceptually invalid, the with-without approach should always be used.

The major stumbling block in the with-without approach is obtaining accurate forecasts. Procedures for deriving such forecasts are discussed in Chapter 9. In general, the best approach for developing an understanding of the cause-effect relationship are statistical models in which historical changes in the economic and technical conditions in the mineral industry are explained as a function of other forces. Whatever the faults of the forecast may be, using the same forecast for the with-without appraisal yields results which possess _relative_ significance for decision purposes.

Time Frame of Analysis

Evaluation of cause-effect models used for the with-without case can be considered from two different time horizons. Policy analysis at the present time is concerned primarily with forecasting the future. Policymakers are thus making decisions regarding the social welfare of a policy without knowing with any degree of certainty the actual impacts that will occur. An alternative procedure is to conduct the evaluation by analyzing a similar policy that has been applied to other mineral industries. If price controls on coal or synthetic gas are being considered, we can evaluate the effect of natural gas price controls in the past to obtain an estimate of their impact.

Forecasts are required in the looking backward (or ex post) approach because the characteristics of the mineral industry without the policy still must be determined. An example of such policies would be the evaluation of prorationing or gas price controls in previous years. By comparing the actual development of the impacted mineral with the growth without such policies, policymakers could better appreciate the implications of policies proposed for other minerals and the accuracy of the forecasts associated with them. Major differences in the impact on desired mineral economic conditions in the ex post study may indicate that adverse costs outweigh positive benefits or vice versa.

Policy evaluation, while providing much needed information about changes in mineral economics, loses this value if improperly prepared and evaluated. No policy evaluation will be totally accurate in its forecasts and its estimates; nor should it be expected to. What is required is as complete an analysis as can reasonably be expected. To fault such studies for failing to be 100% accurate is to demonstrate the unreasonableness of the critic.

Arriving at Conclusions about Mineral Policies

The final decision to accept or reject mineral policies depends upon a multitude of factors, including the availability of funds, the real need for services provided by the policy, etc. Policies should only be accepted when benefits exceed costs. An important point overlooked in some studies is that both the benefits and costs occur over many years and that the proper comparison of a dollar received ten years from now and a dollar today is on a discounted basis, i.e., dollars earned in future years are converted back to the purchasing power of today's dollar (see Chapter 5). The consumption opportunities foregone by today's consumers so that future consumers would have the advantages of the output of the policy are accounted for in this way.

In their simplest form the merit of a policy to society or a firm is the sum of all the costs and benefits as measured in monetary units. Some benefits and costs, such as the cost of environmental damage from strip mining or the benefit of added national security, are difficult if not impossible to monetarize. Though this limitation reduces the value of policy analysis, it does not mean policy analysis cannot be used.

Excluding non-quantifiable elements from policy analysis explicitly recognizes the subjective value of these elements to a society or a corporation. Suppose a policy allowing strip mining on federal lands and one allowing outer continental shelf drilling have been estimated to yield a net gain (the difference in the with and without cases) of $40 million and $100 million, respectively.

If either project is begun there will be some damage to the environment. Strip mining on the designated leases will destroy the nesting area of the last remaining American desert rhinoceros, of which two are known to exist. Offshore drilling, similarly, is anticipated to jeopardize the future of a rare species of milkweed. Rather than place a specific value on the life of the two American desert rhinoceri and the milkweed (which is impossible to do), policymakers in public agencies or management can decide that the continued existence of the American desert rhinoceros exceeds a value of $40 million; whereas, the worth of the rare milkweed is less than $100 million. Strip mining would be stopped and offshore drilling would be permitted.

Whether the American desert rhinoceros, the milkweed, or any other non-quantifiable factor (such as a human life) is worth a specific monetary estimate is a decision which management and public representatives are paid to make, not the mineral professional. The function of policy analysis is the provision of the necessary information, ordered logically, so that policy-makers can reach a decision.

Some individuals may feel that making such decisions is hardhearted. It is--but everyone is required to make them in the course of their existence as human beings, and they cannot be avoided. Public agencies develop policies that may result in adverse health effects (such as the 1976 deaths from a swine flu vaccine designed to protect). The management of mineral companies undertakes policies like offshore drilling where the risk of the loss of human life exists. Individuals also assess these trade-offs when they drink while driving or exceed the speed limit. Virtually every policy decision contains within it the prospect of harmful side effects.

In summary, policy analysis in the mineral industries has three primary functions:

1) Appraisal of a policy's technical, economic, and safety impact as accurately as possible.

2) Development of a logical framework describing the cause-effect relationship among the policy and the impacted technical, economic, and safety elements.

3) Identification of the policy or a variation of a proposed policy that represents the most efficient allocation of mineral resources (allocative efficiency) and that minimizes the impact on specific groups (distributive efficiency).

Selecting the actual formula of a policy that represents a compromise of all possible trade-offs is then the task of management and public agencies.

OUTLINE OF POLICY EVALUATION

Policy analysis is roughly divisible into two distinct sections according to the time period involved--historical and future. The specific steps to be undertaken in a complete policy analysis are:

1. Historical Analysis
   a. Identify mineral industries, users, and any other populations likely to be impacted by a policy. One approach commonly employed is to organize the impacts by geographical region--county, state, nation, hemisphere, and world.
   b. Collect data regarding the historical pattern of change in the mineral supply and demand, including current conditions.
   c. Determine the theoretical causes of historical changes in mineral supply and demand.
   d. Explain statistically the cause-effect relationships hypothesized in (c).

2. Perspectives of the Future
   a. Projections without the policy
      (1) Prepare baseline forecasts assuming a minimum of two scenarios--no change in the structure of the mineral industry, and another policy, either more

restrictive or less restrictive than the proposed policy, may be substituted for the current one.

    (2) Use statistical model from 1.d, for each scenario, to forecast expected future characteristics of mineral supply and demand.

    (3) Compute confidence intervals of forecasts.

  b. Projections with policy

    (1) Identify the economic and technological forces causing historical mineral supply-demand patterns that will be altered by a policy--gas price deregulation obviously will change wellhead gas prices, environmental regulations impact costs and rates of production, etc.

    (2) Quantify changes in the independent variable due to a policy.

    (3) Compute estimates, with confidence intervals, of mineral supply and demand for a given policy.

3. Evaluation of Policy

  a. Subtract the flow of mineral quantity or money volume <u>without</u> the policy from that for the policy case being considered.

  b. Calculate the time value of money.

  c. The policy is acceptable if the time value of money calculations are positive, before consideration of elements which cannot be quantified.

Policy analysis as generally practiced seldom proceeds beyond 1.a, 1.b, and then skipping to 2.b. Analysis of historical conditions in these instances is little more than an inventory process to let policymakers know the composition of mineral supply-demand patterns. Forecasts in such situations follow the before-after approach discussed earlier. Harmful misimpressions created in this way are illustrated in Figure 10.1. Much of the so-called policy evaluation conducted by public and private employees is deficient for these reasons, and often leads to more harm than good.

Each of the steps outlined is discussed with a brief example. Steps 1 represent applications of the mineral economic modelling process presented in Chapters 8 and 9. Forecasting and derivation of the cause-effect relationships in 2.a and 1.d are explained in Chapters 7 and 9. These chapters should be read before continuing. The material that follows draws on the concepts presented therein.

The issue of natural gas pricing in the seventies provides the example that will be used throughout the discussion. To facilitate the discussion, the analysis focuses solely on the impact of controlled versus free market pricing on new additions to natural gas reserves. Important impacts of the policy, like the effect on users of gas of all characteristics, windfall profits for producers, and so on, are ignored. Extending the policy evaluation to consider these elements follows the same approach, however.

<u>Identify Impacted Elements of Mineral Supply and Demand</u>

Proposed policies often affect many areas of society--mineral supply-demand, prices, employment, inflation, balance-of-payments, and profits. An exact listing of such elements varies for each policy proposed and the objective of policy analysis. Regarding natural gas pricing this example limits its objective to just the impact on natural gas reserve additions.

Natural gas reserve additions result from several decisions made by mineral companies. Wells drilled, choice of drilling sites, and discoveries for each well, perhaps, are the most important elements affected by gas pricing.

Studying these three elements is important in evaluation of a gas pricing policy in that their interaction determines new gas additions, excluding extensions and revisions. Actual reserve additions in a year is the sum of discoveries of gas each year. Reserve additions can be expressed mathematically as the product of total wildcat wells drilled (W_i), the success rate which is determined by where the wells are drilled (SR_i), and the average discovery of gas per well (D_i) in year i, or

$$RA_i = (W_i)(SR_i)(D_i)$$

Expressing reserve additions in this way makes the complex drilling patterns in the gas industry amenable to the statistical analysis employed in step 1.d. The diversity of drilling behavior in the gas industry is difficult, if not impossible to analyze otherwise. For the natural gas pricing policy, three elements--wells drilled, success rates, and discovery per well--are analyzed. As in any modelling effort, some simplifications are necessary. Actual evaluation of alternative gas pricing strategies should invoke a more detailed analysis.

Collecting Historical Data

This phase of the policy analysis is simply a "number crunching" effort. Data collection, however, is more than just collecting numbers. Information has to be collected in a manner consistent with the objectives of policy analysis, not in a random manner. At least three decisions are made before data collection begins.

First, there is always the question of how far back in time should data be collected-- 10 years, 20 years, 6 months, 30 months? A purely statistical solution to the right number of observations is to collect at least 30 observations. If the data are measured in yearly units, 30 years of observations are desired, and 30 months if the data exist on a monthly basis. Situations arise when 30 observations cannot be collected. In such circumstances the analyst must make do with available information. Insufficient data bases affect the properties of the econometric estimates, as explained in Chapter 7. But, statistical models with minor problems are generally better than no information at all, or purely subjective analysis of that data. The lesson is--use all available data whenever possible.

From an economic perspective, the lesson is also to use all available information, but for a slightly different reason. Economics does not possess a laboratory in which the interactions of mineral supply and demand can be observed through controlled experiments. Only through careful analysis of actual, observed responses in mineral supply and demand to engineering and economic forces can the underlying structural economic relationships be identified. Economics needs observations on as many types of situations--excess supply, excess demand, price controls, lawsuits, labor strikes, and others--as possible to identify the exact impact of specific economic forces. The situation, at an extreme, is analogous to having more unknowns than equations, yielding an indeterminate solution.

A second decision concerns the level of aggregation of the data. Aggregation refers to the extent to which data are combined together. For example, in analyzing natural gas reserve additions, individual wells, specific fields, geological zones, states, regions, nations, or hemispheres could be used. Combining individual wells to fields, fields into geological zones, zones into states, and so on--all represent increased aggregation. Subject to time, cost, and data availability, it is always best to utilize the most disaggregated data to increase accuracy.

Selection of economic forces affecting mineral economics is the third factor to consider. Collecting data before these forces are identified is obviously an impossible task. A precise research design usually specifies theoretical formulation before data collection. But, because theoretical formulation and data collection occur simultaneously in practice and because supply-demand models were presented in Chapter 7, they are placed second in this discussion.

After specifying the list of economic forces to be considered and the time frame of the policy analysis, data are collected. An example of possible economic forces relating to natural gas supply and demand are presented in Table 10.2 for a 25 year period. The variables and their units of measurement are given below. For example, R,E,NPW signifies revisions, extensions, and new pool wildcats measured in quadrillion cubic feet.

Using the natural gas supply and demand model, evaluate the consequences of alternative natural gas pricing strategies on domestic gas reserves.

Stochastic Equations

1) $R,E,NPW = 4,624 + 39.71\ P_W + 882\ DV - 1,270\ SR - 5.41\ \$/FD$
$\qquad\qquad (3.18)\quad (2.41)\quad\ (1.91)\quad\ (1.26)\quad\quad (2.74)$
$\quad R^2 = 0.787,\quad F_4 - RATIO = 7.94,\quad DW = 2.312$

2) $D_R = 24.20 - 10.72\ P_R + 10.28\ P_E + 4.02\ P_{FO} + 0.013\ RY$
$\qquad\ (12.4)\quad (1.81)\quad\ (20.6)\quad\ (2.25)\quad\ (3.87)$
$\quad R^2 = 0.994,\quad F-RATIO = 1,032,\quad DW = 2.058$

3) $D_C = 73.64 - 242.1\ P_C + 25.8\ P_E - 0.911\ P_{FO} + 2.711\ IIP$
$\qquad\ (3.0)\quad (1.7)\quad\quad (3.0)\quad\ (0.1)\quad\quad (2.3)$
$\quad R^2 = 0.978,\quad F-RATIO = 250,\quad DW = 0.99$

4) $D_I = 36,741 - 1,258\ P_I + 784\ P_E + 3,630\ P_{FO} + 138.7\ IIP$
$\qquad\ (5.73)\quad (0.2)\quad\ (2.2)\quad\ (3.1)\quad\quad (3.8)$
$\quad R^2 = 0.971,\quad F-RATIO = 188,\quad DW = 1.76$

Identities

Proven Reserves (t) = Proven Reserves (t-1) + Additions (t) - Production (t)

Production (t) = Residential (t) + Commercial (t) + Industrial (t)

Residential in TCF (t) = Residential in T Btu (t)*0.001 x # of customers (t)

Definition of Variables

1) R,E,NPW = revisions + extensions + new pool wildcats (trillion CF)
2) $\quad P_W$ = natural gas price at the wellhead (¢)
3) $\qquad DV$ = dummy variable indicating a find of a significant new reservoir, such as North Slope

Table 10.2 Historical Data For The Previous 25 Years

Period	P_W	P_R	P_C	P_I	P_E	P_{FO}	IIP	D_R	D_C	D_I	R,E,NPW	SR	FD	RY
1	12.11	1.58	1.21	0.39	15.73	1.64	47.40	62.50	235.	23187.	11984.	11.2	3679.	2513.
2	12.74	1.43	1.11	0.38	14.38	1.62	51.40	70.30	255.	25268.	15964.	11.1	3870.	2538.
3	13.44	1.27	1.12	0.39	14.00	1.63	53.30	72.70	263.	26912.	14266.	11.1	4084.	2591.
4	15.62	1.47	1.15	0.40	13.63	1.69	47.70	73.20	258.	28384.	20340.	11.2	4034.	2652.
5	16.92	1.45	1.17	0.41	13.25	1.69	54.30	78.89	271.	29548.	9546.	12.2	4060.	2620.
6	17.05	1.47	1.14	0.44	12.74	1.70	61.10	85.20	294.	29166.	21896.	11.3	3964.	2712.
7	17.32	1.44	1.14	0.44	12.16	1.74	63.20	90.50	306.	30751.	24715.	10.0	4021.	2753.
8	17.37	1.41	1.10	0.43	11.53	1.76	63.70	92.50	316.	30662.	20007.	10.9	4053.	2751.
9	18.01	1.42	1.13	0.45	11.33	1.63	59.30	97.70	334.	30374.	18895.	11.3	3960.	2756.
10	19.10	1.43	1.12	0.45	10.90	1.62	66.80	101.00	350.	33477.	20620.	11.0	4041.	2811.
11	20.38	1.45	1.15	0.48	10.54	1.57	68.80	105.00	374.	33494.	13892.	10.2	4078.	2822.
12	21.79	1.47	1.15	0.50	10.36	1.61	69.39	107.00	391.	32620.	17165.	10.8	4018.	2855.
13	21.97	1.44	1.13	0.49	10.00	1.60	74.80	111.00	421.	32712.	19482.	11.6	4299.	2918.
14	22.07	1.42	1.11	0.48	9.70	1.60	78.60	112.00	430.	33629.	18163.	11.7	4222.	2969.
15	21.18	1.38	1.07	0.48	9.31	1.54	83.70	115.00	469.	37088.	20253.	10.6	4197.	3145.
16	20.99	1.35	1.04	0.47	8.86	1.54	90.70	116.00	482.	36981.	21309.	10.3	4379.	3272.
17	20.45	1.30	1.01	0.45	8.40	1.53	98.89	119.00	510.	38258.	20219.	10.3	4369.	3389.
18	20.24	1.26	0.98	0.44	8.04	1.53	100.00	121.00	537.	38709.	21803.	10.3	4291.	3476.
19	19.86	1.21	0.93	0.42	7.52	1.51	105.00	124.00	567.	39413.	13696.	8.5	4626.	3560.
20	19.25	1.16	0.89	0.41	7.06	1.47	111.00	128.00	611.	42219.	8374.	9.0	4823.	3595.
21	18.71	1.16	0.88	0.41	6.73	1.45	107.00	129.00	641.	42385.	37195.	9.7	5036.	3674.
22	18.95	1.16	0.88	0.42	6.68	1.46	107.00	130.00	673.	42120.	9824.	9.7	5077.	3733.
23	18.60	1.19	0.91	0.45	6.71	1.42	115.00	130.00	698.	42952.	9633.	11.1	5125.	3849.
24	20.39	1.18	0.89	0.47	6.59	2.48	126.00	124.00	684.	40036.	6824.	14.1	5206.	4060.
25	26.16	1.22	0.95	0.58	7.13	2.22	125.00	119.00	676.	42005.	8678.	14.2	5298.	4008.
MEAN	18.83	1.35	1.05	0.44	10.13	1.65	81.56	104.58	441.	34494.	16990.	10.9	4353.	3121.
S. DEV.	3.14	0.12	0.10	0.04	2.77	0.23	24.79	21.02	156.	5857.	6649.	1.2	469.	494.

302

4)   SR = success rate of wells drilled

5) $ FD = average cost per well drilled ($)

6) D_R = natural gas demand of residential customers (trillion Btu/customer)

7) D_C = commercial gas demand (trillion Btu/customer)

8) D_I = industrial gas demand (trillion Btu/customer)

9) P_R = real price of residential gas (¢)

10) P_C = real price of commercial gas (¢)

11) P_I = real price of industrial gas (¢)

12)   RY = real per capita disposable income ($)

13)   IIP = index of industrial production (1967 = 100)

14) P_E = real price of electricity (¢)

15) P_{FO} = real price of #2 fuel oil ($)

Many studies designed to evaluate such data merely list or graph these data, and then forecast new reserves by a simple trend analysis. New gas reserves, however, decline in magnitude until year 21 and then increase slightly thereafter, excluding the obvious major find in year 22. The extreme variation in reserve additions is illustrated by the mean reserve additions of 16,990 BCF and its standard deviation of 6,649 in this sample period. Roughly a 40% (6,649/16,990) variation occurred in the sample period. Variations of this magnitude introduce uncertainty into policy evaluation, as explained in Risk Analysis, Chapter 8.

Policy evaluation studies traditionally fail to recognize this uncertainty. Many studies just report the means of the samples, implying that the average size of new discoveries in the population equals the mean. Correct utilization of sample data requires a detailed understanding of the distribution of reserve additions. Figure 10.2 presents two hypothetical distributions of reserve additions. Point X_1 shows reserves added by one well, X_2 by another, and $\bar{X}$ the mean reserve addition per well.

Two distributions are shown in Figure 10.2. Distribution A and B have the same mean, but distribution B has a larger variance. Distribution A, as such, is not equivalent to distribution B, even though analysis of the means indicates the two distributions are equivalent. Such differences would take the form of unequal standard errors. Distribution A corresponds to reserve additions with mean 16,990 BCF and a standard deviation of 6,649 BCF, like that above. Distribution B's mean also equals 16,990 BCF, but its standard deviation is larger, equalling 13,000 BCF. Suppose you were presented with one estimate of reserve additions, but two different standard deviations. Which one would be preferable for forecasting--the smaller variance, of course.

An example of adopting an approach which ignores distributions is illustrated in Figure 10.3. Here actual gas production was plotted by the FPC to 1974, and simple linear trending, with adjustment for new drilling areas, was used to forecast future gas to 1985. Although these forecasts may be correct, knowledge of the changes in the distribution are also important for policy analysis.

Figure 10.4 depicts Curve 1 in Figure 10.3 in association with the standard deviation of production in each year. Note that in the early 60's the dotted lines signifying the standard

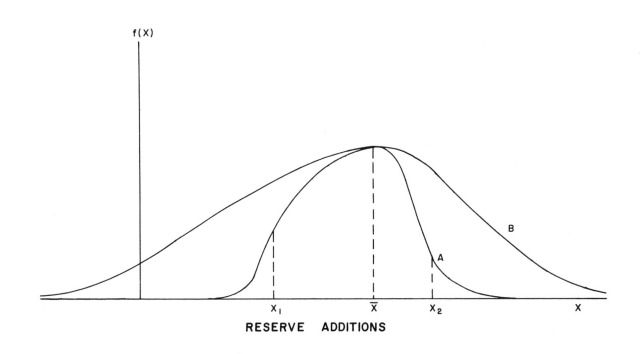

Figure 10.2 Hypothetical distribution of reserve additions

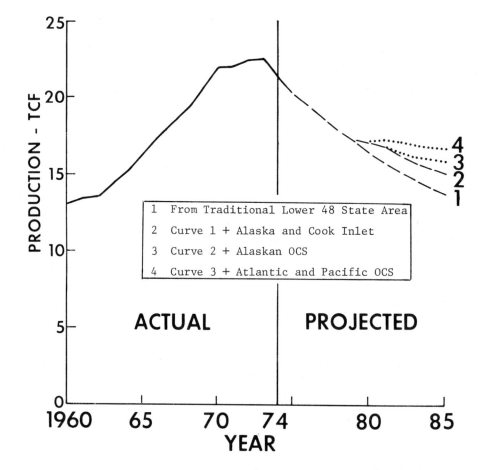

Figure 10.3 Projected productive capability – total U.S. conventional gas supply

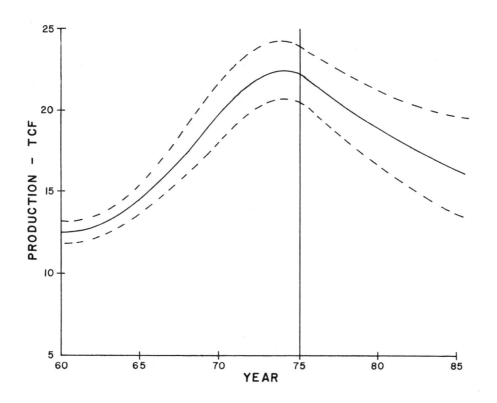

Figure 10.4 Distribution of Projected
Productive Capacity

deviations are close together. But by the early 70's these lines have moved apart, indicating a
larger variation in productive capability. Forecasting needs to recognize the extent to which
risk, as indicated by the larger variation, will occur in the future. For simplicity, the dotted
lines also follow the trend of deviating by larger amounts from the forecast mean. By 1985 the
range of forecast values varies by 10 TCF, certainly a sufficiently large variation to question
any policy based on the mean forecast of productive capacity. What policy analysis must do is
evaluate where actual productive capacity will occur within this range if gas pricing formulas
are changed.

Evaluation of Theoretical Determinants

The theoretical development of models explaining changes in mineral economics hinges
around the expertise of professionals knowledgeable about the subject being studied. Coal supply
and demand models require the inputs of professionals who understand forces operating in that
industry. Basic procedures for specification of mineral models are elaborated on in Chapter 7.

Specification of theoretical models, although a major advancement over most studies where
little or no effort is often made to explain how a policy will affect social welfare, still poses
some problems. Primary among these is that the models are just what the title of this section
indicates--theoretically possible explanations for changes in the elements comprising social welfare. All models, as such, deviate from actual events, and can be criticized for that reason.
Researchers preparing such models and policymakers evaluating them have to decide whether any
perceived deviation from reality significantly affects the estimated impact on social welfare.

A simple example of two alternative approaches that may both be acceptable concerns the estimation of a policy impact on personal income. In a specific approach, personal income (PY) can be taken as the sum of wages and salaries (WS), profit income (PRY), transfer payments (TR), minus contributions to social security (SI). These relationships are shown below:

$$PY = WS + PRY + TR - SI \tag{10.1}$$

A theory explaining the elements of this equation are merely added or subtracted to estimate PY. Equations for two of the component variables might be:

$$WS = F_1 \text{ (Di, EM, BI)} \tag{10.2}$$

$$PRY = F_2 \text{ (Di, RS, BI)} \tag{10.3}$$

where Di are dummy variables corresponding to the political jurisdictions located in the geographical impact area, EM is employment, BI is an index of business activity, and RS is retail sales. Transfer payments and social security contributions are exogenous to this equation, meaning that they are determined by factors not considered by the model.

An alternative model for explaining personal income deals with this value directly with a model of the following type:

$$PY = F_3 \text{ (Di, EM, PEM, GNP)} \tag{10.4}$$

where PEM is employment in the public sector, GNP is gross national product, and the other variables are defined as before.

Empirical estimates of personal income may not be significantly different between the two models, although they will be different. Equations 10.1-10.3 are theoretically better because they identify which segments of the economic environment would benefit from a policy. Suppose that a policy would increase WS income and another policy would increase profit income. Recognition of these differential impacts <u>may</u> be important for distributive evaluations in policy analysis, and if they are, they should be considered. A decision emphasizing just the aggregate effect on personal income would select equation 10.4. Selection of theoretical models affects the analysis.

Another area subject to analysis in the theoretical modeling stage is selection of variables used to explain personal income or its components. Participants in policy analysis must question whether Di, EM, and BI should be used to explain WS and whether other variables should be added. Answers to these questions vary dramatically. Resolution of differential responses revolves around deciding on the importance of such variables. If the developed equation is deemed satisfactory, suggested additions and modifications should not be given undue emphasis.

A similar issue arises in forecasting the effects of alternative gas pricing strategies on natural gas reserve additions. Two types of theoretical models could be employed. One approach, equivalent to Equation 10.4, explains reserve additions (R,E,NPW) directly, or

$$R,E,NPW = f(SR, FD, P_W, P_O, WC) \tag{10.5}$$

where P_O is the free world price of oil, WC is well cost, and the other variables are defined in Table 10.2.

Alternatively, given the approach outlined by the equation $R,E,NPW = (W)(SR)(D)$, we could choose to explain W, SR, and D separately. The equations suggested by Erickson and Spann are

$$W = f_W(P_O, P_G, D_{-1}, R, X_1, X_2, Z_1, Z_2, Z_4, F_{-1})$$
$$S = f_S(P_O, P_G, X_1, X_2, Z_1, Z_1, Z_4)$$
$$D = f_D(P_O, P_G, S_{-1}, X_1, X_2, Z_1, Z_2, Z_4) \tag{10.6}$$

where P_O = the deflated price of crude oil at the wellhead

P_G = the deflated price of gas at the wellhead

F_{-1} = footage drilled a year ago

D_{-1} = discoveries in that district a year ago

R = wildcat wells drilled by major companies

X_1 = number of shutdown days (prorationing) in those years that flow from wells was restricted in Texas

X_2 = same as X_1 for non-Texas areas

$Z_{1,2,3,4}$ = dummy variables distinguishing among petroleum districts

Each explanatory variable is included to capture a specific economic force. Oil and gas prices are included in the three equations to capture the returns to the driller of a successful well. Success rates measure the probability of finding a commercial well in an area. Footage drilled a year ago is a proxy for the cost of drilling.

Faced with increasing prices for these minerals, producers trade off the expected success of their drilling venture and the cost of drilling against higher prices. A 10% increase in price at a time when the success rate declines several percent and costs increase 20% could lead to a reduction in the number of wildcat wells drilled. If wildcat wells drilled are reduced in periods of rising wellhead prices, producers are usually criticized for being non-responsive or of holding their mineral from the market. But, as the equations in 10.6 show, the real cause is the poor economics of drilling. Only by raising the wellhead prices relative to the success rates and cost will drilling increase.

Theoretical models are preferable to simple graphs tracing wildcat drilling, average discovery size, etc. with wellhead price over time because they account for the cost and return to the driller explicitly. Understanding the trade-offs leads to the heart of policy analysis. Any gas pricing policy which does not offset the greater uncertainty from drilling in unknown or higher cost OCS areas will fail. And, this failure cannot be laid at the feet of an unresponsive mineral industry.

The argument that prices are inadequate given the costs and uncertainty facing the developers of almost all mineral forms--oil, gas, coal, solar, geothermal, synthetic hydrocarbons-- are certainly not new. What is gained from these theoretical models is a method for demonstrating that further price increases are not a "rape" of the consuming public. Rather, the price increases are necessary to offset economic and technological forces.

Model Estimation

Converting the implicit functional form of Equation 10.6 to precise relationships proceeds according to the statistical techniques described in Chapter 9. We wish to reiterate here that the best statistical model is the one that explains most of the varation of the dependent variable and wherein the explanatory variables have the correct sign and magnitude.

In the natural gas reserve addition example, the functional form for estimation selected by the authors involved taking logarithms of the following equations.

$$R,E,NPW = (W)(S)(D)$$

$$W = a_o \cdot P_0^{a_1} \cdot P_G^{a_2} \cdot D_{-1}^{a_3} \cdot R^{a_4} \cdot F_{-1}^{a_5} \cdot e^{a_6 \cdot X_i} \cdot e^{a_7 \cdot Z_i}$$

$$S = b_o \cdot P_0^{b_1} \cdot P_G^{b_2} \cdot e^{b_3 \cdot X_i} \cdot e^{b_4 \cdot Z_i}$$

$$D = c_o \cdot P_0^{c_1} \cdot P_G^{c_2} \cdot S_{-1}^{c_3} \cdot e^{c_4 \cdot X_i} \cdot e^{c_5 \cdot X_i}$$

where e is the operator for natural logs; a, b, c are the coefficients measuring the relationship between the explanatory variables and the dependent variable; and the independent variables are defined as before. After taking logs and estimating by ordinary least squares estimation, the equations become

$$\log R,E,NPW = \log W + \log S + \log D$$

$$\log W = -8.21 + 1.48 \log P_0 + 0.352 \log P_G - 0.77 \log D_{-1} + 0.032 \log R \quad (10.7)$$
$$(0.46) \qquad (0.42) \qquad (0.26) \qquad (0.01)$$
$$+ 1.76 \log F_{-1} + 0.0017 X_1 + 0.0013 X_2 - 2 \cdot 30 \cdot Z_1 + 1.55 Z_2 - 0.4 Z_4$$
$$(0.6) \qquad (0.001) \qquad (0.0) \qquad (0.68) \qquad (0.51) \qquad (0.5)$$
$$R^2 = 0.972 , \quad df = 33$$

$$\log S = 2.79 - 0.233 \log P_0 + 0.008 \log P_G + 0.0006 X_1 - 0.0002 X_2 + 3.69 Z_1 \quad (10.8)$$
$$(0.286) \qquad (0.266) \qquad (0.0008) \qquad (0.0001) \qquad (0.23)$$
$$-0.104 Z_2 - 0.33 Z_4$$
$$(0.14) \qquad (0.16) \qquad \qquad R^2 = 0.802 , \quad df = 33$$

$$(10.9)$$
$$\log D = 10.4 - 1.50 \log P_0 + 0.329 \log P_G - 0.241 \log S_{-1} - 0.0016 X_1 - 0.0042 Z_1$$
$$(1.05) \qquad (0.984) \qquad (0.667) \qquad (0.029) \qquad (0.90)$$
$$-1.74 Z_2 - 0.994 Z_4$$
$$(0.54) \qquad (0.61) \qquad \qquad R^2 = 0.599 , \quad df = 31$$

The numbers in the parenthesis below the coefficients are standard errors.

These estimated equations can be evaluated following the procedures outlined in Chapter 9. For the appraisal of natural gas pricing regulations, the main interest is the gas price coefficients in the three stochastic equations. Summing the three coefficients, as suggested by the identity, measures the change in reserve additions for every percentage change in natural gas prices. For example, if a_1 = 0.40, b_1 = 0.10, and c_1 = 0.30, a 10% increase in gas price leads to a 4% increase in wildcat wells drilled, a 1% increase in the success rate, and a 3% increase in average discovery size. Reserve additions rise by the sum of these terms, or 8% = 4% + 1% + 3%.

Actual coefficients for a_1, b_1, c_1 are 0.35, 0.008, and 0.33, respectively. Reserve additions thus increase 6.88% for every 10% increase in natural gas prices. To double reserve additions over a specified period, gas prices have to rise by 145%. From such a model the relationship between gas prices and reserve additions is established. Measuring the expected effect of alternative gas pricing strategies involves plugging alternative values into these equations.

Other economic forces may also be evaluated using the equations. If average discovery size decreased by 10% a year ago, reserve additions would decrease, even with a 10% increase in the price of gas (a 6.88% increase in reserve additions, because the coefficient for D_{-1} is -0.77). A -7.7% decrease in reserve additions more than offsets the 6.88% gas price incentive for adding to reserve additions. Appraisal of the contribution of specific economic forces helps to understand their separate impact on reserve additions.

This example also illustrates the shortcomings of simple time series charts of reserve additions and gas prices. Gas price increases are associated with decreasing reserve additions in this type of analysis, which supports arguments of some consumers that gas prices do not increase reserve additions. The reserve addition model instead indicates that declining discoveries, leading to increased costs and uncertainties, is the real culprit. Gas price increases should be thought of as a positive force inasmuch as they actually reduced the size of reserves depletion. Without a gas price increase, reserve additions would decline by 7.7% instead of 0.9%.

Statistical models are helpful tools in explaining to laymen and decision makers why large increases in mineral prices do not always lead to significant increases in mineral supply. Rapidly rising costs as mineral companies begin to develop more inaccessible areas may offset these price increases. And, as the above example indicates, rising prices may do nothing more than keep reserve additions from declining further. This situation characterizes several mineral industries in the 60's and 70's.

One simple example of this is illustrated by a rule-of-thumb called the daily production cost factor. This is the capital cost of developing an oil production facility divided by the average daily production rate ($/daily barrel). From 1970-78 this number increased from about $2000 to about $8000 per daily barrel for offshore production in one major producing area.

Estimating the Impact of Alternative Policies

Armed with a tool for evaluating changes in mineral supply and demand under different policies, this section describes how the tool is used. Foremost among such evaluations is a prediction of mineral supply and demand without any change in the status quo, or current policy. This provides the benchmark, or baseline forecast against which the proposed policy is judged.

Preparation of Baseline Forecasts

Baseline forecasting might be thought of as an application of trending to mineral supply and demand. The FPC forecast of declining gas productive capacity in Figure 10.3 is the simplest example. But with the more sophisticated statistical tool developed above the capability exists for defining the contribution of individual economic forces to this decline. Without an understanding of the specific causes of the decline, it is obviously impossible to develop remedial policies.

The biggest difficulty in policy analysis is obtaining accurate forecasts of the independent variables. Inaccurate forecasts of these variables, even if the models themselves are accurate, leads to errors in the forecast of the dependent variable. The potential damage to the accuracy of the policy analysis is minimized by the nature of policy. In pure forecasting where the objective is to estimate the future mineral supply or demand that will actually occur, errors in forecasts of the independent variable can lead to disastrously bad forecasts. Policy analysis, however, is concerned with impacts upon the dependent variable "if" a change occurs. The "what if" aspect of policy analysis nullifies most errors in independent variable forecasts, since only the difference in mineral economics resulting from transforming a policy is considered.

Given the equations 10.7-10.9, the independent variables to forecast are P_O, P_G, X, Z, F, and R. Forecasts of the variable S_{-1} are taken from the forecast of the success rate equation in the previous year. For simplicity assume that X, shutdown days, are zero, allowing existing wells to produce at capacity, and that Z, the variable distinguishing among petroleum districts, is also zero. If Z is allowed to take on values other than zero the distributive efficiency of a policy with regard to its impact on specific districts is measurable.

Using the data presented in Table 10.2, assume that the forecast variables increase by 10% in the next three years. Table 10.3 gives the last two years of actual data and the next three years forecasts of the independent variable. These values, plugged into the forecasting equations, yield forecasts of reserve additions for a three year period. These forecasts represent preliminary estimates of reserve additions in the three year period, and are, thus, similar to the types of forecasts discussed in Chapter 9.

Forecasts for year t+1 are found by inserting the appropriate values into the estimating equations, as follows:

$$\log W = -3.77 + 1.48 \log (5.77) + 0.352 \log (48.4) - 0.77 \log (14.2)$$
$$+ 0.032 \log (0.60) + 1.76 \log (5,298) = 3.61$$

$$\log S = 1.19 - 0.233 \log (5.77) + 0.08 \log (48.4) = 1.14$$

$$\log D = 10.9 - 1.50 \log (5.77) + 0.329 \log (48.4) - 0.241 \log (14.2) = 10.03$$

Reserve additions in year t=1 are computed from the anti-logs of 3.61, 1.14, and 10.03.

Table 10.3

Actual and Forecast Values of Independent Economic Forces

Year	Economic Force			
	Wellhead Gas Price (P_W)	Wellhead Oil Price (P_O)	Footage Drilled (F)	% Drilled RY Majors (R)
Actual				
t-1	30.4	$4.50	5,206	0.60
t	44.0	5.25	5,298	0.60
Forecast				
t+1	48.4	5.77	5,826	0.60
t+2	53.2	6.35	6,410	0.60
t+3	58.6	6.98	7,051	0.60

$$4{,}073 = \text{anti-log } 3.61$$

$$13.8 = \text{anti-log } 1.14$$

$$(1.07)(10^{10}) = \text{anti-log } 10.03$$

Estimated reserve additions for year t+1 are

$$RA = 4{,}073 \cdot 0.138 \cdot (1.07 \cdot 10^{10})$$

$$= \underline{6.023 \cdot 10^{12}}$$

where the success rate is converted to the fractional form (13.8 to 0.138). The success rate equation employs 13.8 instead of 0.138 for convenience because the log of a fraction is a negative number, which distorts the coefficients of the model. In year t+2 the estimated reserve addition is $7.76 \cdot 10^{12}$ cubic feet, and $11.35 \cdot 10^{12}$ cubic feet in year t+3. Table 10.4 summarizes the estimated values for the forecast period.

Table 10.4

Summary of Reserve Addition Forecasts

Period	Wells Drilled	Success Rate	Average Discoveries	Reserve Additions
t+1	4,073	0.138	$1.07 \cdot 10^{10}$	$6.02 \cdot 10^{12}$
t+2	5,754	0.138	$9.77 \cdot 10^{9}$	$7.76 \cdot 10^{12}$
t+3	9,591	0.136	$8.70 \cdot 10^{9}$	$11.35 \cdot 10^{12}$

Forecasts such as those presented in Table 10.4 represent the baseline, or benchmark for comparing the effect of gas pricing strategies. Since deviations from these forecasts measure the impact of a policy, it is important to fully understand and accept these figures. The reader should remember that the forecasts presented in Table 10.4 are designed to illustrate the process of policy analysis, and are not intended to reflect past or current events in the mineral industries.

Forecasts with Proposed Policy

The second feature of policy analysis revolves around identifying the likely consequences of a proposed policy. This may be easy or difficult depending on the case. Policies presented in the form of specific, easily recognizable outcomes are rare, however. Mineral pricing policies usually fall into this category, especially natural gas pricing formulas. In 1977, for example, a presidential energy policy specified the exact increments gas prices were to take for several years ahead. An alternative pricing policy supported by industry and many economists proposed merely to discontinue any form of gas price controls.

The difference in the two pricing strategies is significant in policy analysis. Gas prices occurring in the future are known with certainty in the first strategy. It was uncertain what the price of gas would go to if no controls existed. Some experts suggested a price equivalent to oil prices on a Btu basis; others favored prices that existed for the already uncontrolled intrastate markets; and some opted for a several hundred percent increase immediately after decontrol, followed by increases equal to exploration and production costs. One of these pricing scenarios, or a mixture of scenarios, certainly might occur. But specifying the actual pricing situation which occurred was not easy to foresee in advance.

Situations where the actual impact of a policy is highly uncertain is the rule, not the exception, especially if the policy change eminates from legislation. Uncertain policies arise because policy changes originating in law emphasize general guidelines, with few guidelines on how the policy is to be implemented by regulatory agencies. Common examples of the difference between the legal and administrative features of mineral policy revisions include leasing, environmental, development, and pilot programs where the objective passed down to regulatory agencies is to enhance development of mineral resources without placing an undue burden on mineral consumers. Administrators, thus, have the unenviable task of setting specific guidelines that achieve these conflicting objectives.

Unclear or conflicting goals originating by legislation are not the only source of confusion. Public employees frequently help prepare mineral policy legislation. For many subjects there is a real difficulty in reducing the number of available options to a manageable size. Leasing illustrates one such situation. Close to an infinite number of leasing systems and combinations exist. To reduce this number to a size tractable in policy analysis, some cutting would have to be done. Some reductions could be made among leasing strategies which are similar. But many leasing options would be ignored for reasons of incompatibility with the political system, moral and legal conflicts, and technical infeasibility in the sense that the leasing strategy could not be administered effectively.

The exact specification of all policy options is impossible. Every policy analysis represents a subset, or sample, of these options. For this reason designing and implementing a complete policy analysis is a formidable task. Policy analysis design therefore attempts to cover as many possible occurrences as possible.

Natural gas pricing is indicative of evaluation problems in policy analysis. One possible gas price situation under the status quo projections was presented earlier. At the other extreme position is a position allowing gas prices to be set by competitive forces. In between these options is the concept of phased decontrol; whereby gas prices are allowed to return to a free market level at a controlled rate that is slower than immediate decontrol. Immediate decontrol was pushed in the seventies by many producers for obvious reasons. Others favored phased decontrol in order to minimize the potential harmful impact on the economy as a whole. Measuring the impact on inflation, unemployment, industry relocation, consumer costs, and other

312

items could be accomplished within policy analysis. Remember that the example presented here emphasizes solely the effect on reserve additions. To evaluate these impacts would require a more extensive analysis.

Following the two possible pricing strategies--a free market gas price equal to oil on a Btu basis and phased decontrol--hypothetical gas prices for three forecast years might be

Year	Uncontrolled	Phased Decontrol
t+1	191¢	100¢
t+2	210¢	150¢
t+3	231¢	191¢

The distinction between the two pricing formulas is the immediate adjustment to a free market price of $1.91 per thousand cubic feet in year t+1 under free market pricing and a three year gradual adjustment to that level in phased decontrol. To compute the reserve additions in year t+1 with a policy of complete decontrol substitute 191 for 48.8 in the estimating equations. Table 10.5 presents the revised forecasts for higher gas prices.

Table 10.5 Forecasts of Reserve Additions for
Two Gas Pricing Strategies

Year	Pricing Strategy		
	Current	Uncontrolled	Phased Decontrol
t+1	6.02×10^{12}	2.61×10^{14}	1.24×10^{14}
t+2	7.76×10^{12}	4.20×10^{13}	1.28×10^{13}
t+3	11.35×10^{12}	2.92×10^{13}	2.27×10^{13}

Evaluation of Policy Impacts

The measured impact of a policy is the difference between the baseline forecast and the forecasts like Table 10.5. Since the only variation among the forecasts occurs by altering gas prices, this divergence represents likely effects of proposed gas price changes on reserve additions. Quantifying these impacts in any policy analysis perhaps is the easiest part because the results follow simply and directly from the preceding phases. Caution has to be exercised in accepting the conclusions of a policy analysis for this reason. The model and forecast assumptions employed always must be checked, even if the conclusions are acceptable; otherwise, a deficient model might be labelled acceptable when it, in fact, produces serious policy errors. The natural gas model used as an example is an excellent case in point.

Reserve additions increase by the amounts shown in Table 10.6 for each pricing strategy. Column 3 minus column 2 yields column 4, the impact of decontrolled gas prices. In year t+1 the increment in reserve additions resulting from the change in gas pricing formulas is 2.54×10^{14} scf and 1.17×10^{14} scf for decontrol and phased decontrol, respectively. Reserve addition increments decline to 3.42×10^{13} and 5.04×10^{12} in year t+2, and 1.78×10^{13} and 1.13×10^{13} in year t+3. Total additions to reserves resulting from complete gas decontrol is 3.06×10^{14} scf and, for phased decontrol, it is 1.33×10^{13} scf.

313

Table 10.6 Impact of Alternative Gas Pricing Formulas

(1) Year	(2) Baseline Forecast	(3) Decontrol Forecast	(4) Decontrol Difference	(5) Phased Decontrol Forecast	(6) Phased Decontrol Difference
t+1	6.02×10^{12}	2.61×10^{14}	2.54×10^{14}	1.24×10^{14}	1.17×10^{14}
t+2	7.76×10^{12}	4.20×10^{13}	3.42×10^{13}	1.28×10^{13}	5.04×10^{12}
t+3	11.35×10^{12}	2.92×10^{13}	1.78×10^{13}	2.27×10^{13}	1.13×10^{13}

An alternative way of viewing reserve additions corresponds to the monetary value of these mineral policies. Multiplying the incremental reserve additions by the pure differential measures the change in economic value, as shown below.

Economic Value of Mineral Reserves

Year	Decontrol ($)	Phased Decontrol ($)
t+1	3.61×10^{14}	5.99×10^{13}
t+2	5.36×10^{13}	4.87×10^{12}
t+3	3.06×10^{13}	1.50×10^{13}

Subtracting the baseline price (48.8¢) from the decontrolled price (191¢) and multiplying by column 3 in Table 10.6 gives the incremental economic value

$$3.61 \times 10^{14} = (2.54 \times 10^{14})(1.91 - 0.488)$$

Like all proposed policy changes critics would assert that higher gas prices are too expensive for consumers and the economy to absorb. Phased decontrol is favored for this reason. Time value of money concepts (Chapter 5) are useful for evaluating these claims. The present value of the two economic cash flows, discounted at 10% back to year t, are:

Present Value of Policies

Year	Decontrol ($)	Phased Decontrol ($)
t+1	3.28×10^{11}	5.44×10^{13}
t+2	4.42×10^{13}	4.02×10^{12}
t+3	2.30×10^{13}	1.14×10^{13}
Present Value in Year t	$\$3.95 \times 10^{14}$	6.98×10^{13}

where the discount values are 1.10, 1.21, and 1.33. Phased decontrol obviously costs less, but it also produces a smaller quantity of reserve additions.

Policy makers in the public and private sector possess with these estimates the likely cost and benefits of several gas pricing formulas. It is the role of the mineral expert to provide these figures. It is the role of the decision makers to determine which pricing formula is in the best interest of their constituents. After arriving at these estimates the mineral expert cannot offer anything to the decision. To do so moves him from the role as a professional into one of a politician or a member of society.

Recognition of this thin, almost imperceptible line is crucial. Mineral professionals by training and experience are more knowledgeable about the impact of a policy on mineral supply and demand, and should offer that advice, even if it is not requested. But, crossing that thin line and saying that a specific policy is the "best" for a nation's welfare takes the professional out of his area of expertise and into an area where even the least educated layman is as qualified as he. If, after informing as many people as possible of the likely consequences of a proposed policy, it is rejected, the job of the mineral professional is to do the best job that he can in these circumstances, or to try and influence future events in a role as private citizen or publicly elected official.

EXAMPLES OF OTHER TYPES OF POLICY EVALUATION

Gas pricing strategies used as an example in illustrating the technique of policy evaluation represents but one kind of policy analysis. Ex ante policy analysis as explained earlier deals with estimating the impact of events yet to occur using an econometric model. Other kinds of policy analysis include ex post policy evaluation--where the impact of policies implemented in the past are measured, simulation modelling procedures as an alternative to statistical estimation, and optimization in which the selection of a policy hinges upon the maximization of predetermined goals.

All of these different methodologies follow the outline presented in this chapter; in particular, each relies on a model describing the interrelationship among mineral economic forces and assumptions about how a policy changes one or more of these forces. It is important to affirm again that the outcome of the policy analysis is contingent upon the accuracy of the model and the assumptions used. If these items reflect reality, it does not matter how they were arrived at. Each methodology is viewed as being equally valid for this reason.

Three examples of policy evaluation studies are considered. For the ex post study consider the consequences of the U.S. Supreme Court decision ruling that the FPC could control natural gas prices at the wellhead on reserve additions. This is just the reverse view from that taken in the previous example that looked forward. Effects of alternative leasing strategies on outer continental shelf petroleum economics illustrate the simulation procedure. Using policy analysis to design optimal policies is discussed by adding equations to reflect gas demand to a modified version of the gas supply model. These examples should be viewed as alternatives to the specific procedures discussed above.

Ex Post Policy Evaluation

Regulation of gas prices at the wellhead by the FPC was believed to have reduced the new supply of natural gas, while encouraging excess consumption of gas during the sixties and seventies. The net result was excess depletion of an increasingly scarce resource. The model presented in Equations 10.6-10.8 is employed to analyze the effect of allowing gas prices to be set by competitive economic forces. Demand for gas is ignored at this juncture to simplify the analysis (see Erickson and Spann).

The specification of the model in Equations 10.6-10.8 reduces the policy analysis to construction of the baseline forecasts and forecasts of policy additions without court imposed regulation. The major advantage of ex post analysis is the reduction in the uncertainty surrounding the forecasts. Why? Because the baseline forecast is actual reserve additions in the sample period, and is therefore known with certainty. In other words, the baseline forecast represents the effect of gas prices actually mandated which yield specific reserve additions. In ex ante analysis baseline forecasts of reserve additions were determined by expected gas prices, not actual gas prices.

Construction of gas prices under free competition still involves some uncertainty. To specify a gas price in 1967 or any other year without FPC control is difficult. But as a means of beginning the analysis, assume that gas prices are equivalent to oil on a Btu basis in 1961 to 1975. Table 10.7 compares actual gas prices at the wellhead with an oil equivalent gas price. Gas prices without control are approximately four times greater than actual prices in many years, and even more in other years.

Table 10.7 Alternative Gas Prices

	Pricing Formulas (¢ per MCF)	
Year	Average Wellhead Gas Price Cents	Gas Price Equivalent to Oil On A Btu Basis Cents
1961	15.1	63
1962	15.5	63
1963	15.8	63
1964	15.4	62
1965	15.6	62
1966	15.7	63
1967	16.0	65
1968	16.4	65
1969	16.7	71
1970	17.1	75
1971	18.2	81
1972	18.6	84
1973	21.6	95
1974	30.4	148
1975	44.5	174

Following the methodology used earlier and plugging the difference in the two gas prices into the equations yields the incremental gain in reserve additions shown in Table 10.8. The big increase in actual reserve additions in 1970 was North Slope, Alaska. Oil equivalent pricing was assumed to have a zero impact on this level. Had gas prices been free during the sixties, the likely impact could have been to add these reserve additions several years earlier.

The net increase in reserve additions resulting from oil equivalent pricing would have averaged about twice actual reserve additions in each year. Total reserve additions are expected

316

Table 10.8

Past Gas Reserve Additions From
Oil Equivalent Pricing

Year	Actual Reserve Additions	Reserve Additions Resulting From Oil Equivalent Pricing	Total Reserve Additions
1961	17 166	32 615	49 781
1962	19 483	35 848	55 331
1963	18 164	32 513	50 677
1964	20 252	35 846	56 098
1965	21 319	37 095	58 414
1966	20 220	35 585	55 807
1967	21 804	36 447	58 251
1968	13 697	23 695	37 392
1969	8 375	16 163	24 538
1970	37 196	– 0 –	37 196
1971	9 825	20 337	30 162
1972	9 635	20 326	29 961
1973	6 825	13 872	20 679
1974	8 679	20 395	29 074
1975	10 483	17 716	28 199
Total		378 455	

to have increased by 378.4 TCF over the 15 year period between 1961 and 1975, quite a substantial influx to the inventory of minerals. Whether this benefit justifies the higher costs consumers would have to face is a subject for decision makers. But, in the energy crunch of the extremely cold winter in 1977, the extra gas would have saved a lot of lost production and unemployment.

Ex post policy evaluation is obviously just the other side of the coin of ex ante analysis--one looks backwards and the other forward. Looking backward benefits policy analysis in a way that is more than just pointing out the errors or successes of the past. It reminds every policy maker, public or private, that their decisions alter the fundamental characteristics of mineral economics. Such alterations may be so severe that the economy can never recover. Had more flexible gas pricing strategies been employed, the pressing demands for comprehensive energy programs in the mid-seventies would have been lessened. This lesson holds even when it is remembered that the reserve addition estimates in Table 10.8 are biassed upwards due to the failure of the Erickson-Spann model to incorporate technological limitations. The energy crunch's impact would have been minimized even if the reserve addition estimates were cut in half.

Simulation Policy Evaluation

In the examples used heretofore the fundamental feature has been a statistical one based on actual mineral economic forces. Situations arise where a data base sufficient for statistical analysis does not exist. Is policy analysis a viable tool in such instances? Yes, but the degree of uncertainty increases greatly. Without some experience to help analyze a situation, policy analysis, like drilling previously unexplored areas where little geological data is available, is a scientific stab in the dark. We don't know what will actually happen, and any estimate is just a scientific guess.

Developing new leasing policies for the outer continental shelf (OCS) during the late seventies is one such guestimate. Research performed by R. J. Kalter, W. E. Tyner, and D. W. Hughes is the best effort at providing assistance to policy makers in evaluating this complex topic. The basic aim of their work is to examine the effects of alternative leasing systems and schedules on OCS mineral development, made necessary by the high cost and extreme of OCS mineral development. At issue is whether the cash bonus system currently practiced for most U.S. leasing is optimal for orderly OCS mineral reserve development.

Because the model developed by Kalter, et al. is quite complex, the discussion can only highlight certain important areas. Readers should refer to the original article for more specific details.

Alternative OCS leasing systems considered in the policy analysis include:
1) Cash bonus
2) Higher fixed royalty
3) Variable royalty rate
4) Profit sharing on IRS income base
5) Annuity capital recovery profit share
6) British type profit sharing
7) Indonesian type production sharing
8) Variable rate profit sharing
9) Working interest
10) Work programs
11) Royalty bidding
12) Profit share bidding

With these leasing systems and the model discussed below, information on the following items was output.
1) Production time horizon
2) Installed production capacity
3) Present value of payments to government and producer
4) Total and time profile of production and reserve additions
5) Total production cost
6) Percentage dry holes

By minimizing total risk to producer and government, the authors conclude that five leasing systems are better than cash bonus bids. These are: 1) a variable rate British type capital recovery profit sharing system; 2) a fixed rate capital recovery profit share system; 3) variable rate annuity capital recovery profit share system; 4) a fixed rate annuity capital recovery profit share system; and 5) a royalty system with royalty rate varying with yearly production. More importantly, given the controversy that existed over the total volume of mineral reserves that exist in the OCS, total liquid hydrocarbon production is estimated to be 11-13 billion barrels with expected gas production equalling 39-64 billion Mscf.

Model and Assumptions

To arrive at these estimates and policy conclusions, it was necessary to establish at the outset the volume of undiscovered recoverable oil and gas in the OCS area. Beginning any policy analysis built on assumptions of this type exemplifies the extreme uncertainty associated with this kind of policy analysis. Experts have argued for years over estimates of the volume of mineral resources in the OCS area. The only thing that is known for certain is that the exact volume won't be known until the last drop is produced.

Table 10.9 presents estimates developed by the USGS resource appraisal group. Lacking better estimates, these were felt to be the best estimates at the time the model was constructed. A brief word on interpreting these values. Marginal probability estimates reflect the probability that minerals will be found in commercial volumes. One minus the marginal probability is dry hole risk. The Beaufort Sea in Alaska has a 75% chance of finding oil and gas in commercial volumes. The mean and standard deviation measure the true expected value of recoverable reserves (ignoring the probability of finding minerals) and the range, respectively. The 5% and 95% figures are additional estimates of the range of reserves since the mean figure is the 50% figure. Finally, the conditional mean combines the estimate of recoverable reserves with the probability of success by Baye's theorem, or, for the Beaufort Sea again, the conditional mean of 4.37 equals 3.28/0.75.

USGS recoverable reserve estimates combined with other assumptions, like oil price being $11/bbl and gas $0.60/Mscf, and 0.033 barrels of natural gas liquid being produced for each Mscf of gas, has stimulated some criticism by personnel knowledgeable about specific sub-region geological formations that the assumptions are unrealistic. It is safe to say that these criticisms are valid. What is not valid is to make the next step and assert that the model and/or the conclusions are erroneous. Policy analysis' purpose is to build a framework in which rational decisions can be made when faced with unknown and highly uncertain mineral forces, both geological and economic. Since decisions have to be made, is it better to evaluate these subjective elements without recognizing the uncertainty, or to specify assumptions and then analyze their implications as best we can. Certainly the latter approach has more appeal for those professionals desiring to refine policies as more information becomes available to reduce the uncertainty.

Without dwelling excessively on the complex equations utilized in the model, Figure 10.5 gives a flow diagram of the simulation model where installed capacity for each lease is specified by the user. Inserting installed capacity allows the analyst to evaluate the impacts of changing the capacity for producing minerals. One of the major problems facing engineers revolves around the selection of the right size and kind of equipment--too much capacity and the underutilization of equipment, while insufficient capacity retards potential production flows. Costs are incurred in each instance.

Using the USGS estimates and other assumed values, the model begins with baseline exploration costs for oil and natural gas developed by the National Petroleum Council (NPC, 1975). Estimated exploration costs assuming a depth of 200 meters and moderate climate are shown in Table 10.10, where total cost averages $2.714 million per well. Since total cost per well is a function of the location and size of the reservoir to be pumped, Kalter, et al. fit the information in Table 10.11 to the equation $C = aR^b$, where C is cost and R is reservoir size. This provides an equation for estimating development cost. Table 10.12 gives the same information for natural gas.

Table 10.9 USGS Resource Appraisal Group Estimates of Undiscovered Recoverable Oil and Natural Gas Resources for the OSC Areas of the United States

Region, Subregions and Province Names	Oil in Billions of Barrels						Gas in Trillions of Cubic Feet					
	Marg. Prob.	95%	5%	Mean	Std. Dev.	Cond. Mean	Marg. Prob.	95%	5%	Mean	Std. Dev.	Cond. Mean
ALASKA OFFSHORE												
Arctic Ocean												
Beaufort Sea	.75	0	7.60	3.28	3.5800	4.37	.75	0	19.30	8.20	22.380	10.93
North Chukchi	.60	0	6.20	1.89	1.7600	3.15	.60	0	19.50	5.67	15.800	9.45
Central Chukchi	.70	0	11.90	4.41	7.7700	6.30	.70	0	30.00	11.00	47.800	15.71
Hope	.30	0	0.60	0.13	0.0050	0.43	.50	0	3.30	0.86	0.370	1.72
Bering Sea												
Norton	.40	0	2.19	0.54	0.1100	1.35	.60	0	2.80	0.85	0.320	1.42
St. Matthew-Hall				Negligible						Negligible		
Bristol	.50	0	2.50	0.71	0.2100	1.42	.50	0	5.40	1.64	0.930	3.28
Navarin	.30	0	1.90	0.36	0.0500	1.20	.30	0	4.80	0.93	0.330	3.10
Zhemchug-St. George Basin	.50	0	5.10	1.32	0.8300	2.64	.50	0	13.00	3.30	5.210	6.60
Pacific Margin Basins												
Cook Inlet	1.00	0.5	2.30	1.19	0.5900	1.19	1.00	1.0	4.50	2.39	1.170	2.39
Eastern Gulf of Alaska	.70	0	4.40	1.13	1.7000	1.61	.70	0	13.00	3.39	15.310	4.84
Kodiak Tertiary	.40	0	1.00	0.23	0.0800	0.58	.40	0	3.50	0.69	0.700	1.73
Shumagin Shelf	.20	0	0.25	0.04	0.0005	0.20	.20	0	0.50	0.08	0.002	0.40
PACIFIC COAST OFFSHORE												
S. California Borderlands												
Inner Basins	1.00	0.4	2.00	1.01	0.5200	1.01	1.00	0.4	2.00	1.01	0.520	1.01
Outer Basins and Ridges	.40	0	0.24	0.06	0.0010	0.15	.40	0	0.24	0.06	0.001	0.15
Santa Barbara Channel	1.00	0.6	3.00	1.51	0.7900	1.51	1.00	0.7	3.30	1.70	0.850	1.70
Santa Cruz	.50	0	0.37	0.12	0.0050	0.24	.50	0	0.36	0.12	0.005	0.24
Northern Pacific Offshore												
Santa Maria	.60	0	0.28	0.11	0.0040	0.18	.60	0	0.28	0.11	0.004	0.18
Bodega	.40	0	0.53	0.13	0.0070	0.33	.40	0	0.53	0.13	0.007	0.33
Pt. Arena				Negligible						Negligible		
Eel River				Negligible						Negligible		
Oregon-Washington	.30	0	0.72	0.15	0.0070	0.50	.30	0	1.70	0.35	0.040	1.17
GULF OF MEXICO												
Florida Gulf Platform	.70	0	2.80	0.99	0.4500	1.41	.70	0	2.80	0.99	0.4500	1.41
Central and Western Continental Shelf	1.00	2.0	6.50	3.84	1.4200	3.84	1.00	17.0	90.00	49.00	24.0500	49.00
ATLANTIC COAST OFFSHORE												
North Atlantic Shelf	.60	0	2.50	0.89	0.3100	1.48	.60	0	13.20	4.44	7.7700	7.40
Central Atlantic Shelf	.70	0	4.60	1.76	1.2400	2.51	.70	0	14.20	5.29	11.1800	7.56
South Atlantic Shelf	.40	0	1.25	0.34	0.0300	0.85	.40	0	2.50	0.67	0.1400	1.68
Southeast Florida Shelf and Straits				Negligible						Negligible		
TOTALS												
Alaska Offshore	-	3	31	15.23	-	24.44	-	8	80	39.00	-	61.57
Pacific Coast Offshore	-	2	5	3.09	-	3.92	-	2	6	3.48	-	4.78
Gulf of Mexico Offshore	-	3	8	4.83	-	5.25	-	18	91	49.99	-	50.41
Atlantic Coast Offshore	-	2	4	2.99	-	4.84	-	5	14	10.40	-	16.64
TOTAL UNITED STATES OFFSHORE	-	10	49	26.14	-	38.45	-	42	181	102.87	-	133.40

Q_o = installed annual capacity   TAXW = tax write-off available if lease is not developed after exploration
A_TNPV = after-tax net present value   XLOSS = loss incurred from exploration if lease is not developed (exclusive of bonus)

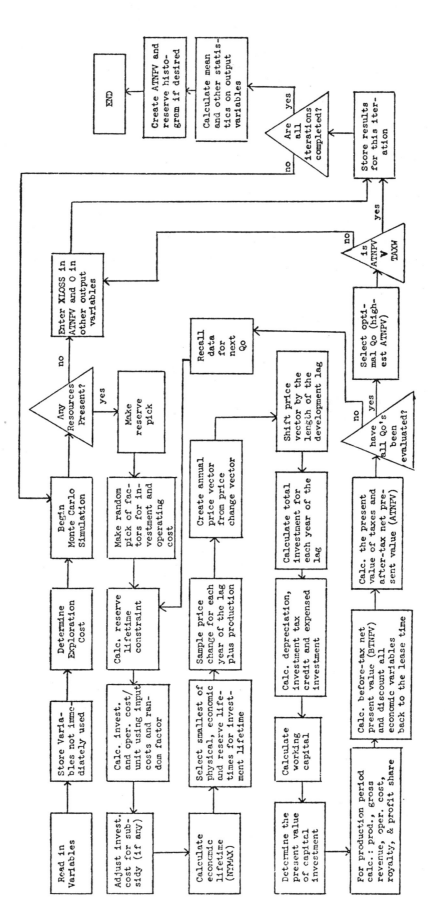

Figure 10.5 Flow Diagram for Simulation Model with Input Q_o

Table 10.10 Base Case Exploratory Drilling Expenditures Per Well

Item	Amount (millions of dollars)
Drilling Expenditures - Day Rate of $27 M/D x 80 Drill Days (10-12,000 Foot Well)	$2.160
Equipping Expenditures - Day Rate of $27 M/D x 7 Equipping Days	.189
Tubular Goods	.264
Wellhead	.050
Testing	.026
Other	.025
Total Per Well Drilling and Equipment Expenditures	$2.714

Note: The Base Case is for 200 meters water depth, moderate climate, expressed in thousands of 1974 constant dollars.

Table 10.11 Development Costs by Oil Reservoir Size

Reservoir Size (million bbls)	Cost Region				
	1 (1975 $)	2	3	4	5
15	$19.45	$36.95	$54.45	$71.95	$89.45
25	16.22	30.81	45.40	60.00	74.59
65	8.11	15.40	22.70	30.00	37.29
175	5.41	10.27	15.13	20.00	24.86
525	2.92	5.55	8.18	10.81	13.44
1050	1.55	2.97	4.38	5.79	7.19

Table 10.12 Gas Investment Costs

Reserves (million Mcf)	Cost Regions ($/Mcf)				
	1	2	3	4	5
90	$3.28	$6.48	$9.39	$12.31	$15.47
150	2.46	4.86	7.05	9.24	11.62
390	1.44	2.85	4.12	5.40	6.79
1050	.83	1.63	2.36	3.10	3.90
3150	.45	.88	1.28	1.67	2.10

322

These inputs, along with others not discussed here, provide the basic information for the Monte Carlo analysis. Monte Carlo analysis selects estimates of reservoir size and cost from predesignated distributions. Repeating the random selection a number of times (200 in this study) yields estimates of the joint probability of an outcome occurring, which is a measure of risk. Some of the distributions used include the following:

Economic Force	Distribution
Investment Cost	Triangular
Operating Cost	Triangular
Occurrence of Commercial Minerals	Uniform (Bernoulli)
Volume of Reserves	Log-normal
Annual Price Changes	Normal

Repeatedly drawing values from these distributions and then calculating the various economic parameters, such as present values of cash flows, royalty payments, taxes, and profits, for alternative types of leasing systems yields estimates that vary by region, reservoir size, and leasing system. Leasing strategies that lead to greater production and cash revenues than the present cash bonus payment system are already preferred.

Table 10.13 repeats Table 29 in Kalter, et al. for medium and large reservoirs in one cost region. Categories under the heading Evaluation Criteria summarize the findings. For medium reservoir sizes the cash bonus system is estimated to yield government revenue of $164.5 million, produce 63.9 million barrels and 14.1 TCF of gas, and have a 12.5% probability of a below normal profit. A royalty which is variable with production improves on all of these values. Government revenue increases to 165.3 million, oil production to 64.1 million barrels, gas production estimates remain about the same at 14.3 TCF, and the probability of below normal profit is reduced to 11.0%. Every criterion for choosing a leasing system is favorable.

For large reservoirs the benefits of royalty variable with production increase even more dramatically. Government revenue increases by $21.1 million from $675.1 to 696.2 million; oil production increases about 2 million barrels, and risk of below normal profit declines 3 percent.

Several leasing systems provide greater benefits than fixed cash bonus or royalty leasing strategies. But what has not been identified are any negative impacts of switching to alternative leasing systems. In all cases the main cost is shifting risk accepted by drillers to the owners of the mineral--the public in this example. The public trade-off is between a cash bonus that is known with certainty, which is smaller in value than revenues from alternative leasing systems, but is uncertain. In return for uncertain cash flows the public also gains extra production. The choice--more revenues and more production at a cost of greater uncertainty. In other words, some of the risk is shifted to the public.

Actually choosing between these alternatives is the role of management. Professionals have identified the risk and returns of alternative leasing strategies as best they can. Further evaluations by changing inputs and assumptions can be performed if these are unacceptable to other professionals. It is unlikely that everyone will agree. But if one leasing strategy is better for a reasonable set of assumptions, that strategy is preferred. By varying these parameters, the critical forces affecting the decision can be identified. At some point someone--the managers--have to say that we are going to accept a specific set of assumptions and the leasing

Table 10.14 Cash Flow in Present Value
Calculations of Mineral Value

Year	Total Revenue (1) (in thousands)	Costs (2) (in thousands)	Net Revenue Less Royalty (3)	Cash Flow (4) Col 3–Col 2
0	0	-30 000	0	-30 000
1	6 000	2 500	5 100	2 600
2	18 000	6 000	15 300	9 300
3	36 000	15 000	30 600	15 600
4	42 000	17 500	35 700	18 200
5	24 000	10 000	20 400	10 400
6	21 000	8 750	17 850	9 100
7	12 000	5 000	10 200	5 200
8	9 600	4 000	8 160	4 160
9	9 000	3 750	7 650	3 900
10	2 400	1 000	2 040	1 040

The project net present value is $34.8 million, an acceptable project as explained in Chapter 5.

Desirability of this project depends on the assumptions of oil price and production costs remaining constant for ten years, a situation which is unlikely to occur. In order to increase the accuracy of the estimated value, the engineering and economics experts agree that the price of oil will increase from $12 to $15 in equal increments over the nine year period ($0.33 per year), producing costs will rise about $0.40 per year, and that higher prices allow the company to engage in secondary recovery, yielding an extra 3 million barrels of production and extend the life of the well for 3 years for a $5 million expenditure in year 8. Such conditions can be based on a supply model or just on the combined expertise of the engineering and economic staff.

In Table 10.15 a new set of cash flows are computed. Total revenue in year 1 is 500 thousand barrels multiplied by $12. Value in year 2 is 1500 thousand barrels times $12.33, and so on. Costs increase by $0.40 per year, except in year 8 when costs rise by $5 million as the result of secondary recovery expenses. With these cash flows the present value of the project is

$$PV = -30\ 000 + (0.953)(2600) + (0.910)(7620) + (0.869)(14\ 883) + (0.831)(16\ 945)$$
$$+ (0.795)(9461) + (0.761)(8069) + (0.729)(4491) + (0.699)(-1176)$$
$$+ (0.671)(5826) + (0.644)(4550) + (0.619)(3112) + (0.595)(3900) + (0.573)(1040)$$
$$= -30\ 000 + 62\ 296 = \$32,296 \quad or \quad \$32,296,000$$

Present value of the project is $32.29 million, which is a $2.5 million decrease from the baseline valuation of $34.79 million.

The differences in these values appear trivial. Chapter 5 explains that any investment with a positive net present value is acceptable. But to maximize treasury growth when faced with limited capital, investments selected should be those with the highest PV/I ratios. In the baseline case the PV/I ratio is 34.79/30 = 1.15, and 32.29/30 = 1.07 in the second. Faced with other investments, including this one without secondary recovery, in which PV/I exceeds 1.07 and which use up the entire investment budget, secondary recovery would be rejected at this time.

Table 10.15 Project Cash Flows Under Alternative Conditions

Year	Production (1)	Total Revenue (2)	Cost (3)	Net Revenue Less Royalty (4)	Cash Flow (5)
0	0	0	-30 000	0	-30 000
1	500	6 000	2 500	5 100	2 600
2	1500	18 495	8 100	15 720	7 620
3	3000	37 980	17 400	32 283	14 883
4	3500	45 465	21 700	38 645	16 945
5	2000	26 660	13 200	22 661	9 461
6	1750	23 905	12 250	20 319	8 069
7	1000	13 990	7 400	11 891	4 491
8	800	11 464	10 920	9 744	- 1 176
9	1250	18 325	9 750	15 576	5 826
10	1000	15 000	8 200	12 750	4 550
11	750	11 250	6 450	9 562	3 112
12	550	8 250	4 730	7 012	2 282
13	200	3 000	1 720	2 550	830

Other goals can offset the results of the valuation process. Mineral companies are in the inventory business, building up their stock of minerals and then selling from that stock. Depletion of inventories requires additional efforts to replace each sale. The alternatives facing management are simple--replace the stock or find a more profitable line of business with higher PV/I ratios. A management decision to remain in the inventory business could lead to accepting the secondary recovery project because it augments their inventory by 3 million barrels and is profitable. It is just less profitable than other, non-mineral investments. Acquisitions by many mineral companies of non-mineral industries in the sixties and seventies are explained by a similar appraisal of mineral investment profitability relative to non-mineral investments.

SUMMARY

Policy analysis represents a valuable guide for making mineral decisions. Policies, whether formulated by public or private institutions, too often fail to properly account for the trade-offs associated with the policy. Oftentimes everyone can agree on the general direction of the impact--reducing reserves, increasing demand, changing prices, lowering profitability, and so on. What is never clear is the importance of one impact relative to another and the relative magnitude of a policy's benefits and cost. And, unfortunately, decision makers seldom possess the training or experience to fully evaluate these impacts. Possession of these necessary credentials is also not a license to make these decisions for society. In a world moving toward a greater interference in mineral activity, practical policy analysis offers the best hope for realizing sound policies for developing and operating mineral resources.

Policy analysis is not a panacea against bad policies, but it is better than the alternatives of special interest groups in governments, in companies, or in consumer groups manipulating statistics to benefit them at the expense of others.

REFERENCES

1. Newendorp, P.D., _Decision Analysis for Petroleum Exploration_, Tulsa: Petroleum Publishing Co., 1975.

2. Kalter, R.J., Tyner, W.E., and D.W. Hughes, _Alternative Leasing Strategies and Schedules for the Outer Continental Shelf_, Cornell University, 1976.

3. Campbell, R. and J. Campbell, Jr., "Optimizing Treasury Growth with Proper Evaluation of Long-Term Projects," SPE 6332, 1977.

4. Campbell, J. and R.A. Campbell, "Effects of Alternative Gas Pricing Strategies on Natural Gas Reserve Additions: Past, Present, and Future," SPE 6318, 1977.

5. Uhler, R.S. and P.G. Bradley, "A Stochastic Model for Determining the Economic Prospects of Petroleum Exploration Over Large Regions," _Journal of the American Statistical Association_, June 1970.

6. National Petroleum Council, _Ocean Petroleum Resources_, 1975.

7. National Petroleum Council, _U.S. Energy Outlook Oil and Gas Availability_, 1972.

8. Hubbert, M.E., _Energy Resources_, National Academy of Sciences, 1962.

9. Miller, B.M., et al., "Geological Estimates of Undiscovered Recoverable Oil and Gas Resources in the United States," _United States Geological Survey Circular_ 725, 1975.

10. Erickson, E. and R. Spann, "Supply Response in a Regulated Industry: The Case of Natural Gas," _Bell Journal of Economics and Management Science_, Spring 1971.

11. McAvoy, P. and P. Pindyck, "Alternative Regulatory Policies for Dealing with the Natural Gas Shortage," _Bell Journal of Economics and Management Science_, Autumn 1973.

11

OVERVIEW OF EVALUATION TOOLS

In the previous chapters we have briefly reviewed the principles governing mineral economics. The purpose has been to make the minerals specialist aware of the many factors which affect the value of the mineral property output. The total evaluation consists of two separate parts:

1. Determination of the characteristics, size and productive capacity of the minerals reserves.
2. Forecasting of the value of the reserves produced.

These are not independent activities. One is not more important than the other. These are two separate skills areas and no one human being can be totally competent in both. This means that a team approach of some sort must be involved. The minimum team must of necessity contain persons with expertise in geology, engineering and economics. These functions do not have to occur simultaneously or in the same locality necessarily, but intimate communication is needed. It is unfeasible for each professional to do his part completely independent of the other.

In Volume 2 which follows we will review the physical characteristics of energy mineral reservoirs. Petroleum reservoirs are emphasized. But, many of the principles for petroleum apply for other minerals.

Extent of Reserves

The first step in any exploration effort, regardless of the mineral, is to establish the size of the reserve. Is there enough mineral to justify the anticipated capital cost to bring it to the surface? If the answer is yes, the second step is to appraise downstream facilities needed to get the raw product to the marketplace in a salable form. What is the anticipated cost to modify existing facilities or construct new downstream facilities? These total capital costs must be compared to forecast revenues in some sort of value system before the decision is made to proceed with exploitation of the reserve.

The determination of reserve size is essentially a combined geological-engineering function. A series of core holes or wells are drilled. But, the sample obtained is too small statistically to fix all pertinent variables. The subjective input of the geologist is necessary to "extrapolate" the limited hard data to provide sufficient input for decision purposes.

Chapters 12-15 present an overview of mineral reservoirs to outline the factors affecting overall characteristics that influence this early appraisal. Chapter 17 is a more specific chapter for determining petroleum reservoir size.

The accuracy of reservoir size estimations obviously depends on the amount of hard data available. In the very early stages when only geophysical-geological data are available, unconfirmed by sampling, the risk in any estimate is large. But the decision at this point involves only the expenditure necessary to confirm early interpretations. For coal or oil shale relatively close to the surface a large amount of sampling via core holes can be obtained at a relatively low cost. For petroleum reserves offshore or in remote locations and/or occurring at great depths, sampling via wells is very expensive. Thus, petroleum sampling tends to be less extensive than that for coal and oil shale. The net result of this is a greater inherent risk in the petroleum decision to develop than for the solid fossil fuels.

On the other hand, the downstream processing capital needs for the solid minerals may be higher because of the lack of existing facilities. There is also more uncertainty for said facilities if they are to convert the solid to gas or liquid because of less proven technology.

Consideration of all variables leads to the minimum size reserve that is economically viable under the economic constraints prevailing. A reserve less than this will not be developed until the economics become favorable. For example, you have found a reserve of oil expected to be about 200 million barrels under water about 500 feet deep in a country where the political climate is unstable. After calculating capital costs, debt service costs, taxes, operating costs, etc. the economics are marginal (if not unfavorable). In view of the political situation the normal decision would be to forego current development. Some form of incentive is necessary if the host country wishes to promote such development.

As a practical matter about 40% of mineral exploitation capital investments are externally financed. The reserve is the collateral supporting said financing. This early calculation is therefore primarily for this purpose.

Producing Rate

Time value of money concepts (Chapter 5) dictate required rates of production from the reserve. In the beginning this must be deduced from the reservoir characteristics and past experience. In the case of petroleum it might dictate the number of wells needed; for coal the number of shafts or strip mining units might be fixed. This in turn fixes the downstream capacity needed.

Can this necessary capacity be sold at a favorable price? This is where supply-demand forecasts come into play. It is this aspect of evaluation where the mineral industry has a rather unimpressive record, particularly in the special product area. Historically prices of fertilizers, sulfur, ethylene, some petrochemicals, and the like have gone through pronounced price swings as demand affected price elasticity. This also has been true of hard rock minerals such as copper. Too often the evaluator has assumed that anything produced could be sold automatically at the current price.

Production Strategies

Once production has been initiated the beginning strategy is usually the one presumed during planning. Now the emphasis shifts to altering this strategy as more hard information accumulates. This information is applied in two basic categories of decisions:

1. Future exploitation of the reserve
2. Developing more specific strategies for producing the reserve as development continues

Necessarily broad, general goals and conclusions increasingly become more specific.

The sampling and testing of the first production refines the raw knowledge presumed available to that time. In petroleum, for example, fluid sampling, production tests, pressure tests and the like enable us to modify earlier interpretations of reservoir size, characteristics and the like. This should alter the exploration program visualized initially. The techniques for doing this are outlined in Volume 2. Comparable techniques exist for other mineral areas.

Chapter 18 covers material balances based on production data. Such balances can be used to calculate reserves, but the accuracy is often unsatisfactory until about 10% of the reserves have been produced. The same equations can be used to predict future performance--a very useful calculation early in the reservoir life. Chapter 19 on decline curves accomplishes a similar goal. These are merely plots of production data versus some variable dependent on time or time itself. Said curves may be used to predict future production or reserves. Early in the life of a given reserve there is insufficient data to extrapolate the curve meaningfully. The usual procedure is to use a more complete curve from an analogous reservoir to accomplish an extrapolation that is at least consistent with the comparable geological properties.

Chapters 18 and 19 are thus similar approaches which vary only in complexity. Many feel the decline curve approach is more reliable if judicious choice is made of the curves used to determine analogous performance.

Modeling (Chapter 20) is really nothing more than a more detailed, complex version of Chapter 18 feasible with machine computation. Regardless of the mathematical complexity, it is nothing more than a process of matching existing production and pressure data as a means for predicting future performance.

The remainder of the chapters outline techniques used primarily as input for other evaluation calculations. This input can be used for two purposes: (1) reservoir development and (2) production strategies.

The need for the analysis of well logs and cores is obvious. Pressure data may be less obvious but equally important. Proper pressure tests can locate geological faults and other production barriers not evident from geophysical and geological data. Such pressure interpretations must be included from each new well as reservoir development proceeds. This can be accomplished only if effective communication is established between geologists and engineers.

As noted in other sections of the book, the proper interpretation of pressure test data may be the most positive indicator of what is happening in the reservoir.

Large reservoirs. - The evaluation procedure outlined herein is particularly important for large reservoirs producing large amounts of fluids. Because of their high front-end capital cost it is customary to produce at maximum feasible rates in order to reduce debt service, improve money liquidity and achieve comparable economic benefits. This maximum feasible rate usually will be higher than the maximum efficient rate. The word "efficient" refers to that rate which maximizes recovery efficiency while maintaining satisfactory economics.

If excess rates are carried on too long irreparable reservoir damage may result. By-passing oil, water production, sand production and excessive pressure draw-down are among the symptoms of trouble. The various ramifications are too numerous to cover herein so one example must suffice.

In many large reservoirs pressure maintenance is instituted early to increase recoverable oil. Water injection is the most common fluid used. It is often injected around the perimeter of the field and flows toward the middle as oil is displaced. In theory this water moves through the reservoir like a piston and it takes one barrel of water to fill the void created by one barrel of oil being produced. In practice, the water moves faster through some parts of the formation than others and "fingers" of water extend beyond the main water front. As a finger reaches a producing well the water cut of the production increases.

One practice is to close in such a well when no water handling facilities are available. New, or other, wells are used to maintain the desired production level. Closing in a well in this situation can cause the entrapment of oil with by-passed water. Substantial quantities of oil may be lost for future recovery by ordinary means. The problem can be particularly acute with fractured reservoir systems.

Reservoir modeling is not adequate to quantitatively handle the problem. It is by necessity a somewhat idealized analysis of the reservoir that is concerned with "average" performance. Production monitoring is the basic tool although modeling can supplement it.

In this example, monitoring may dictate a change in the production system to handle water and/or a change in injection rate or pattern. In early planning injection facilities are normally designed for an injection rate higher than that needed in order to provide a reasonable safety factor. In the early stages the full capacity of the injection system is utilized somewhat by default. Production monitoring provides the proper rate--the one which provides an intelligent compromise between pressure decline and the amount of water breakthrough in producing wells. This rate normally will be less than system capacity in most cases.

SUMMARY

One must view mineral property evaluation as a multi-faceted exercise which continues over time, the goals of which change with time. Early evaluations stress equity value for exploration-exploitation decisions. Continuing evaluations serve to guide such decisions until that phase is completed. From that point on the sole purpose is to govern operational decisions.

Historically, one must note that most evaluations have suffered because each specialist tends to stress those calculations he knows how to make, not necessarily the ones that should be made. Evaluation at best involves much uncertainty. The only manner in which risk can be reduced to an acceptable level is to use all information and combine it in a meaningful manner, consistent with the real world as we interpret it. This means that technology alone is insufficient to yield satisfactory answers.

APPENDIX A

This appendix contains three tables whose primary purpose is to simplify the arithmetic of calculations in Chapters 5 and 9.

TABLE A.1

This table applies for the assumption that annual cash returns from an investment are received in a lump sum at the mid-point of the year.

Column 2, *Factor for Year*, is the single discount factor for the year in question expressed in Equ. 5.2. It is $1/(1 + i)^n$. If $i = 10\%$, the present value of $1000 received in Year 5 is found by simple multiplication

$$C = (0.65123)(1000) = \$651.23$$

Once one knows the successive annual cash flows on a project, each such flow may be multiplied by its discount factor and said discounted amounts summed to find net present value (NPV).

Column 3, *Equal Payments*, may be used for a series of equal payments. It is a *composite* or *lump sum* factor. If $1000 is to be received for each of 5 years, at $i = 10\%$, the total payment is $5000. In Column 3 for Year 5, the equal payment discount factor is 0.79516.

$$C = (5000)(0.79516) = \$3975.80$$

The same result could be obtained by multiplying $1000 by its year factor from Column 2 for the years 1-5 and summing them.

Columns 4 and higher are for declining income presumed to follow *exponential decline* as described in Chapter 9. The *percent annual decline* is the value of d, expressed by Equ. 9.6 as a fraction. Please note in Chapter 9, before using this table, that d is not the true slope of the exponential decline curve. It is related to a, the true slope, by the table and equation shown just below Equ. 9.6.

Suppose $20,000 is the *gross amount* to be received *on the decline* during a ten year period for $i_o = 10\%$ and $d = 0.20$. The present value of this amount *as of the time at which decline began* is

$$(20,000)(0.75089) = \$15,017.80$$

If this decline had been preceded by a series of equal payments or a time delay, the number $15,017.80$ would have to be discounted further to true time zero by a single payment discount factor. Suppose a two year time delay occurs before the above declining production begins. From Column 2 for Year 2 at 10%, the factor equals 0.86678. The NPV of the declining production as of true time zero (not at beginning of decline) is

$$(0.86678)(15,017.80) = \$13,017.13$$

Suppose instead of delay that during the two years *constant production before decline* that the revenue is $4,000 per year. From Column 3 for Year 2 at $i_0 = 10\%$, the factor is 0.91012. Present value of constant production is $(\$8000)(0.91012) = \$7,280.96$.

Total present value for this example is the sum of constant and declining revenue – $13,017.13 + \$7,280.96 = \$20,298.09$.

TABLE A.2

This table of factors is for *continuous compounding with annual cash flows occurring uniformly* throughout the year. This table contains the value of

$$(e^{-jn}) \text{ in the equation } C = S(e^{-jn}) \tag{A.1}$$

as noted in Equations 5.1-5.4.

For $j = 0.10$, what is the present value of $1000 received uniformly during Year 5 (Year 4-5 in table)? At 10% interest, the factor is 0.6379. So,

$$C = (\$1000)(0.6379) = \$637.90$$

This is about 2% different from the amount found using Table A.1.

Comparison of Table A.2 with the "Factor for Year" column of Table A.1, at the same interest rate, shows that the discount factors diverge with increasing time. During the 20th year at $j = 0.10$, they vary by about 8-9%. In terms of the practical decisions that are made using such tables, the choice is somewhat arbitrary and likely will not affect the decision.

TABLE A.3

This table is for *continuous compounding with an annual payment being received any time during the year*. This table may be used to relate the time value of single payments using Equ. A.1. For this use $x = jn$. If $j = 0.10$ and $n = 5$, $jn = x = 0.5$. For $x = 0.5$ (Column 1), $e^{-x} = 0.607$ or

$$C = (\$1,000)(0.607) = \$607.00$$

We could reverse the calculation to find the value of $607.00 compounded continuously for 5 years at 10%. In this case x *(jn)* would again be *0.5*. For $x = 0.5$, $e^x = 1.649$ or

$$S = (\$607)(1.649) = \$1,000.00$$

Column (4) of this table may be used to find a *composite discount factor* for a series of equal payments as expressed by Equ. 5.8. If $1000 is to be received during each of 5 years, at $j = 0.10$, what is the present value? Once again, $x = 0.5$. From Column (4), the composite factor is 0.787. Or,

$$C = (\$5000)(0.787) = \$3935$$

One of the best uses of this table is in conjunction with declining revenues which follow exponential decline. The details of this are shown in Chapter 9.

TABLE A.1 Present Worth Factors for Income Received in <u>One</u>
<u>Payment</u> at The <u>Middle of The Year</u>

6%

Year	Factor for Year	Equal Payments	5% Per Yr. Decline	10% Per Yr. Decline	15% Per Yr. Decline	20% Per Yr. Decline	30% Per Yr. Decline
1	.97129	.97129	.97129	.97129	.97129	.97129	.97129
2	.91631	.94380	.94450	.94524	.94603	.94685	.94865
3	.86444	.91734	.91917	.92109	.92311	.92524	.92981
4	.81551	.89189	.89521	.89871	.90238	.90620	.91433
5	.76935	.86738	.87256	.87799	.88365	.88953	.90178
6	.72580	.84378	.85112	.85881	.86678	.87499	.89172
7	.68472	.82106	.83085	.84107	.85162	.86236	.88376
8	.64596	.79917	.81167	.82469	.83803	.85146	.87752
9	.60940	.77809	.79352	.80956	.82587	.84208	.87269
10	.57490	.75777	.77635	.79560	.81501	.83404	.86898
11	.54236	.73818	.76011	.78273	.80534	.82719	.86616
12	.51166	.71931	.74473	.77088	.79675	.82137	.86403
13	.48270	.70111	.73019	.75996	.78913	.81644	.86243
14	.45538	.68356	.71642	.74993	.78238	.81229	.86124
15	.42960	.66662	.70339	.74070	.77642	.80880	.86035
16	.40528	.65029	.69105	.73222	.77117	.80588	.85970
17	.38234	.63453	.67938	.72444	.76655	.80344	.85923
18	.36070	.61932	.66833	.71730	.76249	.80141	.85888
19	.34028	.60463	.65786	.71076	.75893	.79972	.85862
20	.32102	.59045	.64796	.70477	.75581	.79833	.85844
21	.30285	.57676	.63858	.69928	.75309	.79718	.85831
22	.28571	.56353	.62969	.69426	.75071	.79622	.85821
23	.26954	.55074	.62128	.68967	.74864	.79544	.85814
24	.25428	.53839	.61332	.68548	.74684	.79480	.85809
25	.23989	.52645	.60577	.68165	.74527	.79428	.85806
26	.22631	.51491	.59863	.67815	.74391	.79384	.85803
27	.21350	.50374	.59186	.67497	.74274	.79349	.85801
28	.20141	.49295	.58544	.67206	.74172	.79321	.85800
29	.19001	.48250	.57937	.66941	.74083	.79297	.85799
30	.17926	.47239	.57361	.66700	.74007	.79278	.85799
31	.16911	.46261	.56816	.66481	.73941	.79263	.85798
32	.15954	.45314	.56299	.66281	.73885	.79250	.85798
33	.15051	.44397	.55810	.66100	.73836	.79240	.85797
34	.14199	.43509	.55346	.65935	.73794	.79232	.85797
35	.13395	.42648	.54906	.65785	.73758	.79225	.85797
36	.12637	.41815	.54489	.65649	.73726	.79220	.85797
37	.11922	.41007	.54094	.65525	.73700	.79215	.85797
38	.11247	.40223	.53720	.65413	.73677	.79212	.85797
39	.10610	.39464	.53365	.65311	.73657	.79209	.85797
40	.10010	.38728	.53028	.65219	.73640	.79207	.85797

TABLE A.1   Present Worth Factors for Income Received in <u>One</u>
Payment at The <u>Middle of The Year</u>

			8%				

Year	Factor for Year	Equal Payments	5% Per Yr. Decline	10% Per Yr. Decline	15% Per Yr. Decline	20% Per Yr. Decline	30% Per Yr. Decline
1	.96225	.96225	.96225	.96225	.96225	.96225	.96225
2	.89097	.92661	.92753	.92849	.92950	.93057	.93290
3	.82497	.89273	.89508	.89755	.90014	.90287	.90875
4	.76387	.86052	.86476	.86921	.87388	.87876	.88913
5	.70728	.82987	.83641	.84327	.85043	.85787	.87339
6	.65489	.80071	.80990	.81953	.82954	.83984	.86090
7	.60638	.77295	.78510	.79781	.81096	.82435	.85111
8	.56146	.74651	.76190	.77797	.79446	.81110	.84352
9	.51987	.72133	.74019	.75983	.77986	.79982	.83769
10	.48136	.69733	.71986	.74327	.76695	.79024	.83325
11	.44571	.67446	.70083	.72815	.75555	.78215	.82990
12	.41269	.65264	.68300	.71435	.74552	.77533	.82739
13	.38212	.63183	.66630	.70177	.73670	.76961	.82552
14	.35382	.61198	.65064	.69030	.72896	.76483	.82414
15	.32761	.59302	.63596	.67985	.72218	.76084	.82313
16	.30334	.57491	.62220	.67034	.71625	.75753	.82238
17	.28087	.55762	.60929	.66168	.71108	.75479	.82184
18	.26007	.54109	.59718	.65380	.70656	.75252	.82145
19	.24080	.52528	.58581	.64663	.70264	.75065	.82116
20	.22297	.51017	.57514	.64011	.69922	.74911	.82096
21	.20645	.49570	.56512	.63419	.69626	.74785	.82081
22	.19116	.48186	.55571	.62882	.69369	.74681	.82071
23	.17700	.46861	.54686	.62393	.69147	.74596	.82063
24	.16389	.45591	.53855	.61950	.68955	.74527	.82058
25	.15175	.44374	.53073	.61548	.68789	.74471	.82054
26	.14051	.43208	.52339	.61184	.68645	.74425	.82051
27	.13010	.42089	.51647	.60853	.68522	.74388	.82049
28	.12046	.41016	.50997	.60554	.68416	.74358	.82048
29	.11154	.39987	.50385	.60283	.68324	.74333	.82047
30	.10328	.38998	.49809	.60037	.68245	.74314	.82046
31	.09563	.38049	.49266	.59814	.68178	.74298	.82046
32	.08854	.37136	.48755	.59613	.68120	.74285	.82045
33	.08198	.36259	.48274	.59431	.68070	.74274	.82045
34	.07591	.35416	.47820	.59266	.68027	.74266	.82045
35	.07029	.34605	.47393	.59117	.67991	.74259	.82045
36	.06508	.33825	.46989	.58982	.67959	.74253	.82045
37	.06026	.33073	.46609	.58861	.67932	.74249	.82045
38	.05580	.32350	.46251	.58751	.67910	.74245	.82045
39	.05166	.31653	.45912	.58651	.67890	.74242	.82045
40	.04784	.30981	.45593	.58561	.67873	.74240	.82045

TABLE A.1 Present Worth Factors for Income Received in <u>One</u> <u>Payment</u> at The <u>Middle of The Year</u>

10%

Year	Factor for Year	Equal Payments	5% Per Yr. Decline	10% Per Yr. Decline	15% Per Yr. Decline	20% Per Yr. Decline	30% Per Yr. Decline
1	.95346	.95346	.95346	.95346	.95346	.95346	.95346
2	.86678	.91012	.91123	.91240	.91364	.91494	.91777
3	.78799	.86941	.87224	.87522	.87835	.88164	.88873
4	.71635	.83115	.83621	.84154	.84713	.85297	.86539
5	.65123	.79516	.80291	.81105	.81955	.82839	.84685
6	.59203	.76131	.77211	.78345	.79524	.80740	.83229
7	.53820	.72943	.74361	.75847	.77384	.78954	.82097
8	.48928	.69941	.71723	.73586	.75503	.77440	.81228
9	.44480	.67112	.69279	.71541	.73853	.76163	.80566
10	.40436	.64445	.67014	.69690	.72407	.75089	.80066
11	.36760	.61928	.64913	.68017	.71143	.74188	.79691
12	.33418	.59552	.62964	.66504	.70039	.73436	.79413
13	.30380	.57308	.61155	.65136	.69076	.72810	.79207
14	.27618	.55187	.59475	.63899	.68238	.72290	.79056
15	.25108	.53182	.57914	.62782	.67510	.71860	.78946
16	.22825	.51285	.56462	.61772	.66877	.71505	.78866
17	.20750	.49489	.55112	.60860	.66329	.71213	.78808
18	.18864	.47787	.53854	.60036	.65854	.70973	.78766
19	.17149	.46175	.52684	.59292	.65444	.70776	.78736
20	.15590	.44646	.51593	.58620	.65089	.70615	.78714
21	.14173	.43194	.50575	.58013	.64783	.70484	.78699
22	.12884	.41817	.49627	.57465	.64519	.70377	.78688
23	.11713	.40508	.48741	.56971	.64292	.70289	.78680
24	.10648	.39264	.47914	.56525	.64096	.70219	.78674
25	.09680	.38080	.47142	.56122	.63928	.70161	.78670
26	.08800	.36954	.46420	.55759	.63784	.70115	.78667
27	.08000	.35882	.45745	.55431	.63660	.70077	.78665
28	.07273	.34860	.45113	.55136	.63554	.70047	.78664
29	.06612	.33886	.44521	.54869	.63463	.70022	.78663
30	.06011	.32957	.43967	.54629	.63385	.70002	.78662
31	.05464	.32070	.43448	.54412	.63318	.69986	.78662
32	.04967	.31223	.42962	.54217	.63261	.69973	.78661
33	.04516	.30414	.42505	.54041	.63212	.69963	.78661
34	.04105	.29640	.42077	.53882	.63170	.69955	.78661
35	.03732	.28900	.41675	.53739	.63135	.69948	.78661
36	.03393	.28191	.41298	.53610	.63104	.69943	.78661
37	.03084	.27513	.40943	.53494	.63078	.69938	.78661
38	.02804	.26862	.40610	.53389	.63056	.69935	.78661
39	.02549	.26239	.40296	.53295	.63037	.69932	.78661
40	.02317	.25641	.40002	.53210	.63021	.69930	.78661

TABLE A.1   Present Worth Factors for Income Received in One
Payment at The Middle of The Year

$$12\%$$

Year	Factor for Year	Equal Payments	5% Per Yr. Decline	10% Per Yr. Decline	15% Per Yr. Decline	20% Per Yr. Decline	30% Per Yr. Decline
1	.94491	.94491	.94491	.94491	.94491	.94491	.94491
2	.84367	.89429	.89559	.89696	.89840	.89992	.90322
3	.75328	.84729	.85056	.85401	.85764	.86145	.86967
4	.67257	.80361	.80943	.81555	.82197	.82869	.84298
5	.60051	.76299	.77182	.78110	.79080	.80089	.82199
6	.53617	.72518	.73740	.75023	.76359	.77738	.80566
7	.47872	.68998	.70588	.72257	.73987	.75756	.79308
8	.42743	.65716	.67700	.69779	.71922	.74093	.78350
9	.38163	.62654	.65050	.67557	.70126	.72700	.77625
10	.34074	.59796	.62617	.65565	.68567	.71538	.77083
11	.30424	.57126	.60382	.63780	.67214	.70573	.76679
12	.27164	.54629	.58327	.62178	.66042	.69772	.76382
13	.24254	.52293	.56435	.60742	.65028	.69110	.76163
14	.21655	.50104	.54693	.59454	.64151	.68564	.76003
15	.19335	.48053	.53086	.58298	.63394	.68115	.75888
16	.17263	.46129	.51604	.57261	.62741	.67747	.75804
17	.15414	.44322	.50235	.56330	.62178	.67446	.75743
18	.13762	.42624	.48970	.55495	.61694	.67199	.75700
19	.12288	.41027	.47800	.54745	.61277	.66999	.75669
20	.10971	.39525	.46717	.54072	.60919	.66835	.75647
21	.09796	.38109	.45714	.53468	.60612	.66703	.75631
22	.08746	.36774	.44783	.52925	.60348	.66595	.75620
23	.07809	.35515	.43920	.52437	.60122	.66508	.75612
24	.06972	.34326	.43118	.52000	.59929	.66437	.75606
25	.06225	.33202	.42372	.51606	.59763	.66380	.75602
26	.05558	.32138	.41679	.51253	.59621	.66334	.75600
27	.04963	.31132	.41034	.50935	.59500	.66297	.75598
28	.04431	.30178	.40432	.50650	.59396	.66267	.75596
29	.03956	.29274	.39872	.50393	.59307	.66242	.75595
30	.03532	.28416	.39349	.50163	.59231	.66223	.75595
31	.03154	.27601	.38861	.49956	.59167	.66207	.75594
32	.02816	.26827	.38406	.49770	.59112	.66195	.75594
33	.02514	.26090	.37980	.49602	.59065	.66185	.75593
34	.02245	.25389	.37581	.49452	.59025	.66177	.75593
35	.02004	.24720	.37208	.49316	.58990	.66170	.75593
36	.01790	.24083	.36859	.49195	.58961	.66165	.75593
37	.01598	.23476	.36532	.49085	.58936	.66161	.75593
38	.01427	.22895	.36225	.48987	.58915	.66157	.75593
39	.01274	.22341	.35937	.48898	.58897	.66155	.75593
40	.01137	.21811	.35667	.48819	.58882	.66152	.75593

TABLE A.1 Present Worth Factors for Income Received in <u>One</u>
Payment at The <u>Middle of The Year</u>

Year	Factor for Year	Equal Payments	5% Per Yr. Decline	10% Per Yr. Decline	15% Per Yr. Decline	20% Per Yr. Decline	30% Per Yr. Decline	
1	.94072	.94072	.94072	.94072	.94072	.94072	.94072	
2	.83250	.88661	.88800	.88946	.89100	.89262	.89616	
3	.73672	.83665	.84013	.84381	.84767	.85173	.86049	
4	.65197	.79048	.79665	.80314	.80995	.81708	.83225	
5	.57696	.74777	.75710	.76690	.77716	.78782	.81015	
6	.51059	.70824	.72110	.73460	.74867	.76320	.79303	
7	.45185	.67161	.68829	.70580	.72396	.74254	.77991	
8	.39986	.63765	.65837	.68010	.70254	.72527	.76994	
9	.35386	.60611	.63105	.65718	.68399	.71088	.76244	
10	.31315	.57682	.60609	.63672	.66795	.69892	.75685	13%
11	.27713	.54957	.58325	.61844	.65409	.68901	.75270	
12	.24524	.52421	.56233	.60212	.64212	.68082	.74965	
13	.21703	.50058	.54316	.58754	.63180	.67408	.74741	
14	.19206	.47855	.52557	.57451	.62291	.66853	.74579	
15	.16997	.45797	.50942	.56285	.61526	.66399	.74461	
16	.15041	.43875	.49456	.55243	.60868	.66027	.74376	
17	.13311	.42077	.48089	.54310	.60303	.65723	.74315	
18	.11780	.40394	.46830	.53476	.59818	.65476	.74271	
19	.10424	.38817	.45669	.52729	.59401	.65275	.74240	
20	.09225	.37337	.44597	.52060	.59044	.65111	.74218	

Year	Factor for Year	Equal Payments	5% Per Yr. Decline	10% Per Yr. Decline	15% Per Yr. Decline	20% Per Yr. Decline	30% Per Yr. Decline	
1	.93250	.93250	.93250	.93250	.93250	.93250	.93250	
2	.81087	.87169	.87325	.87489	.87662	.87845	.88242	
3	.70511	.81616	.82005	.82414	.82845	.83298	.84275	
4	.61314	.76541	.77223	.77941	.78696	.79485	.81166	
5	.53316	.71896	.72919	.73996	.75123	.76296	.78754	
6	.46362	.67640	.69041	.70514	.72050	.73638	.76903	
7	.40315	.63736	.65541	.67437	.69407	.71427	.75496	
8	.35056	.60151	.62378	.64718	.67137	.69594	.74436	15%
9	.30484	.56855	.59517	.62312	.65187	.68078	.73644	
10	.26508	.53820	.56925	.60183	.63514	.66828	.73057	
11	.23050	.51023	.54573	.58296	.62079	.65800	.72624	
12	.20044	.48441	.52436	.56623	.60849	.64956	.72308	
13	.17429	.46056	.50493	.55139	.59795	.64264	.72078	
14	.15156	.43849	.48722	.53821	.58893	.63700	.71911	
15	.13179	.41804	.47108	.52650	.58120	.63239	.71791	
16	.11460	.39908	.45633	.51609	.57460	.62864	.71705	
17	.09965	.38146	.44284	.50683	.56895	.62559	.71643	
18	.08665	.36508	.43048	.49858	.56413	.62312	.71599	
19	.07535	.34983	.41916	.49124	.56001	.62112	.71568	
20	.06552	.33562	.40876	.48469	.55649	.61950	.71546	

TABLE A.1 Present Worth Factors for Income Received in One
Payment at The Middle of The Year

Year	Factor for Year	Equal Payments	5% Per Yr. Decline	10% Per Yr. Decline	15% Per Yr. Decline	20% Per Yr. Decline	30% Per Yr. Decline	
1	.92057	.92057	.92057	.92057	.92057	.92057	.92057	
2	.78015	.85036	.85216	.85406	.85605	.85816	.86275	
3.	.66114	.78729	.79173	.79640	.80131	.80649	.81764	
4	.56029	.73054	.73824	.74635	.75486	.76378	.78279	
5	.47482	.67940	.69082	.70284	.71545	.72858	.75613	
6	.40239	.63323	.64869	.66498	.68199	.69960	.73592	
7	.34101	.59148	.61121	.63198	.65360	.67581	.72073	
8	.28899	.55367	.57778	.60317	.62950	.65632	.70941	
9	.24491	.51936	.54792	.57800	.60904	.64037	.70104	
10	.20755	.48818	.52119	.55596	.59168	.62736	.69489	18%
11	.17589	.45979	.49721	.53665	.57693	.61675	.69040	
12	.14906	.43390	.47567	.51970	.56441	.60812	.68715	
13	.12632	.41024	.45628	.50480	.55377	.60112	.68479	
14	.10705	.38858	.43878	.49169	.54475	.59543	.68310	
15	.09072	.36872	.42297	.48014	.53708	.59083	.68189	
16	.07688	.35048	.40865	.46995	.53056	.58711	.68103	
17	.06515	.33370	.39566	.46095	.52503	.58411	.68041	
18	.05522	.31823	.38385	.45299	.52033	.58168	.67997	
19	.04679	.30394	.37310	.44594	.51634	.57973	.67967	
20	.03966	.29073	.36329	.43969	.51295	.57815	.67945	

Year	Factor for Year	Equal Payments	5% Per Yr. Decline	10% Per Yr. Decline	15% Per Yr. Decline	20% Per Yr. Decline	30% Per Yr. Decline	
1	.91287	.91287	.91287	.91287	.91287	.91287	.91287	
2	.76073	.83680	.83875	.84080	.84297	.84525	.85022	
3	.63394	.76918	.77395	.77897	.78426	.78982	.80183	
4	.52828	.70895	.71717	.72583	.73493	.74446	.76479	
5	.44023	.65521	.66732	.68007	.69345	.70739	.73669	
6	.36686	.60715	.62344	.64060	.65855	.67715	.71555	
7	.30572	.56409	.58473	.60649	.62917	.65251	.69979	
8	.25477	.52542	.55050	.57695	.60442	.63246	.68812	
9	.21230	.49063	.52016	.55133	.58357	.61618	.67955	
10	.17692	.45926	.49320	.52906	.56597	.60297	.67329	20%
11	.14743	.43091	.46920	.50966	.55113	.59227	.66874	
12	.12286	.40524	.44777	.49275	.53860	.58361	.66546	
13	.10238	.38195	.42859	.47797	.52801	.57661	.66310	
14	.08532	.36076	.41139	.46502	.51906	.57096	.66141	
15	.07110	.34145	.39593	.45368	.51150	.56640	.66020	
16	.05925	.32381	.38200	.44371	.50509	.56273	.65934	
17	.04938	.30767	.36942	.43494	.49968	.55977	.65873	
18	.04115	.29286	.35804	.42721	.49509	.55739	.65830	
19	.03429	.27925	.34771	.42039	.49121	.55548	.65799	
20	.02857	.26672	.33833	.41437	.48791	.55394	.65778	

Year	Factor for Year	Equal Payments	5% Per Yr. Decline	10% Per Yr. Decline	15% Per Yr. Decline	20% Per Yr. Decline	30% Per Yr. Decline	
1	.89443	.89443	.89443	.89443	.89443	.89443	.89443	
2	.71554	.80498	.80728	.80969	.81224	.81492	.82077	
3	.57243	.72747	.73298	.73878	.74489	.75132	.76520	
4	.45795	.66009	.66941	.67925	.68959	.70044	.72360	
5	.36636	.60134	.61486	.62912	.64409	.65973	.69267	
6	.29309	.54997	.56786	.58677	.60658	.62716	.66983	
7	.23447	.50489	.52724	.55088	.57560	.60111	.65309	
8	.18757	.46523	.49200	.52037	.54995	.58027	.64088	
9	.15006	.43021	.46133	.49435	.52868	.56360	.63204	25%
10	.12005	.39919	.43453	.47208	.51100	.55026	.62566	
11	.09604	.37163	.41103	.45298	.49629	.53959	.62108	
12	.07683	.34707	.39035	.43653	.48401	.53105	.61781	
13	.06146	.32510	.37209	.42232	.47376	.52422	.61548	
14	.04917	.30539	.35591	.41002	.46517	.51876	.61382	
15	.03934	.28765	.34153	.39935	.45798	.51439	.61264	
16	.03147	.27164	.32870	.39005	.45194	.51089	.61181	
17	.02518	.25714	.31722	.38193	.44687	.50809	.61123	
18	.02014	.24398	.30692	.37484	.44260	.50586	.61081	
19	.01611	.23198	.29764	.36861	.43901	.50407	.61052	
20	.01289	.22103	.28927	.36314	.43597	.50263	.61032	

Year	Factor for Year	Equal Payments	5% Per Yr. Decline	10% Per Yr. Decline	15% Per Yr. Decline	20% Per Yr. Decline	30% Per Yr. Decline	
1	.87706	.87706	.87706	.87706	.87706	.87706	.87706	
2	.67466	.77586	.77845	.78119	.78406	.78710	.79372	
3	.51897	.69023	.69636	.70281	.70961	.71677	.73224	
4	.39921	.61747	.62768	.63845	.64979	.66169	.68715	
5	.30708	.55540	.56997	.58536	.60155	.61849	.65424	
6	.23622	.50220	.52122	.54136	.56251	.58453	.63035	
7	.18171	.45641	.47986	.50472	.53081	.55781	.61310	
8	.13977	.41683	.44458	.47408	.50496	.53674	.60069	
9	.10752	.38247	.41434	.44832	.48382	.52011	.59180	30%
10	.08271	.35249	.38829	.42657	.46647	.50695	.58546	
11	.06362	.32623	.36575	.40813	.45218	.49654	.58095	
12	.04894	.30312	.34615	.39242	.44038	.48828	.57775	
13	.03765	.28270	.32902	.37898	.43060	.48173	.57548	
14	.02896	.26458	.31399	.36745	.42249	.47652	.57388	
15	.02228	.24842	.30074	.35750	.41573	.47238	.57275	
16	.01713	.23397	.28900	.34890	.41008	.46908	.57196	
17	.01318	.22098	.27857	.34144	.40537	.46646	.57140	
18	.01014	.20927	.26926	.33494	.40141	.46436	.57101	
19	.00780	.19866	.26092	.32926	.39809	.46269	.57073	
20	.00600	.18903	.25342	.32429	.39530	.46136	.57054	

Year	Factor for Year	Equal Payments	5% Per Yr. Decline	10% Per Yr. Decline	15% Per Yr. Decline	20% Per Yr. Decline	30% Per Yr. Decline	
1	.84515	.84515	.84515	.84515	.84515	.84515	.84515	
2	.60368	.72442	.72751	.73077	.73421	.73783	.74572	
3	.43120	.62668	.63376	.64123	.64911	.65741	.67535	
4	.30800	.54701	.55848	.57059	.58337	.59680	.62561	
5	.22000	.48161	.49754	.51442	.53222	.55089	.59049	
6	.15714	.42753	.44783	.46940	.49214	.51592	.56573	
7	.11225	.38249	.40694	.43302	.46051	.48914	.54828	
8	.08018	.34470	.37304	.40338	.43537	.46853	.53601	
9	.05727	.31276	.34471	.37906	.41526	.45259	.52738	
10	.04091	.28558	.32085	.35895	.39906	.44021	.52132	40%
11	.02922	.26227	.30061	.34219	.38595	.43055	.51707	
12	.02087	.24216	.28330	.32814	.37526	.42300	.51408	
13	.01491	.22467	.26840	.31628	.36652	.41706	.51199	
14	.01065	.20939	.25549	.30621	.35932	.41239	.51052	
15	.00761	.19594	.24422	.29760	.35338	.40870	.50950	
16	.00543	.18403	.23434	.29022	.34846	.40578	.50878	
17	.00388	.17343	.22563	.28385	.34436	.40346	.50827	
18	.00277	.16395	.21790	.27834	.34095	.40162	.50792	
19	.00198	.15543	.21101	.27354	.33809	.40016	.50767	
20	.00141	.14773	.20485	.26936	.33569	.39900	.50750	

Year	Factor for Year	Equal Payments	5% Per Yr. Decline	10% Per Yr. Decline	15% Per Yr. Decline	20% Per Yr. Decline	30% Per Yr. Decline	
1	.81650	.81650	.81650	.81650	.81650	.81650	.81650	
2	.54433	.68041	.68390	.68758	.69145	.69553	.70443	
3	.36289	.57457	.58234	.59053	.59917	.60828	.62801	
4	.24192	.49141	.50367	.51663	.53032	.54474	.57573	
5	.16128	.42538	.44203	.45970	.47838	.49802	.53985	
6	.10752	.37241	.39317	.41532	.43875	.46333	.51514	
7	.07168	.32945	.35400	.38031	.40819	.43735	.49809	
8	.04779	.29424	.32224	.35239	.38436	.41772	.48628	
9	.03186	.26509	.29619	.32986	.36561	.40276	.47809	
10	.02124	.24070	.27459	.31150	.35072	.39129	.47240	50%
11	.01416	.22011	.25651	.29640	.33878	.38243	.46844	
12	.00944	.20255	.24123	.28385	.32914	.37555	.46568	
13	.00629	.18745	.22818	.27334	.32131	.37018	.46375	
14	.00420	.17436	.21696	.26447	.31490	.36597	.46240	
15	.00280	.16293	.20723	.25693	.30962	.36266	.46147	
16	.00186	.15286	.19873	.25048	.30527	.36005	.46081	
17	.00124	.14394	.19126	.24494	.30165	.35798	.46035	
18	.00083	.13599	.18466	.24015	.29864	.35634	.46003	
19	.00055	.12886	.17879	.23599	.29613	.35504	.45980	
20	.00037	.12244	.17354	.23237	.29403	.35401	.45965	

Table A.2

Discount Factors for Continuous Compounding and Annual Cash Flow

Received Uniformly Throughout the Year

Year	1	2	3	4	5	6	7	8	9	10	11	12	13	14	15	16	17	18	19	20
0-1	.9950	.9901	.9851	.9803	.9754	.9706	.9658	.9610	.9563	.9516	.9470	.9423	.9377	.9332	.9286	.9241	.9196	.9152	.9107	.9063
1-2	.9851	.9705	.9560	.9418	.9278	.9141	.9005	.8872	.8740	.8611	.8483	.8358	.8234	.8112	.7993	.7875	.7759	.7644	.7531	.7421
2-3	.9753	.9512	.9278	.9049	.8826	.8608	.8396	.8189	.7988	.7791	.7600	.7413	.7230	.7053	.6879	.6710	.6546	.6385	.6228	.6075
3-4	.9656	.9324	.9004	.8694	.8395	.8107	.7829	.7560	.7300	.7050	.6808	.6574	.6349	.6131	.5921	.5718	.5522	.5333	.5150	.4974
4-5	.9560	.9140	.8737	.8353	.7986	.7635	.7299	.6979	.6672	.6379	.6099	.5831	.5575	.5330	.5096	.4873	.4659	.4455	.4259	.4072
5-6	.9465	.8959	.8479	.8026	.7596	.7190	.6806	.6442	.6098	.5772	.5463	.5172	.4895	.4634	.4386	.4152	.3931	.3721	.3522	.3334
6-7	.9371	.8781	.8229	.7711	.7226	.6772	.6346	.5947	.5573	.5223	.4894	.4588	.4299	.4029	.3775	.3538	.3316	.3108	.2913	.2730
7-8	.9278	.8607	.7985	.7409	.6874	.6377	.5917	.5490	.5093	.4726	.4385	.4069	.3775	.3502	.3250	.3015	.2798	.2596	.2409	.2235
8-9	.9185	.8437	.7749	.7118	.6538	.6006	.5517	.5068	.4655	.4276	.3928	.3609	.3314	.3045	.2797	.2569	.2360	.2168	.1992	.1830
9-10	.9094	.8270	.7520	.6839	.6219	.5656	.5144	.4678	.4254	.3869	.3519	.3201	.2910	.2647	.2407	.2189	.1991	.1811	.1647	.1498
10-11	.9003	.8106	.7298	.6571	.5916	.5327	.4796	.4318	.3888	.3501	.3152	.2839	.2556	.2301	.2072	.1866	.1680	.1513	.1362	.1227
11-12	.8914	.7946	.7082	.6312	.5628	.5016	.4472	.3986	.3553	.3168	.2824	.2518	.2244	.2000	.1783	.1590	.1417	.1264	.1126	.1004
12-13	.8825	.7788	.6873	.6065	.5353	.4724	.4169	.3680	.3248	.2866	.2530	.2233	.1970	.1739	.1535	.1355	.1196	.1055	.0932	.0822
13-14	.8737	.7634	.6670	.5827	.5092	.4449	.3888	.3397	.2968	.2593	.2266	.1981	.1730	.1512	.1321	.1154	.1009	.0882	.0770	.0673
14-15	.8650	.7483	.6473	.5599	.4844	.4190	.3625	.3136	.2713	.2347	.2030	.1757	.1519	.1314	.1137	.0984	.0851	.0736	.0637	.0551
15-16	.8564	.7335	.6282	.5380	.4608	.3946	.3380	.2895	.2479	.2123	.1819	.1558	.1334	.1143	.0979	.0838	.0718	.0615	.0527	.0451
16-17	.8479	.7139	.6096	.5169	.4383	.3716	.3151	.2672	.2266	.1921	.1629	.1382	.1172	.0993	.0842	.0714	.0606	.0514	.0436	.0369
17-18	.8395	.7047	.5916	.4966	.4169	.3500	.2938	.2467	.2071	.1739	.1460	.1225	.1029	.0864	.0725	.0609	.0511	.0429	.0360	.0303
18-19	.8311	.6908	.5741	.4772	.3966	.3296	.2740	.2277	.1893	.1573	.1308	.1087	.0903	.0751	.0624	.0519	.0431	.0358	.0298	.0248
19-20	.8228	.6771	.5571	.4584	.3772	.3104	.2554	.2102	.1730	.1423	.1171	.0964	.0793	.0653	.0537	.0442	.0364	.0299	.0246	.0203
20-21	.8147	.6637	.5407	.4405	.3588	.2923	.2382	.1940	.1581	.1288	.1049	.0855	.0697	.0568	.0462	.0377	.0307	.0250	.0204	.0166
21-22	.8065	.6505	.5247	.4232	.3413	.2753	.2221	.1791	.1445	.1165	.0940	.0758	.0612	.0493	.0398	.0321	.0259	.0209	.0169	.0136
22-23	.7985	.6376	.5092	.4066	.3247	.2593	.2071	.1653	.1320	.1054	.0842	.0673	.0537	.0429	.0343	.0274	.0218	.0175	.0139	.0111
23-24	.7906	.6250	.4941	.3907	.3089	.2442	.1931	.1526	.1207	.0954	.0754	.0596	.0472	.0373	.0295	.0233	.0184	.0146	.0115	.0091
24-25	.7827	.6126	.4795	.3753	.2938	.2300	.1800	.1409	.1103	.0863	.0676	.0529	.0414	.0324	.0254	.0199	.0156	.0122	.0095	.0075

Table A.3

Convenient Summary of Natural Logarithmic Functions

In addition to the specific applications of this table noted in Chapters 5 and 9, and in the introduction to Appendix A, this may be regarded as a basic natural log table.

$$\ln e^x = x$$

In any calculation when it is desired to take the natural log of a number, locate that number in the e^x column. The natural log of the number is read from the x column. Also, a log to the base 10 (log) = 0.434 ln of the same number.

Conversely, knowing the natural log of a number x, the number is found from e^x.

This table may be used also to find numbers to a power. Suppose that you need to find the answer for a number like $(3.39)^4$. Let $y = (3.39)^4$. Equivalently, $\ln y = 4 \ln 3.39$

$$\ln 3.39 = 1.22 \qquad (4)(1.22) = 4.88$$

$$\ln y = 4.88 \quad , \quad y = 131.63 = (3.39)^4$$

x	e^x	e^{-x}	$\dfrac{1-e^{-x}}{x}$	x	e^x	e^{-x}	$\dfrac{1-e^{-x}}{x}$
0.00	1.000	1.000	0.000	0.35	1.419	0.705	0.844
0.01	1.010	0.990	0.995	0.36	1.433	0.698	0.840
0.02	1.020	0.980	0.990	0.37	1.448	0.691	0.836
0.03	1.031	0.970	0.985	0.38	1.462	0.684	0.832
0.04	1.041	0.961	0.980	0.39	1.477	0.677	0.828
0.05	1.051	0.951	0.975	0.40	1.492	0.670	0.824
0.06	1.062	0.942	0.971	0.41	1.507	0.664	0.820
0.07	1.073	0.932	0.966	0.42	1.522	0.657	0.817
0.08	1.083	0.923	0.961	0.43	1.537	0.651	0.813
0.09	1.094	0.914	0.956	0.44	1.553	0.644	0.809
0.10	1.105	0.905	0.952	0.45	1.568	0.638	0.805
0.11	1.116	0.896	0.947	0.46	1.584	0.631	0.802
0.12	1.128	0.887	0.942	0.47	1.600	0.625	0.798
0.13	1.139	0.878	0.938	0.48	1.616	0.619	0.794
0.14	1.150	0.869	0.933	0.49	1.632	0.613	0.791
0.15	1.162	0.861	0.929	0.50	1.649	0.607	0.787
0.16	1.174	0.852	0.924	0.51	1.665	0.600	0.783
0.17	1.185	0.844	0.920	0.52	1.682	0.595	0.780
0.18	1.197	0.835	0.915	0.53	1.699	0.589	0.776
0.19	1.209	0.827	0.911	0.54	1.716	0.583	0.773
0.20	1.221	0.819	0.906	0.55	1.733	0.577	0.769
0.21	1.234	0.811	0.902	0.56	1.751	0.571	0.766
0.22	1.246	0.803	0.898	0.57	1.768	0.566	0.762
0.23	1.259	0.795	0.893	0.58	1.786	0.560	0.759
0.24	1.271	0.787	0.889	0.59	1.804	0.554	0.755
0.25	1.284	0.779	0.885	0.60	1.822	0.549	0.752
0.26	1.297	0.771	0.881	0.61	1.840	0.543	0.749
0.27	1.310	0.763	0.876	0.62	1.859	0.538	0.745
0.28	1.323	0.756	0.872	0.63	1.878	0.533	0.742
0.29	1.336	0.748	0.868	0.64	1.897	0.527	0.739
0.30	1.350	0.741	0.864	0.65	1.916	0.522	0.735
0.31	1.363	0.733	0.860	0.66	1.935	0.517	0.732
0.32	1.377	0.726	0.856	0.67	1.954	0.512	0.729
0.33	1.391	0.719	0.852	0.68	1.974	0.507	0.726
0.34	1.405	0.712	0.848	0.69	1.994	0.502	0.722

Table A.3 – Exponential Functions

x	e^x	e^{-x}	$\frac{1-e^{-x}}{x}$	x	e^x	e^{-x}	$\frac{1-e^{-x}}{x}$	x	e^x	e^{-x}	$\frac{1-e^{-x}}{x}$	x	e^x	e^{-x}	$\frac{1-e^{-x}}{x}$
0.70	2.014	0.497	0.719	1.05	2.858	0.350	0.619	1.40	4.055	0.247	0.538	1.75	5.755	0.174	0.472
0.71	2.034	0.492	0.716	1.06	2.886	0.346	0.617	1.41	4.096	0.244	0.536	1.76	5.812	0.172	0.470
0.72	2.054	0.487	0.713	1.07	2.915	0.343	0.614	1.42	4.137	0.242	0.534	1.77	5.871	0.170	0.469
0.73	2.075	0.482	0.710	1.08	2.945	0.340	0.611	1.43	4.179	0.239	0.532	1.78	5.930	0.169	0.467
0.74	2.096	0.477	0.707	1.09	2.974	0.336	0.609	1.44	4.221	0.237	0.530	1.79	5.989	0.167	0.465
0.75	2.117	0.472	0.704	1.10	3.004	0.333	0.606	1.45	4.263	0.235	0.528	1.80	6.050	0.165	0.464
0.76	2.138	0.468	0.700	1.11	3.034	0.330	0.604	1.46	4.306	0.232	0.526	1.81	6.110	0.164	0.462
0.77	2.160	0.463	0.697	1.12	3.065	0.326	0.602	1.47	4.349	0.230	0.524	1.82	6.172	0.162	0.460
0.78	2.182	0.458	0.694	1.13	3.096	0.323	0.599	1.48	4.393	0.228	0.522	1.83	6.234	0.160	0.459
0.79	2.203	0.454	0.691	1.14	3.127	0.320	0.597	1.49	4.437	0.225	0.520	1.84	6.297	0.159	0.457
0.80	2.226	0.449	0.688	1.15	3.158	0.317	0.594	1.50	4.482	0.223	0.518	1.85	6.360	0.157	0.456
0.81	2.248	0.445	0.685	1.16	3.190	0.313	0.592	1.51	4.527	0.221	0.516	1.86	6.424	0.156	0.454
0.82	2.271	0.440	0.682	1.17	3.222	0.310	0.589	1.52	4.572	0.219	0.514	1.87	6.488	0.154	0.452
0.83	2.293	0.436	0.679	1.18	3.254	0.307	0.587	1.53	4.618	0.217	0.512	1.88	6.554	0.153	0.451
0.84	2.316	0.432	0.677	1.19	3.287	0.304	0.585	1.54	4.665	0.214	0.510	1.89	6.619	0.151	0.449
0.85	2.340	0.427	0.674	1.20	3.320	0.301	0.582	1.55	4.712	0.212	0.508	1.90	6.686	0.150	0.448
0.86	2.363	0.423	0.671	1.21	3.354	0.298	0.580	1.56	4.759	0.210	0.506	1.91	6.753	0.148	0.446
0.87	2.387	0.419	0.668	1.22	3.387	0.295	0.578	1.57	4.807	0.208	0.504	1.92	6.821	0.147	0.444
0.88	2.411	0.415	0.665	1.23	3.421	0.292	0.575	1.58	4.855	0.206	0.503	1.93	6.890	0.145	0.443
0.89	2.435	0.411	0.662	1.24	3.456	0.289	0.573	1.59	4.904	0.204	0.501	1.94	6.959	0.144	0.441
0.90	2.460	0.407	0.659	1.25	3.490	0.287	0.571	1.60	4.953	0.202	0.499	1.95	7.029	0.142	0.440
0.91	2.484	0.403	0.657	1.26	3.525	0.284	0.569	1.61	5.003	0.200	0.497	1.96	7.099	0.141	0.438
0.92	2.509	0.399	0.654	1.27	3.561	0.281	0.566	1.62	5.053	0.198	0.495	1.97	7.171	0.139	0.437
0.93	2.535	0.395	0.651	1.28	3.597	0.278	0.564	1.63	5.104	0.196	0.493	1.98	7.243	0.138	0.435
0.94	2.560	0.391	0.648	1.29	3.633	0.275	0.562	1.64	5.155	0.194	0.491	1.99	7.316	0.137	0.434
0.95	2.586	0.387	0.646	1.30	3.669	0.273	0.560	1.65	5.207	0.192	0.490	2.00	7.389	0.135	0.432
0.96	2.612	0.383	0.643	1.31	3.706	0.270	0.557	1.66	5.259	0.190	0.488	2.01	7.463	0.134	0.431
0.97	2.638	0.379	0.640	1.32	3.743	0.267	0.555	1.67	5.312	0.188	0.486	2.02	7.538	0.133	0.429
0.98	2.665	0.375	0.637	1.33	3.781	0.264	0.553	1.68	5.366	0.186	0.484	2.03	7.614	0.131	0.428
0.99	2.691	0.372	0.635	1.34	3.819	0.262	0.551	1.69	5.420	0.185	0.483	2.04	7.691	0.130	0.426
1.00	2.718	0.368	0.632	1.35	3.857	0.259	0.549	1.70	5.474	0.183	0.481	2.05	7.768	0.129	0.425
1.01	2.746	0.364	0.629	1.36	3.896	0.257	0.547	1.71	5.529	0.181	0.479	2.06	7.846	0.127	0.424
1.02	2.773	0.361	0.627	1.37	3.935	0.254	0.544	1.72	5.585	0.179	0.477	2.07	7.925	0.126	0.422
1.03	2.801	0.357	0.624	1.38	3.975	0.252	0.542	1.73	5.641	0.177	0.476	2.08	8.005	0.125	0.421
1.04	2.829	0.353	0.622	1.39	4.015	0.249	0.540	1.74	5.697	0.176	0.474	2.09	8.085	0.124	0.419

Table A.3 – Exponential Functions

x	e^x	e^{-x}	$\dfrac{1-e^{-x}}{x}$
2.10	8.166	0.122	0.418
2.11	8.248	0.121	0.416
2.12	8.331	0.120	0.415
2.13	8.415	0.119	0.414
2.14	8.499	0.118	0.412
2.15	8.585	0.116	0.411
2.16	8.671	0.115	0.410
2.17	8.758	0.114	0.408
2.18	8.846	0.113	0.407
2.19	8.935	0.112	0.406
2.20	9.025	0.111	0.404
2.21	9.116	0.110	0.403
2.22	9.207	0.109	0.402
2.23	9.300	0.108	0.400
2.24	9.393	0.106	0.399
2.25	9.488	0.105	0.398
2.26	9.583	0.104	0.396
2.27	9.679	0.103	0.395
2.28	9.777	0.102	0.394
2.29	9.875	0.101	0.392
2.30	9.974	0.100	0.391
2.31	10.074	0.099	0.390
2.32	10.176	0.098	0.389
2.33	10.278	0.097	0.387
2.34	10.381	0.096	0.386
2.35	10.486	0.095	0.385
2.36	10.591	0.094	0.384
2.37	10.697	0.093	0.382
2.38	10.805	0.093	0.381
2.39	10.914	0.092	0.380
2.40	11.023	0.091	0.379
2.41	11.134	0.090	0.378
2.42	11.246	0.089	0.376
2.43	11.359	0.088	0.375
2.44	11.473	0.087	0.374

x	e^x	e^{-x}	$\dfrac{1-e^{-x}}{x}$
2.45	11.588	0.086	0.373
2.46	11.705	0.085	0.372
2.47	11.822	0.085	0.371
2.48	11.941	0.084	0.369
2.49	12.061	0.083	0.368
2.50	12.183	0.082	0.367
2.51	12.305	0.081	0.366
2.52	12.429	0.080	0.365
2.53	12.554	0.080	0.364
2.54	12.680	0.079	0.363
2.55	12.807	0.078	0.362
2.56	12.936	0.077	0.360
2.57	13.066	0.077	0.359
2.58	13.197	0.076	0.358
2.59	13.330	0.075	0.357
2.60	13.464	0.074	0.356
2.61	13.599	0.074	0.355
2.62	13.736	0.073	0.354
2.63	13.874	0.072	0.353
2.64	14.013	0.071	0.352
2.65	14.154	0.071	0.351
2.66	14.296	0.070	0.350
2.67	14.440	0.069	0.349
2.68	14.585	0.069	0.348
2.69	14.732	0.068	0.347
2.70	14.880	0.067	0.345
2.71	15.029	0.067	0.344
2.72	15.180	0.066	0.343
2.73	15.333	0.065	0.342
2.74	15.487	0.065	0.341
2.75	15.643	0.064	0.340
2.76	15.800	0.063	0.339
2.77	15.959	0.063	0.338
2.78	16.119	0.062	0.337
2.79	16.281	0.061	0.336

x	e^x	e^{-x}	$\dfrac{1-e^{-x}}{x}$
2.80	16.445	0.061	0.335
2.81	16.610	0.060	0.334
2.82	16.777	0.060	0.333
2.83	16.946	0.059	0.333
2.84	17.116	0.058	0.332
2.85	17.288	0.058	0.331
2.86	17.462	0.057	0.330
2.87	17.637	0.057	0.329
2.88	17.814	0.056	0.328
2.89	17.993	0.056	0.327
2.90	18.174	0.055	0.326
2.91	18.357	0.054	0.325
2.92	18.541	0.054	0.324
2.93	18.728	0.053	0.323
2.94	18.916	0.053	0.322
2.95	19.106	0.052	0.321
2.96	19.298	0.052	0.320
2.97	19.492	0.051	0.319
2.98	19.688	0.051	0.319
2.99	19.886	0.050	0.318
3.00	20.086	0.050	0.317
3.01	20.287	0.049	0.316
3.02	20.491	0.049	0.315
3.03	20.697	0.048	0.314
3.04	20.905	0.048	0.313
3.05	21.115	0.047	0.312
3.06	21.328	0.047	0.311
3.07	21.542	0.046	0.311
3.08	21.758	0.046	0.310
3.09	21.977	0.046	0.309
3.10	22.198	0.045	0.308
3.11	22.421	0.045	0.307
3.12	22.646	0.044	0.306
3.13	22.874	0.044	0.306
3.14	23.104	0.043	0.305

x	e^x	e^{-x}	$\dfrac{1-e^{-x}}{x}$
3.15	23.336	0.043	0.304
3.16	23.571	0.042	0.303
3.17	23.808	0.042	0.302
3.18	24.047	0.042	0.301
3.19	24.288	0.041	0.301
3.20	24.533	0.041	0.300
3.21	24.779	0.040	0.299
3.22	25.028	0.040	0.298
3.23	25.280	0.040	0.297
3.24	25.534	0.039	0.297
3.25	25.790	0.039	0.296
3.26	26.050	0.038	0.295
3.27	26.311	0.038	0.294
3.28	26.576	0.038	0.293
3.29	26.843	0.037	0.293
3.30	27.113	0.037	0.292
3.31	27.385	0.037	0.291
3.32	27.660	0.036	0.290
3.33	27.938	0.036	0.290
3.34	28.219	0.035	0.289
3.35	28.503	0.035	0.288
3.36	28.789	0.035	0.287
3.37	29.079	0.034	0.287
3.38	29.371	0.034	0.286
3.39	29.666	0.034	0.285
3.40	29.964	0.033	0.284
3.41	30.265	0.033	0.284
3.42	30.569	0.033	0.283
3.43	30.877	0.032	0.282
3.44	31.187	0.032	0.281
3.45	31.500	0.032	0.281
3.46	31.817	0.031	0.280
3.47	32.137	0.031	0.279
3.48	32.460	0.031	0.279
3.49	32.786	0.031	0.278

Table A.3 — Exponential Functions

x	e^x	e^{-x}	$\frac{1-e^{-x}}{x}$	x	e^x	e^{-x}	$\frac{1-e^{-x}}{x}$	x	e^x	e^{-x}	$\frac{1-e^{-x}}{x}$	x	e^x	e^{-x}	$\frac{1-e^{-x}}{x}$
3.50	33.115	0.030	0.277	3.85	46.993	0.021	0.254	4.20	66.686	0.015	0.235	4.55	94.632	0.011	0.217
3.51	33.448	0.030	0.276	3.86	47.465	0.021	0.254	4.21	67.357	0.015	0.234	4.56	95.584	0.010	0.217
3.52	33.784	0.029	0.276	3.87	47.942	0.021	0.053	4.22	68.034	0.015	0.233	4.57	96.544	0.010	0.217
3.53	34.124	0.029	0.275	3.88	48.424	0.021	0.252	4.23	68.717	0.015	0.233	4.58	97.514	0.010	0.216
3.54	34.467	0.029	0.274	3.89	48.911	0.020	0.252	4.24	69.408	0.014	0.232	4.59	98.494	0.010	0.216
3.55	34.813	0.029	0.274	3.90	49.402	0.020	0.251	4.25	70.106	0.014	0.232	4.60	99.484		0.215
3.56	35.163	0.028	0.273	3.91	49.899	0.020	0.251	4.26	70.810	0.014	0.231	4.61	100.484		0.215
3.57	35.517	0.028	0.272	3.92	50.400	0.020	0.250	4.27	71.522	0.014	0.231	4.62	101.494		0.214
3.58	35.874	0.028	0.272	3.93	50.907	0.020	0.249	4.28	72.241	0.014	0.230	4.63	102.514		0.214
3.59	36.234	0.028	0.271	3.94	51.419	0.019	0.249	4.29	72.967	0.014	0.230	4.64	103.544		0.213
3.60	36.598	0.027	0.270	3.95	51.935	0.019	0.248	4.30	73.700	0.014	0.229	4.65	104.585		0.213
3.61	36.966	0.027	0.270	3.96	52.457	0.019	0.248	4.31	74.441	0.013	0.229	4.66	105.636		0.213
3.62	37.338	0.027	0.269	3.97	52.985	0.019	0.247	4.32	75.189	0.013	0.228	4.67	106.698		0.212
3.63	37.713	0.027	0.268	3.98	53.517	0.019	0.247	4.33	75.944	0.013	0.228	4.68	107.770		0.212
3.64	38.092	0.026	0.268	3.99	54.055	0.019	0.246	4.34	76.708	0.013	0.227	4.69	108.853		0.211
3.65	38.475	0.026	0.267	4.00	54.598	0.018	0.245	4.35	77.479	0.013	0.227	4.70	109.947		0.211
3.66	38.861	0.026	0.266	4.01	55.147	0.018	0.245	4.36	78.257	0.013	0.226	4.71	111.052		0.210
3.67	39.252	0.025	0.266	4.02	55.701	0.018	0.244	4.37	79.044	0.013	0.226	4.72	112.168		0.210
3.68	39.646	0.025	0.265	4.03	56.261	0.018	0.244	4.38	79.838	0.013	0.225	4.73	113.296		0.210
3.69	40.045	0.025	0.264	4.04	56.826	0.018	0.243	4.39	80.641	0.012	0.225	4.74	114.434		0.209
3.70	40.447	0.025	0.264	4.05	57.397	0.017	0.243	4.40	81.451	0.012	0.224	4.75	115.584		0.209
3.71	40.854	0.024	0.263	4.06	57.974	0.017	0.242	4.41	82.270	0.012	0.224	4.76	116.746		0.208
3.72	41.264	0.024	0.262	4.07	58.557	0.017	0.242	4.42	83.096	0.012	0.224	4.77	117.919		0.208
3.73	41.679	0.024	0.262	4.08	59.146	0.017	0.241	4.43	83.932	0.012	0.223	4.78	119.104		0.207
3.74	42.098	0.024	0.261	4.09	59.740	0.017	0.240	4.44	84.775	0.012	0.223	4.79	120.302		0.207
3.75	42.521	0.024	0.260	4.10	60.340	0.017	0.240	4.45	85.627	0.012	0.222	4.80	121.511		0.207
3.76	42.948	0.023	0.260	4.11	60.947	0.016	0.239	4.46	86.488	0.012	0.222	4.81	122.732		0.206
3.77	43.380	0.023	0.259	4.12	61.559	0.016	0.239	4.47	87.357	0.011	0.221	4.82	123.965		0.206
3.78	43.816	0.023	0.259	4.13	62.178	0.016	0.238	4.48	88.235	0.011	0.221	4.83	125.211		0.205
3.79	44.256	0.023	0.258	4.14	62.803	0.016	0.238	4.49	89.122	0.011	0.220	4.84	126.470		0.205
3.80	44.701	0.022	0.257	4.15	63.434	0.016	0.237	4.50	90.017	0.011	0.220	4.85	127.741		0.205
3.81	45.150	0.022	0.257	4.16	64.072	0.016	0.237	4.51	90.922	0.011	0.219	4.86	129.024		0.204
3.82	45.604	0.022	0.256	4.17	64.716	0.015	0.236	4.52	91.836	0.011	0.219	4.87	130.321		0.204
3.83	46.063	0.022	0.255	4.18	65.366	0.015	0.236	4.53	92.759	0.011	0.218	4.88	131.631		0.203
3.84	46.526	0.021	0.255	4.19	66.023	0.015	0.235	4.54	93.691	0.011	0.218	4.89	132.954		0.203

Table A.3 – Exponential Functions

x	e^x	e^{-x}	$\frac{1-e^{-x}}{x}$
4.90	134.290		0.203
4.91	135.640		0.202
4.92	137.003		0.202
4.93	138.380		0.201
4.94	139.771		0.201
4.95	141.175		0.201
4.96	142.594		0.200
4.97	144.027		0.200
4.98	145.475		0.199
4.99	146.937		0.199
5.00	148.413		0.199
5.01	149.905		0.198
5.02	151.412		0.198
5.03	152.933		0.198
5.04	154.470		0.197
5.05	156.023		0.197
5.06	157.591		0.196
5.07	159.175		0.196
5.08	160.774		0.196
5.09	162.390		0.195
5.10	164.022		0.195
5.11	165.671		0.195
5.12	167.336		0.194
5.13	169.018		0.194
5.14	170.716		0.193
5.15	172.432		0.193
5.16	174.165		0.193
5.17	175.915		0.192
5.18	177.683		0.192
5.19	179.469		0.192
5.20	181.273		0.191
5.21	183.095		0.191
5.22	184.935		0.191
5.23	186.793		0.190
5.24	188.671		0.190

x	e^x	e^{-x}	$\frac{1-e^{-x}}{x}$
5.25	190.567		0.189
5.26	192.482		0.189
5.27	194.417		0.189
5.28	196.371		0.188
5.29	198.344		0.188
5.30	200.338		0.188
5.31	202.351		0.187
5.32	204.385		0.187
5.33	206.439		0.187
5.34	208.513		0.186
5.35	210.609		0.186
5.36	212.726		0.186
5.37	214.864		0.185
5.38	217.023		0.185
5.39	219.204		0.185
5.40	221.407		0.184
5.41	223.633		0.184
5.42	225.880		0.184
5.43	228.150		0.183
5.44	230.443		0.183
5.45	232.759		0.183
5.46	235.098		0.182
5.47	237.461		0.182
5.48	239.848		0.182
5.49	242.258		0.181
5.50	244.692		0.181
5.51	247.152		0.181
5.52	249.636		0.180
5.53	252.144		0.180
5.54	254.679		0.180
5.55	257.238		0.179
5.56	259.823		0.179
5.57	262.435		0.179
5.58	265.072		0.179
5.59	267.736		0.178

x	e^x	e^{-x}	$\frac{1-e^{-x}}{x}$
5.60	270.427		0.178
5.61	273.145		0.178
5.62	275.890		0.177
5.63	278.663		0.177
5.64	281.463		0.177
5.65	284.292		0.176
5.66	287.149		0.176
5.67	290.035		0.176
5.68	292.950		0.175
5.69	295.894		0.175
5.70	298.868		0.175
5.71	301.872		0.175
5.72	304.906		0.174
5.73	307.970		0.174
5.74	311.065		0.174
5.75	314.192		0.173
5.76	317.349		0.173
5.77	320.539		0.173
5.78	323.760		0.172
5.79	327.014		0.172
5.80	330.301		0.172
5.81	333.620		0.172
5.82	336.973		0.171
5.83	340.360		0.171
5.84	343.781		0.171
5.85	347.236		0.170
5.86	350.725		0.170
5.87	354.250		0.170
5.88	357.811		0.170
5.89	361.407		0.169
5.90	365.039		0.169
5.91	368.708		0.169
5.92	372.413		0.168
5.93	376.156		0.168
5.94	379.937		0.168

x	e^x	e^{-x}	$\frac{1-e^{-x}}{x}$
5.95	383.755		0.168
5.96	387.612		0.167
5.97	391.507		0.167
5.98	395.442		0.167
5.99	399.416		0.167
6.00	403.431		0.166
6.01	407.485		0.166
6.02	411.580		0.166
6.03	415.717		0.165
6.04	419.895		0.165
6.05	424.115		0.165
6.06	428.377		0.165
6.07	432.683		0.164
6.08	437.031		0.164
6.09	441.424		0.164
6.10	445.860		0.164
6.11	450.341		0.163
6.12	454.867		0.163
6.13	459.439		0.163
6.14	464.056		0.163
6.15	468.720		0.162
6.16	473.431		0.162
6.17	478.189		0.162
6.18	482.995		0.161
6.19	487.849		0.161
6.20	492.752		0.161
6.21	497.704		0.161
6.22	502.706		0.160
6.23	507.758		0.160
6.24	512.861		0.160
6.25	518.016		0.160
6.26	523.222		0.159
6.27	528.480		0.159
6.28	533.792		0.159
6.29	539.156		0.159

Table A.3 – Exponential Functions

x	e^x	e^{-x}	$\frac{1-e^{-x}}{x}$
6.30	544.575		0.158
6.31	550.048		0.158
6.32	555.576		0.158
6.33	561.160		0.158
6.34	566.800		0.157
6.35	572.496		0.157
6.36	578.250		0.157
6.37	584.061		0.157
6.38	589.931		0.156
6.39	595.860		0.156
6.40	601.849		0.156
6.41	607.898		0.156
6.42	614.007		0.156
6.43	620.178		0.155
6.44	626.411		0.155
6.45	632.706		0.155
6.46	639.065		0.155
6.47	645.488		0.154
6.48	651.975		0.154
6.49	658.528		0.154
6.50	665.145		0.154
6.51	671.829		0.153
6.52	678.581		0.153
6.53	685.401		0.153
6.54	692.290		0.153
6.55	699.247		0.152
6.56	706.275		0.152
6.57	713.373		0.152
6.58	720.543		0.152
6.59	727.784		0.152
6.60	735.099		0.151
6.61	742.487		0.151
6.62	749.949		0.151
6.63	757.486		0.151
6.64	765.099		0.150

x	e^x	e^{-x}	$\frac{1-e^{-x}}{x}$
6.65	772.788		0.150
6.66	780.555		0.150
6.67	788.400		0.150
6.68	796.323		0.150
6.69	804.327		0.149
6.70	812.410		0.149
6.71	820.575		0.149
6.72	828.822		0.149
6.73	837.152		0.148
6.74	845.565		0.148
6.75	854.064		0.148
6.76	862.647		0.148
6.77	871.317		0.148
6.78	880.074		0.147
6.79	888.919		0.147
6.80	897.853		0.147
6.81	906.876		0.147
6.82	915.990		0.146
6.83	925.196		0.146
6.84	934.495		0.146
6.85	943.887		0.146
6.86	953.373		0.146
6.87	962.955		0.145
6.88	972.632		0.145
6.89	982.408		0.145
6.90	992.281		0.145
6.91	1002.254		0.145
6.92	1012.327		0.144
6.93	1022.501		0.144
6.94	1032.777		0.144
6.95	1043.157		0.144
6.96	1053.641		0.144
6.97	1064.230		0.143
6.98	1074.926		0.143
6.99	1085.729		0.143

x	e^x	e^{-x}	$\frac{1-e^{-x}}{x}$
7.00	1096.641		0.143
7.01	1107.662		0.143
7.02	1118.794		0.142
7.03	1130.038		0.142
7.04	1141.395		0.142
7.05	1152.867		0.142
7.06	1164.453		0.142
7.07	1176.156		0.141
7.08	1187.977		0.141
7.09	1199.916		0.141
7.10	1211.975		0.141
7.11	1224.156		0.141
7.12	1236.459		0.140
7.13	1248.885		0.140
7.14	1261.437		0.140
7.15	1274.114		0.140
7.16	1286.920		0.140
7.17	1299.853		0.139
7.18	1312.917		0.139
7.19	1326.112		0.139
7.20	1339.440		0.139
7.21	1352.901		0.139
7.22	1366.498		0.138
7.23	1380.232		0.138
7.24	1394.103		0.138
7.25	1408.114		0.138
7.26	1422.266		0.138
7.27	1436.560		0.137
7.28	1450.998		0.137
7.29	1465.580		0.137
7.30	1480.310		0.137
7.31	1495.187		0.137
7.32	1510.214		0.137
7.33	1525.392		0.136
7.34	1540.722		0.136

x	e^x	e^{-x}	$\frac{1-e^{-x}}{x}$
7.35	1556.207		0.136
7.36	1571.847		0.136
7.37	1587.644		0.136
7.38	1603.600		0.135
7.39	1619.717		0.135
7.40	1635.995		0.135
7.41	1652.437		0.135
7.42	1669.044		0.135
7.43	1685.818		0.135
7.44	1702.761		0.134
7.45	1719.874		0.134
7.46	1737.159		0.134
7.47	1754.618		0.134
7.48	1772.252		0.134
7.49	1790.063		0.133
7.50	1808.055		0.133
7.51	1826.226		0.133
7.52	1844.580		0.133
7.53	1863.118		0.133
7.54	1881.843		0.133
7.55	1900.756		0.132
7.56	1919.858		0.132
7.57	1939.153		0.132
7.58	1958.642		0.132
7.59	1978.327		0.132
7.60	1998.209		0.132
7.61	2018.292		0.131
7.62	2038.576		0.131
7.63	2059.064		0.131
7.64	2079.758		0.131
7.65	2100.660		0.131
7.66	2121.772		0.130
7.67	2143.096		0.130
7.68	2164.634		0.130
7.69	2186.389		0.130

Table A.3 - Exponential Functions

x	e^x	e^{-x}	$\frac{1-e^{-x}}{x}$	x	e^x	e^{-x}	$\frac{1-e^{-x}}{x}$	x	e^x	e^{-x}	$\frac{1-e^{-x}}{x}$	x	e^x	e^{-x}	$\frac{1-e^{-x}}{x}$
7.70	2208.363		0.130	8.05	3133.816		0.124	8.40	4447.114		0.119	8.75	6310.744		0.114
7.71	2230.557		0.130	8.06	3165.312		0.124	8.41	4491.809		0.119	8.76	6374.169		0.114
7.72	2252.974		0.129	8.07	3197.124		0.124	8.42	4536.952		0.119	8.77	6438.231		0.114
7.73	2275.617		0.129	8.08	3229.256		0.124	8.43	4582.550		0.119	8.78	6502.937		0.114
7.74	2298.487		0.129	8.09	3261.711		0.124	8.44	4628.606		0.118	8.79	6568.294		0.114
7.75	2321.588		0.129	8.10	3294.492		0.123	8.45	4675.125		0.118	8.80	6634.307		0.114
7.76	2344.920		0.129	8.11	3327.603		0.123	8.46	4722.111		0.118	8.81	6700.984		0.113
7.77	2368.487		0.129	8.12	3361.046		0.123	8.47	4769.570		0.118	8.82	6768.330		0.113
7.78	2392.290		0.128	8.13	3394.825		0.123	8.48	4817.505		0.118	8.83	6836.354		0.113
7.79	2416.333		0.128	8.14	3428.944		0.123	8.49	4865.922		0.118	8.84	6905.061		0.113
7.80	2440.618		0.128	8.15	3463.406		0.123	8.50	4914.799		0.118	8.85	6974.459		0.113
7.81	2465.146		0.128	8.16	3498.214		0.123	8.51	4964.194		0.117	8.86	7044.554		0.113
7.82	2489.922		0.128	8.17	3533.372		0.122	8.52	5014.086		0.117	8.87	7115.354		0.113
7.83	2514.946		0.128	8.18	3568.884		0.122	8.53	5064.479		0.117	8.88	7186.865		0.113
7.84	2540.221		0.128	8.19	3604.752		0.122	8.54	5115.378		0.117	8.89	7259.095		0.112
7.85	2565.751		0.127	8.20	3640.981		0.122	8.55	5166.789		0.117	8.90	7332.051		0.112
7.86	2591.537		0.127	8.21	3677.574		0.122	8.56	5218.717		0.117	8.91	7405.740		0.112
7.87	2617.582		0.127	8.22	3714.534		0.122	8.57	5271.166		0.117	8.92	7480.170		0.112
7.88	2643.890		0.127	8.23	3751.867		0.121	8.58	5324.143		0.117	8.93	7555.348		0.112
7.89	2670.461		0.127	8.24	3789.574		0.121	8.59	5377.652		0.116	8.94	7631.281		0.112
7.90	2697.300		0.127	8.25	3827.660		0.121	8.60	5431.699		0.116	8.95	7707.978		0.112
7.91	2724.408		0.126	8.26	3866.129		0.121	8.61	5486.289		0.116	8.96	7785.445		0.112
7.92	2751.789		0.126	8.27	3904.985		0.121	8.62	5541.428		0.116	8.97	7863.691		0.111
7.93	2779.445		0.126	8.28	3944.231		0.121	8.63	5597.121		0.116	8.98	7942.723		0.111
7.94	2807.378		0.126	8.29	3983.872		0.121	8.64	5653.374		0.116	8.99	8022.550		0.111
7.95	2835.593		0.126	8.30	4023.911		0.120	8.65	5710.191		0.116	9.00	8103.179		0.111
7.96	2864.091		0.126	8.31	4064.352		0.120	8.66	5767.580		0.115				
7.97	2892.876		0.125	8.32	4105.200		0.120	8.67	5825.546		0.115				
7.98	2921.950		0.125	8.33	4146.458		0.120	8.68	5884.095		0.115				
7.99	2951.316		0.125	8.34	4188.131		0.120	8.69	5943.231		0.115				
8.00	2980.976		0.125	8.35	4230.223		0.120	8.70	6002.963		0.115				
8.01	3010.936		0.125	8.36	4272.738		0.120	8.71	6063.294		0.115				
8.02	3041.197		0.125	8.37	4315.680		0.119	8.72	6124.232		0.115				
8.03	3071.762		0.124	8.38	4359.054		0.119	8.73	6185.782		0.115				
8.04	3102.634		0.124	8.39	4402.864		0.119	8.74	6247.951		0.114				

SUBJECT INDEX

In searching for a given item in this book please also refer to the TABLE OF CONTENTS, LIST OF FIGURES and LIST OF TABLES in the front of the book. The latter two are particularly useful since they show the content of the accompanying text.